Ocean Dumping
and
Marine Pollution

Ocean Dumping and Marine Pollution

Geological Aspects of Waste Disposal

Edited by

Harold D. Palmer
Dames & Moore
Washington, D. C.

and

M. Grant Gross
Chesapeake Bay Institute
The Johns Hopkins University
Baltimore, Maryland

Dowden, Hutchinson & Ross, Inc.
Stroudsburg Pennsylvania

Library of Congress Catalog Card Number: 78-10436
ISBN: 0-87933-343-X

81 80 79 1 2 3 4 5
Manufactured in the United States of America.

Library of Congress Cataloging in Publication Data

Main entry under title:
Ocean dumping and marine pollution.
Includes bibliographies and index.
1.Submarine geology—Congresses. 2. Waste disposal in the ocean—Congresses. 3. Marine pollution—Congresses. 4. Marine sediments—Congresses. I. Palmer, Harold Dean, 1934- II. Gross, Meredith Grant, 1933-
QE39.025 551.4′6 78-10436
ISBN 0-87933-343-X

Distributed world wide by Academic Press,
a subsidiary of Harcourt Brace Jovanovich,
Publishers.

Contents

List of Contributors vii

Preface xi

Waste Disposal and Dredging Activities: The Geological Perspective 1
M. Grant Gross and Harold D. Palmer

Shelf-Sediment Dynamics and Solid-Waste Disposal 9
D. S. Gorsline

Stability of Dredged Material Deposited Seaward of the Columbia River Mouth 17
R. W. Sternberg, J. S. Creager, J. Johnson, and W. Glassley

Geologic Effects of Ocean Dumping on the New York Bight Inner Shelf 51
S. Jeffress Williams

Mud Deposits near the New York Bight Dumpsites: Origin and Behavior 73
George L. Freeland, Donald J. P. Swift, and Robert A. Young

Depositional Characteristics of Sediments at a Low Energy Ocean Disposal Site, Savannah, Georgia 97
George F. Oertel

Containment of Particulate Wastes at Open-Water Disposal Sites 109
H. J. Bokuniewicz and R. B. Gordon

Dredging and Disposal in Chesapeake Bay, 1975–2025 131
M. Grant Gross and W. B. Cronin

The Problem of Misplaced Sediment 147
Maynard M. Nichols

Two Waste Disposal Sites on the Continental Shelf off the Middle Atlantic States: Observations Made from Submersibles 163
D. W. Folger, H. D. Palmer, and R. A. Slater

Mathematical Modeling Predictions of the Geological Effects of Sewage Sludge Dumping on the Continental Shelf 185
David O. Cook

Distribution of Suspended Particulate Matter near Sewage Outfalls in Santa Monica Bay, California 205
Ronald L. Kolpack

Dredged Material, Ocean Disposal, and the Regulatory Maze 241
David D. Smith

Index 263

About the Editors 269

List of Contributors

H. J. Bokuniewicz	Department of Geology and Geophysics Yale University New Haven, Connecticut
David O. Cook	Environmental Research Laboratory Raytheon Company Portsmouth, Rhode Island
J. S. Creager	Department of Oceanography University of Washington Seattle, Washington
W. B. Cronin	Chesapeake Bay Institute The Johns Hopkins University Baltimore, Maryland
D. W. Folger	U. S. Geological Survey Woods Hole, Massachusetts
George L. Freeland	Atlantic Oceanographic and Meteorological Laboratories National Oceanic and Atmospheric Administration Miami, Florida
W. Glassley	Department of Geology Middlebury College Middlebury, Vermont
R. B. Gordon	Department of Geology and Geophysics Yale University New Haven, Connecticut

D. S. Gorsline — Department of Geological Sciences, University of Southern California, Los Angeles, California

M. Grant Gross — Chesapeake Bay Institute, The Johns Hopkins University, Baltimore, Maryland

J. Johnson — Department of Oceanography, University of Washington, Seattle, Washington

Ronald L. Kolpack — Department of Geological Sciences, University of Southern California, Los Angeles, California

Maynard M. Nichols — Virginia Institute of Marine Sciences, Gloucester Point, Virginia

George F. Oertel — Skidaway Institute of Oceanography, Savannah, Georgia

Harold D. Palmer — Dames & Moore, Washington, D. C.

R. A. Slater — Department of Earth Sciences, University of Northern Colorado, Greeley, Colorado

David D. Smith — David D. Smith Associates, San Diego, California

R. W. Sternberg — Department of Oceanography, University of Washington, Seattle, Washington

Donald J. P. Swift — Atlantic Oceanographic and Meteorological Laboratories, National Oceanic and Atmospheric Administration, Miami, Florida

S. Jeffress Wiliams

Geotechnical Engineering Branch
Coastal Engineering Branch
U. S. Army, Corps of Engineers
Fort Belvoir, Virginia

Robert A. Young

Atlantic Oceanographic and Meteorological Laboratories
National Oceanic and Atmospheric Administration
Miami, Florida

Preface

The majority of the papers contained in this book were presented in a symposium convened during the 51st Annual meeting of the Society of Economc Paleontologists and Mineralogists in Washington, D. C., in June, 1977. Over the past decade, the consensus among many of the marine geologists with whom we had discussed the geological aspects of waste disposal in the sea was that this discipline, more than others in the marine sciences, provides a perspective of time scales that is necessary in considering the fate of wastes released in the ocean. For that reason, and since much of the policy and regulatory activity dealing with ocean dumping originates in federal agencies located in Washington, it was appropriate to hold a symposium addressing this subject during a major geological meeting in that city.

The contributors come from a variety of institutions, agencies, and private firms involved in the study of the fate of wastes dumped at sea. By way of an introduction, we have attempted to set the theme in a brief review of the source of materials dumped at sea, their fates, and some statistics on volumes and the nature of wastes. This is followed by a review paper provided by Donn Gorsline, President of the SEPM, that examines the dynamics of solid waste disposal. In this, and many subsequent chapters, emphasis has been placed on dredge spoil—that sedimentary material excavated from channels, harbors, and other navigable waterways—a nuisance that has persisted since the earliest days of waterborne commerce. Inasmuch as the great majority of materials disposed of at sea is a product of dredging activities, and since many dollars in research funds have addressed the fates and effects of spoil disposal, it is not surprising that this substance becomes the theme of many papers in this volume. Again, it is appropriate to point out that it is the marine geologist, or more specifically the sedimentologist, who is best equipped to determine the *rates* and *scales* of dredge spoil transport.

These two parameters are the critical factors in assessing the impact of spoil disposal on the marine environment. How quickly are these materials dispersed by marine processes? What are the paths of transport? Are there harmful components in the materials and if so, what is their availability to the marine or estuarine biota? These are some of the questions addressed by many of our contributors.

Dick Sternberg and co-workers have examined a high-energy shelf site, while Jeff Williams documents studies in a moderate energy environment, stressing site specific effects. George Freeland and colleagues examine the regional aspects of waste disposal in an intensely utilized moderate energy environment. George Oertel deals with conditions in a low energy environment, somewhat similar to that examined by Hank Bokuniewicz and Bob Gordon, the latter, however, dealing with intense utilization in an estuarine region, Long Island Sound, as opposed to the Georgia inner shelf. And for our purposes, estuaries are considered part of the marine environment, since the exchange of marine and freshwater watermasses is significant, and the agents and processes acting to disperse and dilute wastes are essentially the same. Grant Gross, Bill Cronin, and Maynard Nichols have examined aspects of disposal in the Chesapeake Bay region—a most productive estuary with a history of several centuries of conflicts in use.

Direct observations of materials other than dredge spoils are included in chapters by Dave Folger and colleagues and Ron Kolpack. The former describe conditions at two mid-shelf dumpsites where industrial wastes and municipal sludge have been released in large quantities, while the latter examines the fate of effluents piped into Santa Monica Bay off southern California. A treatment of modeling of waste discharges by Dave Cook and associates and David Smith's analysis of the complex route required for permitting of ocean disposal round out the volume.

Our purpose has been to bring geological perspective to a serious environmental concern, and if our readers share the knowledge generously provided by our contributors and perhaps apply some of their insights into problems elsewhere, then the editors will feel that this volume has served the purpose intended—namely, providing the experience of marine scientists, engineers, planners, and others associated with managing the by-products of our civilization to the consideration of an obvious, but certainly not limitless, sink for waste products. By law, ocean dumping of all but dredge spoil must cease (or be severly restricted and controlled) by 1981. Yet the cumulative effects of years of dumping at sea remain, to a large degree, unknown and unmeasured. Our collective concern over the overall health of the oceans, however large they may seem, is not misdirected, for as Edward Goldberg says in his volume *Health of the Oceans*,

> The slow but continuous alteration of the open-ocean waters can offer future generations the legacy of a poisonous ocean. It is most unreasonable to titrate the

seas with man's wastes to the endpoint of a world-wide mass mortality of organisms.[1]

To our contributors, who share a mutual concern for the future, we are thankful for their interest and participation in this summary of geological aspects of waste disposal in the ocean. The patient staff at Dowden, Hutchinson & Ross coped with numerous problems, and we are grateful to Shirley End and others who helped bring this volume to you, our readers.

Harold D. Palmer
M. Grant Gross

1. E. D. Goldberg. 1976. *The Health of the Oceans*. UNESCO Press, Paris, p. 166.

Ocean Dumping and Marine Pollution

Waste Disposal and Dredging Activities: The Geological Perspective

M. Grant Gross
Harold D. Palmer

INTRODUCTION

From prehistoric times to the present, the coastal margins of continents and islands have offered man a variety of natural resources. At first, they were sites where food could be found in abundance, and where trade or conquest could be pursued by ship much more easily than by overland traffic. The first sites were within estuaries or in protected coastal reaches, but as commerce expanded, enlargement of such facilities required man's intervention into natural processes to ensure continuing and uninterrupted use of ports and harbors. Natural channels were widened and deepened, and the practice of dredging was initiated. We know that by Roman times coastal engineering had become a profession, as witnessed by the carefully laid out harbors at Tyre, Sidon, and other ports which included jetties, breakwaters, and dredged channels. The excavated material, or spoil, was undoubtedly dumped outside the harbor, but traces of such disposal efforts have long been erased by natural submarine processes. By at least 1806 dredging engines were in use to create and/or maintain shipping channels, and in that year we know that a large steam bucket was employed to maintain the East India Docks in London (Kemp, 1976).

Although, today dredge spoil remains as the greatest single substance to be disposed of at sea, other "superfluous" materials have been dumped in coastal waters. For centuries, sailing ships visiting ports would discharge ballast stone after sea voyages, and in certain cases some careful detective work has revealed the sources of such transposed materials (Emery et al., 1968). Other solid jetsam includes excavated stone, cellar dirt (mainly building demolition debris), fly ash and old vessels dumped in coastal regions such as the New York Bight. Obviously, dumping of such materials in navigable waterways produced a conflict in use and federal regulation of waste disposal in the ocean began nearly a century ago when Congress designated the supervisor of New York Harbor as responsible for establishing disposal sites clear of the busy channels. This law, under Section 1 of the Act of Congress approved June 29, 1888, also authorized the supervisors of Baltimore and Hampton Roads Harbors to likewise designate disposal sites clear of navigable waters so we may be sure the problem was not confined to New York.

With the majority of the world's population inhabiting the coastal zones of continents and islands, it is inevitable that sooner or later conflicts arise in the use of the nearshore marine environment. For the "developed," higher industrialized nations coastal conflict has long since arrived, and in countries where intensive multiple use of wastes has continued for decades (or centuries in some cases) the inhabitants may be victims of poor planning and overuse. As Goldberg (1976) points out, it is not surprising that prominent environmental diseases have Japanese names--Minamata Disease (methyl mercury poisoning), Yusho (PCB poisoning) and Itai-itai (cadmium poisoning). His "population ratio" comparing Gross National Product to national area, population and waste flux provides an interesting if not alarming perspective on local and regional impacts in marine environments.

The coastal zones are, for many countries, their prime natural resource as well as their portal to world commerce. In these shelf areas, waste deposits of various types are common features. Because of the importance of these activities, we attempted in this volume to focus attention on what we know and--equally important--what we do not know. Our purpose is to bring to the attention of scientists, planners, regulatory staff members, and legislators the importance of the activities and the technical principles that must be considered in understanding, planning, regulating, and controlling the processes and events.

Waste Control

Dredging and disposal of dredged materials are among the most important activities affecting estuaries and the coastal ocean in the United States. Both dredging and disposal involve movements of large volumes of materials (Boyd et al., 1972). In the United States, about 2×10^8 m^3 of sediment is removed from navigation channels (maintenance dredging) while nearly $6 \times 10^7 \times m^3$ is dredged during construction of new facilities (new work) at an annual cost of $150,000,000 (1970 dollars). In most cases, the materials dredged (about $1.90 \times 10^8 m^3 yr.^{-1}$) are discharged to open-water sites in bays, estuaries, rivers, and lakes. Of the remainder, nearly equal amounts are discharged in freshwater (primarily rivers) and marine sites. Maintenance dredging accounts for 80 percent of dredging activity. Half the annual maintenance dredging is done in the Gulf of Mexico with the Southern Atlantic region of the U.S. and Delaware Bay also contributing significant quantities (Boyd et al., 1972).

Materials removed during dredging in urban areas come from the natural sediment load of rivers and from littoral drift. Part of this sediment load is polluted because of waste discharges from up-river cities or runoff from agriculture or mine drainage. Some sediment, such as beach sand, is initially clean but mixes with other wastes when deposited in industrial harbors or in dredged navigation channels. Such deposits are eventually incorporated into the urban waste stream and require disposal (Gross, 1972).

Waste Sources

Agriculture and mining are major sediment contributors to rivers in the United States (Kenahan, 1971). In urban areas however, the principal sediment source is erosion of construction sites left bare during construction (Wolman, 1967; Wolman and Schick, 1967). Erosion of construction sites yields 10 to 100 times more sediment per unit area than either mining or agriculture. Sediment from erosion of cropland in the United States has decreased markedly from its peak in 1900-1920 owing to a reduction in land areas farmed and to better conservation practices. For example, sediment yield from an area in the Southern Piedmont (South Carolina, Alabama, Georgia) decreased from about 200 t $km^{-2}yr.^{-1}$ in 1910-1934 to about 30 t $km^{-2}yr.^{-1}$ in 1967-1972 (Meade and Trimble, 1974).

Despite recently decreased sediment yield to rivers, soils eroded from farm lands in past decades no doubt reach urban estuaries today, especially during floods. However, only about 5 percent of the soil eroded from upland slopes in the southeastern United States since European settlement has reached the ocean (Trimble, 1975). The remainder is deposited temporarily in stream channel and bank sediments (Wolman, 1967; Wolman and Schick, 1967) or impounded by dams (Meade and Trimble, 1974). These deposits are scoured during high river flow and carried downstream, and some of this sediment load is deposited in urbanized estuaries.

Erosion accompanying conversion of croplands or woodlands to housing developments is another major sediment source in urban areas (Wolman, 1967). The sediment eroded from a small construction site can equal 28,000 t $km^{-2}yr.^{-1}$ (Wolman and Schick, 1967). For the Baltimore-Washington metropolitan area, Wolman and Schick estimated that 630-1600t of sediment are mobilized per 1000 increase in population. Assuming comparable erosion in the New York-New Jersey metropolitan region we can estimate that as much as 2×10^5 t of sediment were mobilized by the increase in 200,000 persons between 1971 and 1972 as approximately 75 km^2 of agricultural land were converted to suburbs (Tri-State Regional Planning Commission 1973). Such large sediment yields cause extensive alteration of stream channels because of temporary sediment deposition.

Another natural source of deposits removed during dredging is found in the littoral drift along ocean beaches. Such drift may be reduced slightly by human intervention through seawalls and jetties (Caldwell, 1966; Yasso and Hartman, 1975) or because of dredging and removal of sand deposited in inlets. In areas where beaches are restored by adding sand (beach replenishment), there may be slight increase in littoral drift but the amounts of sand involved usually are small.

Sediment moved during maintenance operations is poorly characterized. About half ($1.2 \times 10^8 m^3$) is mixed silt and sand. Another $6 \times 10^7 m^3 yr.^{-1}$ is finer-grained mud, clay and silt, while $4 \times 10^7 m^3$ $yr.^{-1}$ is coarse grained sand, gravel, and shell (Boyd et al., 1972). While it is even more difficult to determine the concentration of pollutants in the dredged materials, about one third of the materials moved during maintenance dredging were considered polluted (Boyd et al., 1972). Discharge of materials dredged from heavily polluted areas like New York Harbor is a major source of sediment, metals, and nutrients to the coastal ocean (Mueller et al., 1976).

Waste Deposits

Identification and mapping of waste deposits usually involve two approaches. The simplest is to identify artifacts or physical properties. Human artifacts, such as building debris, fly ash, urban refuse, and so forth, are often found in dredged materials. In the case of sewage sludge, persistant materials that defy digestion in treatment plants (hair, tomato seeds, cellulose fibers, etc.) have been used as tracers of treated effluent.

The second approach is to compare the chemical composition of wastes (e.g., dredged sediment and sewage sludges) with that of natural substances likely to accumulate in estuaries or on continental shelves. Such analyses indicate that physical properties, such as grain size, and the abundance of certain chemical constituents from industrial or domestic wastes, such as carbon and metals like lead and silver, are useful tags for some waste deposits (Gross, 1970a, 1970b; Carmody et al., 1973).

Changed chemical characteristics aid in mapping waste deposits in areas of sandy sediment (Gross, 1976). They have less application in areas where the sediment deposits naturally covering the bottom are similar in carbon content and physical properties to many wastes. Presence of sewage solids is still a useful indicator of some waste deposits, regardless of when they are deposited. Using these criteria to map waste accumulation, one can easily show that deposits of carbon-rich, metal-rich wastes are widespread in industrialized harbors, such as New York (Gross, 1972) and Baltimore (Villa and Johnson, 1974).

In shallow waters, waste solids accumulate in the disposal sites unless their physical characteristics--low density, fine particle size--permit them to be transported by currents beyond the disposal site before settling to the bottom. For example, in the New York Bight, waste deposits are most common in the disposal site near the head of Hudson Channel (Gross, 1976; Carmody et al., 1973), over 15m (50 ft) deep in the axis (Williams and Duane, 1974), where waste disposal has been carried out since the late nineteenth century.

Effects of Dredged Material Disposal

Effects of dredged material disposal on continental shelves can be appreciated only when we realize that, except for a few major rivers, nearly all sediment transported by rivers is trapped and normally deposited in estuaries, bays, and harbors (Emery, 1965; Meade, 1969). Consequently, dredging operations are not only supplying a large quantity of waste solids but are moving them to the continental shelf where little other sediment is being deposited to dilute or bury the wastes (Gross, 1970c; 1972; this volume).

Disposal operations are localized and usually involve releasing many tons of solids whenever a barge or dredge is emptied. Thus, the ocean bottom in the disposal area may receive a relatively thick layer of waste solids almost instantaneously. Designated disposal grounds are usually small (a few square kilometers), although it is probable that dumping actually occurs over much larger areas owing to navigational errors, adverse weather conditions, and illegal dumping. Numerous studies of the fate and distribution of waste materials such as sludge, packaged radioactive substances, and industrial byproducts have been carried out by means of small submersibles (see Folger et al., this volume). Such surveys, providing direct observation of the effects of dumping practices, are summarized by Palmer (1977).

Studies of the behavior of sewage sludges discharged to the marine environment have pinpointed some of the probable effects of fine-grained, low-density wastes in coastal ocean waters (NOAA, 1975; 1976). Fine-grained low-density particles remain suspended and when discharged in near-surface waters, form visible plumes of discolored water which move with the local surface currents. Waves and currents scouring the ocean bottom can resuspend and move deposits in shallow waters (Harris, 1976; Swift et al., 1976).

Dredged material deposits are often rich in organic carbon and metal content (Gross, 1976; Segar and Cantillo, 1976; Thomas et al., 1976), but the effect of metal enrichment on the marine ecosystem is poorly understood. There is evidence that metals accumulate in bottom-dwelling organisms. The public health implications of metal transfers through seafood to man have not been widely studied (Verber, 1976).

In areas where large volumes of materials have been dumped for an extended period, bottom-dwelling communities have been transformed substantially. The abundance and number of different types of marine organisms was substantially reduced in the waste disposal areas of the New York Bight, for example (Pearce, 1972). Observed changes could have been caused by altered physical properties of the bottom, by toxic metals and hydrocarbons associated with waste solids, and by reduction in dissolved oxygen concentrations of near-bottom waters. Contaminated deposits have also been strongly implicated in the occurrence of diseases in marine organisms such as finrot--the erosion of fishes' fin tissue (Murchelano and Ziskowski, 1976)--and shell erosion in crabs, lobsters, and other crustacea (Rosenfeld, 1976).

The presence of human pathogens (disease-causing agents) in the sewage-derived constituents in many dredged materials makes ocean bottoms used for waste disposal unsuitable for production of shellfish for human consumption (Verber, 1976) and less attractive for recreational fishing.

Disposal of waste solids in continental shelf areas can alter bottom topography and sediment composition and texture. Dredged wastes deposited at the head of the Hudson Shelf Valley have partially filled one of the tributary arms (Williams, 1975; Freeland et al., 1976). Between surveys made in 1936 and 1973, deposits of dredged material more than 10m thick accumulated at the site, amounting to 87 percent of the volume reportedly discharged (Freeland et al., 1976). Previous disposal operations have formed hills that stand 8m above the surrounding shelf (Williams, 1975).

Some changes in sediment texture are quite distinctive. Artifact gravels--mixtures of bricks, concrete fragments, and rock fragments--are conspicuous in the surficial sediments accumulating in the area used for disposal of construction and demolition debris in the Hudson Shelf Valley (Freeland et al., 1976). The fate of the fine-grained carbon-rich materials is less obvious. The deepest portions of the Hudson Shelf Valley are floored by fine-grained sediment (Freeland et al., 1976). These deposits are enriched in metals typical of industrial wastes (Gross, 1972, 1976; Harris, 1976) and in organic carbon-total carbohydrates (Hatcher and Keister, 1976) suggesting some deposition of fine-grained wastes. None of the available tests are adequate to indicate the amount of waste material mixed with sediments derived from natural sources.

Regardless of how we examine dredging and disposal operations, it is clear they cannot be ignored. The amounts of material involved are large and the costs are increasing rapidly. The materials can cause problems in the disposal areas, and there are unknown environmental risks and possibly public health hazards as well. Studies such as those described in this volume are essential steps in our understanding of the multiple interactions between natural agents and processes and the by-products of "civilization."

REFERENCES

Biscaye, P. E., and C. R. Olsen. 1976. Suspended particulate concentrations and compositions in the New York Bight. *Am. Soc. Limnol. Oceanogr. Spec. Symp.* 2:124-37.

Boyd, M. G., R. T. Saucier, J. W. Keeley, R. L. Montgomery, R. D. Brown, D. B. Mathis, and C. J. Guice. 1972. Disposal of dredge spoil. U. S. Army Waterw. Exp. Stn. Tech. Rep. H-72-8, Vicksburg, Mississippi, 121 pp.

Caldwell, J. M. 1966. Coastal process and beach erosion. *J. Soc. Civ. Eng.* 53(2):142-57.

Carmody, D. J., J. B. Pearce, and W. E. Yasso. 1973. Trace metals in sediments of New York Bight. *Mar. Pollut. Bull.* 4(9):132-35.

Emery, K. P. 1965. Geology of the continental margin off eastern United States. In *Submarine Geology and Geophysics,* W. F. Whittard and R. Bradshaw, eds., pp. 1-20. Butterworth, London.

Emery, K. O., Kaye Loring, and Nota Djg. 1968. European Cretaceous flints on the coast of North America. *Science* 160:1112-25.

Freeland, G. L., D. J. P. Swift, W. L. Stubblefield, and A. E. Cook. 1976. Surficial sediments of the NOAA-MESA areas in the New York Bight. In Biscaye and Olsen, 1972, pp. 99-101.

Goldberg, E. D. 1976. *The Health of the Oceans.* UNESCO Press, Paris, 172 pp.

Gross, M. G. 1970a. New York metropolitan region--A major sediment source. *Water Resource Res.* 6:927-31.

Gross, M. G. 1970b. Preliminary analysis of urban wastes, New York metropolitan region. Mar. Sci. Res. Cent. Tech. Rep. 5. State University of New York, Stony Brook, 35 pp. Also in *Congr. Rec.* 116(31):S2885-90.

Gross, M. G. 1970c. Analyses of dredged wastes, fly ash and waste chemicals, New York metropolitan region. Mar. Sci. Res. Cent. Tech. Rep. 7. State University of New York, Stony Brook, 33 pp.

Gross, M. G. 1972. Geologic aspects of waste solids and marine waste deposits, New York metropolitan region. *Geol. Soc. Am. Bull.* 83: 3163-76.

Gross, M. G. 1976. Waste disposal. MESA New York Bight Atlas, Monogr. 26. New York Sea Grant Inst., Albany, 32 pp.

Harris, W. H. 1976. Spatial and temporal variation in sedimentary grain-size facies and sediment heavy metal ratios in the New York Bight apex. In Biscaye and Olsen, 1976, pp. 102-23.

Hatcher, P. G., and L. E. Keister. 1976. Carbohydrates and organic carbon in New York Bight sediments as possible indicators of sewage contamination. In Biscaye and Olsen, 1976, pp. 240-48.

Kemp, P. 1976. *The Oxford Companion to Ships and the Sea.* Oxford University Press, New York, 971 pp.

Kenahan, C. B. 1971. Solid wastes--resources out of place. *Environ. Sci. Technol.* 5:594-600.

Meade, R. H. 1969. Landward transport of bottom sediment in estuaries of the Atlantic Coastal Plain. *J. Sediment. Petrol.* 29:222-34.

Meade, R. H., and S. W. Trimble. 1974. Changes in sediment loads in rivers of the Atlantic drainage of the United States since 1900. *Int. Assoc. Sci. Hydrol. Publ.* 113:100-104.

Mueller, J. A., A. R. Anderson, and J. S. Jeris. 1976. Contaminants entering the New York Bight: Sources, mass loads, significance. In Biscaye and Olsen, 1976, pp. 162-70.
Murchenalo, R. A., and J. Ziskowski. 1976. Fin-rot disease studies in the New York Bight. In Biscaye and Olsen, 1976, pp. 329-36.
National Oceanic and Atmospheric Administration (NOAA). 1975. Ocean dumping in the New York Bight. NOAA Tech. Rep. ERL-321-MESA, 278 pp.
National Oceanic and Atmospheric Administration (NOAA). 1976. Evaluation of proposed sewage sludge dumpsite areas in the New York Bight. NOAA Tech. Rep. ERL-MESA-11, 212 pp.
Palmer, H. D. 1977. The use of manned submersibles in the study of ocean waste disposal. In *Submersibles and Their Use in Oceanography and Ocean Engineering*, R. A. Geyer, ed., Elsevier Oceanography Series No. 17, Elsevier Science Publ. Co., Amsterdam, 383 pp.
Pararas-Carayannis, G. 1973. Ocean dumping in the New York Bight: An assumption of environmental studies. Tech. Memo. 39, Coastal Engineering Research Center, U. S. Army Corps of Engineers, 159 pp.
Pearce, J. B. 1972. The effects of solid waste disposal on benthic communities in the New York Bight. In *Marine Pollution and Sea Life*, M. Raivo, ed., pp. 404-11. Fishing News (Books) Ltd, London.
Rosenfeld, A. 1976. Infectious diseases in commercial shellfish on the Middle Atlantic coast. In Biscaye and Olsen, 1976, pp. 414-23.
Segar, D. A., and A. Y. Cantillo. 1976. Trace metals in the New York Bight. In Biscaye and Olsen, pp. 171-98.
Swift, D. J. P., G. L. Freeland, P. E. Gadd, G. Han, J. W. Lavelle, and W. L. Stubblefield. 1976. Morphologic evolution and coastal sand transport New York-New Jersey shelf. In Biscaye and Olsen, 1976, pp. 69-89.
Thomas, J. P., W. C. Phoel, F. W. Steimle, J. E. O'Reilly, and C. A. Evans. 1976. Seabed oxygen consumption--New York Bight apex. In Biscaye and Olsen, 1976, pp. 354-69.
Trimble, S. W. 1975. Denudation studies: Can we assume steady state? *Science* 188:1207-08.
Verber, J. L. 1976. Safe shellfish from the sea. In Biscaye and Olsen, 1976, pp. 433-41.
Villa, O., Jr., and P. G. Johnson. 1974. Distribution of metals in Baltimore Harbor sediments. Tech. Rep. 59, Annapolis Field Off. Reg. III, EPA-90319-74-012.
Williams, S. J. 1975. Anthropogenic filling of the Hudson River (shelf) Channel. *Geology* 3:597-600.
Williams, S. J., and D. B. Duane. 1974. Geomorphology and sediments of the inner New York Bight continental shelf. U. S. Army Corps of Eng. Tech. Mem. 45. Coastal Eng. Res. Cent., Fort Belvoir, Va.
Wolman, M. G. 1967. A cycle of sedimentation and erosion in urban river channels. *Geografiska Annaler* 49A:385-95.
Wolman, M. G., and A. P. Schick. 1967. Effects of construction on fluvial sediment, urban and suburban areas of Maryland. *Water Resource Res.* 3:451-64.
Yasso, W. E., and E. M. Hartman, Jr. 1975. Beach forms and coastal processes. MESA New York Bight Atlas, Monogr. 11, 51 pp.

Shelf-Sediment Dynamics and Solid-Waste Disposal

D. S. Gorsline

ABSTRACT

Dredged material is commonly natural material that has been deposited under riverine, estuarine or deltaic conditions. It includes natural terrigenous and biogenic detritus, organic matter and debris and adsorbed metal ions and organic compounds of natural or man-generated origins. Because the great bulk of the material is natural particles, its dispersion is identical to dispersion for naturally contributed sedimentary materials and the particles will follow the same paths. Other than the effect of over whelming mass dumped in a locality the major problem environmentally is the smaller fraction of dredge spoil that contains pollutants. Since a large proportion of these contaminants is adsorbed on the clay fraction which moves as suspended load, the major environmental impact is the disposal of the fine fraction.

Most harbors are located in estuaries or the lower reaches of major rivers, and so the most complex problems of control of spoil disposal are due to the inherent characteristics of estuarine circulation and particularly to the fact that esturies are sediment traps. They typically draw from both river input and from adjacent shelves and thus the problem of finding suitable dump sites outside a given system may be economically and technically difficult.

We need more information about such problems as the rates of benthic boundary layer processes, the paths of suspended sediment transport over shelves and the net contributions of estuaries to shelves in normal and storm periods.

Deep water sites should be avoided until we know more about the biology of those environments.

INTRODUCTION

This symposium volume includes papers that deal with several specific examples of dredge disposal systems (see Sternberg et al.; Williams, this volume), discuss the magnitude of the dredge spoil contribution (Smith and Peguegnat; Gross and Cronin, this volume), and review processes and controls (Kolpack; Nichols and Faas; Oertel, all in this volume). I will therefore restrict this introductory paper to a discussion of the sedimentologic aspects of the problem of dredge spoil disposal.

To do this, it is useful to examine the basic elements involved. Most dredge spoil is natural particulate material derived from river, coastal or shallow marine deposits. It includes terrigenous detritals, biogenic clasts, organic material and varying amounts of adsorbed metal ions and organic compounds of both natural and man-generated origins (National Research Council, 1976).

Since harbors and navigation channels are the primary sources of such materials, the typical sedimentary materials are commonly estuarine or deltaic in origin. The sediments will therefore have large contributions of natural organic and biogenic components and fine silt-clays will be a typical element. Sands are common in these environments and will usually be a less critical element in the spoil due to their rapid settling rates and relative freedom from adsorbed contaminants. The major sources of trouble will be the fine sediments.

In dynamic terms the estuary circulation system contains elements inimical to effective dredge spoil disposal as will be noted below. Estuaries are sediment sinks for a coastal segment and draw sediment both from the tributary rivers and from the open shelf adjacent to the estuary (Wiley, 1976). Therein lies a major disposal problem.

Clays are chemically active because of their high ion exchange capacity which in turn is related to their chemical structure (Grim, 1953). Rates of exchange and the ions or molecules involved are functions of ion concentrations and of the dominant elements present in solution. Oxidation state and pH are also important.

Thus as clays move from fresh to salt water and from reduced to oxidized environments the particles release some organics and metals and take up others (DeGroot, et al., 1973; Forstner and Muller, 1974). The large surface area of these fine particles and their common association in low density aggregates facilitates the exchanges. These properties are the major source of the pollution problems associated with contaminated dredge spoil (International Atomic Energy Agency, 1973).

Most historic disposal sites are on the continental margins and the majority are in depths of less than 30 m (Kirby et al., 1975). Within those depths the major agents of sediment dispersal are surface waves of all periods including tides, wind-generated and sometimes thermohaline currents, edge waves and internal waves. The application of this spectrum of forces depends on latitude, shelf relief, climatic regime and the ratio of normal wave impact to storm wave impact and of storm frequency (Booth and Gorsline, 1973).

The resulting pattern of energy application together with the location of sediment sources, rates of input and nature of the sediment input determines the areas of erosion and deposition and produces a surface distribution of sediment types. Obviously, the best locale for solid waste disposal if dispersal is desired would be regions of high energy systems and nondeposition. If a fixed or nondispersive mode is the desired objective then low energy systems and deposition would be best. In either example, pretreatment might be required to enhance the response of the waste to the shelf energy regime.

Shelf dispersal systems must obviously be quantitatively described to determine the acceptability of such disposal. Since sediments in nature move by reason of their hydraulic characteristics (simplified to size and relative density), a necessary prerequisite for disposal is to determine the hydraulic equivalence of the solid wastes. If these two are known then the capacity of the given shelf can readily be determined. The definition of capacity in turn will depend on the acceptable levels of such factors as water turbidity, sediment type changes and resulting bioeffects.

The comparative effects of normal energy levels and of storm levels must be known. Particles move by floating, suspension and bottom traction and of these processes the most critical are probably floating and suspension since these represent the modes of fastest transport. These processes are amplified many times under extreme conditions. We know least about bedload and floating transport processes.

DISCUSSION

Although precise figures for the annual dumping of solid wastes into the ocean are hard to come by and can vary by a factor of 10 depending on definitions and the sources of information, the rough magnitude of the volume dumped every year in rivers and coastal areas is of the order of 200 x 10^6 tons per year (see National Research Council, 1976; Smith and Peguegnat, this volume; Kirby et al., 1975) of which 75-80% is from dredging. Of this dredge contribution about one-third is polluted material, or about 1-2 times the total solid contribution from all other waste sources (NRC, 1976). Kirby and his associates (1975) estimate that 80% of dredging is for maintenance of existing channels and 20% is to open new channels. Thus much dredged material is reworked from potentially contaminated sites.

On some coasts and for some large urbanized areas this form of discharge can be larger than the natural stream contribution of solid particles to coastal areas (Gross, 1972). For better comparison the average solids discharged by the Mississippi River have been estimated at about 280-300 x 10^6 tons per year (Leopold and others, 1964), the Columbia discharges about 30 x 10^6 tons per year and the St. Lawrence delivers perhaps 2-4 x 10^6 tons per year (Strakov, 1967; Leopold and others, 1964). Thus solids discharged to the U. S. coastal waters from dredging alone may deliver from 1 to 10% of the total solids delivered by streams and rivers. It compares in mass to a major river system sediment discharge.

Were it not for the contaminants carried with these solids the additional discharges from this source would pose environmental problems only in terms of the volume of deposits laid down near the points of discharge, in the additional turbidity of the coastal waters near the discharge points, and in the possible establishment of unstable deposits on the sea floor. The transport of these solids would be by the same agents that distribute the natural sediment contributions in coastal waters. However, since a proportion of the dredged material and most of the industrial, seqage and other solid wastes contain materials that are detrimental to life in the sea, the problem becomes a critical one. Although Congressional action in the 1970's (NRC, 1976) has established controls over waste disposal in the oceans and has curtailed or stopped much of the non-dredge waste contributions, the dredge spoil input continues to be large and growing. Certainly dredging is a vital requirement for maintenance of the nation's ports and shipping channels and cannot be easily controlled for one aspect without creating problems in other aspects.

Once solids have been introduced into ocean waters the dispersal and sedimentation processes are the same ones that affect natural particles. Since the dominant agents and processes change from one environment to the next, planning for the dispersal of waste solids requires a thorough knowledge of the natural dynamic equilibrium of the shelf discharge area.

If we examine the dredged material problem we find that it exists because some minimum water depth must be maintained in waterways of commercial interest. The locations of these economically useful waterways are responses to many factors that are historic, socio-political, geomorphologic and economic. The ideal harbor is an area of quiet water which has easily navigable entrances from the sea and which serves a large hinterland. It is likely that all such locales are now in use and have been for at least a century and that few remain to be exploited barring some large shift in populations in the future. We can therefore identify accurately where dredging problems will be most critical.

Because of the historic requirement that harbors be centers that serve large interior areas, most harbors are located in estuaries. The entering rivers are or were the access routes to the interior. Thus dredge spoil operations are much involved with the mechanics of estuarine systems. There are, of course, some exceptions to this rule. Los Angeles Harbor has been built by man in a coastal area that provided only partial shelter and had no large continuous stream associated with the anchorage. Some harbors are located in lagoons rather than true estuaries. An extreme example would be the wartime harbor at Eniwetok in the southwest Pacific. Others are located within major rivers as in the example of New Orleans or some of the ports in the lower reaches of other great rivers of the world.

We may generalize by saying that the great majority of the world's large harbors are estuarine or have been developed from natural estuarine systems. These alterations include artificial deepening of original channels and secondarily the opening of new channels or turning basins by dredging filled land areas within the old system. Marshes and deltaic areas are typical sites.

Much work has been done on the sedimentation processes in estuaries (see Allen et al., 1973; Cronin, 1975; Lauf, 1967; Wiley, 1976) and these studies show that the estuary is a sink for sediments moving from both land and sea. Although major floods can sweep material out of an estuary, these materials often return from the adjacent shelf in intervening periods of normal river flow (Allen et al., 1976). A typical feature of these systems is the turbid wedges associated with the strong salinity gradients in the estuarine circulation system. The simple model for circulation is the classic salt wedge in which the outflow of fresh water and entrainment of underlying salt water causes an inflow of salt water at depth. At the salinity gradient between the superposed fresh water and the underlying saline water a turbidity maxima occurs as a result of sedimentation from the fresh water lense and inflow of suspended sediment in the underflow. This turbid maxima migrates back and forth with the salinity gradient in response to tide. In some estuaries of high sedimentation rates this zone also defines a bottom low density mud deposit which can be reworked and swept out by strong floods (see Allen et al., 1973; Allen, et al., 1976).

Since the interface between fresh and salt water is also a chemical "fence" both in terms of ion type and concentration and often of oxidation-reduction potential and pH, considerable ion exchange takes place as supended material enters the zone and is temporarily deposited on the estuary floor below this migrating zone. After sedimentation, sediments will become reduced.

When estuary sediments are transported out to the adjacent shelf by strong floods or exceptional tides or wind surges the fine sediments are suspended and move into the shelf suspended sediment system.

Typically the suspended load in nearshore waters moves in layers or plumes (Drake and Gorsline, 1973; Karl, 1976, McCave, 1972; Rodolfo, 1964) controlled by the density distribution in the coastal waters. Nearshore waters above the local thermocline depth areusually homogenized by wave stirring and with the suspended load from runoff or bottom erosion turbid plumes form the surface layer and move in response to winds and tides. These are strongly limited at their base by the density change at the thermocline and form the surface turbid layer. Below this layer sedimenting particles collect at deeper density discontinuities within the shelf water column and ultimately collect at the base of the column where turbulence is high at the benthic boundary layer. Here a second strong turbidity maximum typically is found. Thus most shelf water columns have turbid surface layer, varying numbers of less concentrated plumes or zones of turbid water in intermediate depths and a well defined bottom turbid layer. These bottom plumes commonly detach from the bottom as they move out over the shelf edge following density surfaces and then disperse and diffuse in the current systems of the shelf edge and upper slope (Karl, 1976).

On narrow Pacific shelves where submarine canyons are common, the turbid plumes often show a strong relationship to the canyon heads (Drake and Gorsline, 1973), and the canyon axis circulations (Shepard and Marshall, 1969, 1973) pump the suspended materials to depth. This phenomenon is found to be typical of most submarine canyons and even on broad shelves these relief features probably exert some influence on shelf water circulation and turbidity plume motion.

Where turbid, nepheloid transport is strong, a fine sediment bottom deposit may be developed even in relatively high energy areas if supply exceeds capacity for transport (McCave, 1972). Such deposits may be heavily eroded by occasional exceptional storm wave action at shelf depths normally below the usual wave influence. Again, the problem of reworking of sediments that may have become reduced after burial and are then reintroduced into a well oxygenated water medium is one of ion exchange and the possible release of stored contaminants to the water column.

With regard to the disposal problem, all of the above operate. If dredged materials are introduced to the shelf or nearshore by discharge from pipes, by barge dumping or from dredge disturbance of the bottom, the sand fraction probably settles rapidly to the bottom while the silts and clays will enter the suspended transport system (IAEA, 1973). Diffusion by wave stirring and current flow will be the same as for natural suspended sediment transport and will follow the local circulation pattern. Thus it is evident that a knowledge of the surface and subsurface flow patterns at various seasons and under storm and normal conditions is a prerequisite for predicting the paths of dispersion of dumped material. Recall that most dumping is in waters less than 30 m deep.

Inman and Brush (1973) have discussed the effects of dumping suspensates or any contaminant into the nearshore surf zone. In that zone the circulation patterns may concentrate the pollutant within the coastal cells. This applies to dispersal of fine components of dredge spoil that may be dumped directly into the nearshore zone. Certainly this is to be avoided for both aesthetic reasons and health safety.

Passage down submarine canyons to the deep ocean floor or to marginal deep basins (Drake and Gorsline, 1973) is possible for dredge spoil fines dumped near such features or in central shelf areas within the circulation system of canyons. In this movement sediments may be deposited and resuspended many times in transit.

Limited data would suggest that deep water biological systems are sensitive to contaminants and so this form of possible contaminant transfer should be avoided in spoil disposal. Until we know more about deep water biological systems input to these regions should be controlled and perhaps excluded.

A major problem involved in dumping within the zone of influence of major estuaries has been indicated earlier. Suspended and bedload materials can rapidly move back into the estuary (harbor) under the influence of tidal circulation. In this instance, dredge spoil may rapidly return to the place from which it was dredged.

Ludwick (1975) has shown through ingenious use of current meter observations that in a given estuarine system there are locales of deposition and erosion that are relatively stable over periods of years. A knowledge of these patterns can save money in harbor maintenance program in that use may be made of such naturally defined zones in planning the positioning of turning basins and channels. It should be remembered that harbor improvement means changing the volume of the natural water body with attendant response by the forces acting on that water body. In Panama City, Florida, a study of bathymetric surveys before and after the construction of an artificial inlet showed that the areas of deposition and erosion in the natural bay rapidly changed as a result of the alterations of channel positions and depths (T. L. Hopkins, personal communication). Modelling of proposed harbor alterations is therefore a necessary preliminary study before dredging operations are commenced. Prediction of frequency of dredging and the proper disposal sites should be based on such studies and upon studies of the coastal and shelf circulation systems.

CONCLUSION

It is evident that effective improvement of harbors and the least environmental impact of the disposal of the resulting dredge spoil requires several preliminary studies. Modelling of the harbor area under present and proposed conditions is a necessity. This in turn will require a thorough study of the circulation in the harbor and the adjacent coastal waters to provide dimensional data for the model and as a means of identifying sites of deposition and erosion in the existin system.

Much needs to be known about suspended load transport mechanisms and patterns in shelf areas. Much research is required on the processes and rates of transport in the benthic boundary layer of the shelf. The frequency and effects of storms and floods must be examined since these may rework deposits that in normal times are stable.

The release of fine sediments from polluted deposits into aerated waters and different chemical environments will release contaminants to the water and thus the movement of the generated turbidity plumes may not be indicative of the path of the contaminants that may pass into solution in the water and separate from the particulates (De Groot et al., 1973). Dredging of such deposits must be carefully controlled and the method of dredging and the mode of disposal carefully defined and monitored.

Harbors will always require some maintenance and coastal construction will always be with us. Thus a full understanding of shelf and coastal dynamics is of high national priority. Without this basic knowledge, efficient planning of dredged material disposal cannot be done.

REFERENCES

Allen, G. P., P. Castaing, and A. Klingebiel. 1973. Suspended sediment transport in the Gironde Estuary and adjacent shelf. In Proc. Intern. Symp. on Interrelationships of Estuarine and Continental Shelf Sedimentation, Bordeaux, pp. 27-37.

Allen, G. P., G. Sauzay, and P. Castaing. 1976. Transport and deposition of suspended sediment in the Gironde Estuary, France. In Estuarine Processes, Vol. II,Circulation Sediments, and Transfer of Material in the Estuary. Academic Press, New York, pp. 63-81.

Booth, J. S., and D. S. Gorsline. 1973. Thoughts on new areas for research in shelf sediment transport studies. In Proc. Intern. Symp. on Interrelationships of Estuarine and Continental Shelf Sedimentation, Bordeaux, pp. 145-148.

Cronin, L. E., ed. 1975. Estuarine Research, Vol. II, Geology and Engineering. Academic Press, New York, 587 pp.

De Groot, A. J., E. Allersma, N. de Bruin, and J. P. W. Houtman. 1973. Use of activatable tracers. In Tracer Techniques in Sediment Transport, Intern. Atomic Energy Agency, Vienna, Tech. Report 145, pp. 151-168.

Drake, D. E., and D. S. Gorsline. 1973. Distribution and transport of suspended particulate matter in Hueneme, Redondo, Newport and La Jolla Submarine Canyons, California. Geol. Soc. America Bull., 84:3949-3968.

Förstner, U., and G. Müller. 1974. Schwermetalle in Flüssen und Seen. Springer-Verlag, Berlin, 225 pp.

Grim, R. E. 1953. Clay Mineralogy. McGraw-Hill, New York, 384 pp.

Gross, M. G. 1972. Oceanography: A View of the Earth. Prentice-Hall, Englewood Cliffs, 581 pp.

Gross, M. G., and W. B. Cronin. 1978. Dredging and disposal in Chesapeake Bay, 1975-2025. In this volume.

Inman, D. L., and B. M. Brush. 1973. The Coastal Challenge. Science 181:20-32.

International Atomic Energy Agency. 1973. Tracer Techniques in Sediment Transport. Intern. Atomic Energy Agency, Vienna, Tech. Report 145, 234 pp.

Karl, H. A. 1976. Processes influencing transportation and deposition of sediment on the continental shelf, southern California. Ph.D. dissertation, University of Southern California, Los Angeles, 331 pp.

Lorbu. C. J., J. W. Keeley, and J. Hauison. 1975. An overview of the technical aspects of the Corps of Engineers National Dredged Material Research Program. In Estuarine Research, Vol. II, Geology and Engineering. Academic Press, New York, pp. 523-536.

Kolpack. R. L. 1978. Distribution of suspended particulate matter near sewage outfalls in Santa Monica Bay, California. In this volume.
Lauf, G. H., ed. 1967. Estuaries. Am. Assoc. Adv. Sci., Publication 83, 757 pp.
Ludwick, J. C. 1975. Tidal currents, sediment transport and sand banks in Chesapeake Bay entrance, Virginia. In Estuarine Research, Vol. II, Geology and Engineering. Academic Press, New York, pp. 365-380.
Leopold, L. B., M. G. Wolman, and J. P. Miller. 1964. Fluvial Processes in Geomorphology. W. H. Freeman, San Francisco, 523 pp.
McCave, I. N. 1972. Transport and escape of fine-grained sediment from shelf areas. In Shelf Sediment Transport. Dowden, Hutchinson & Ross, Stroudsburg, Pa., pp. 225-248.
Nichols, M. M. 1978. The problem of misplaced sediment. In this volume.
National Research Council. 1976. Disposal in the marine environment. National Academy of Sciences, Washington, D. C., 76 pp.
Oertel, G. F. 1978. Depositional characteristics of sediments at a low energy Ocean Disposal Site, Savannah, Georgia. In this volume.
Rodolfo, K. S. 1964. Suspended sediment in southern California waters. M.S. Thesis, University of Southern California, Los Angeles, 135 pp.
Sternberg, R. W., J. S. Creager, J. Johnson, and W. Glassley. 1978. Stability of dredged material deposited seaward of the Columbia River mouth. In this volume.
Smith, D. D. 1978. Dredged material, ocean disposal, and the regulatory maze. In this volume.
Strakov, N. M. 1967. Principles of Lithogenesis. Consultants Bureau. Oliver and Boyd, Edinburgh, 245 pp.
Shepart, F. P., and N. F. Marshall. 1969. Currents in La Jolla and Scripps Submarine Canyons. Science 165:177-178.
Shepard, F. P., and N. F. Marshall. 1973. Currents along the floors of submarine canyons. Am. Assoc. Petroleum Geologists Bull., 57:244-264.
Wiley, M., ed. 1976. Estuarine Processes, Vol. II, Circulation, sediments, and transfer of material in the estuary. Academic Press, New York, 428 pp.
Williams, S. J. 1978. Geologic effects of ocean dumping on the New York Bight inner shelf. In this volume.

Stability of Dredged Material Deposited Seaward of the Columbia River Mouth

R. W. Sternberg
J. S. Creager
J. Johnson
W. Glassley

ABSTRACT

A two-part study was conducted in a region seaward of the Columbia River where disposal of large quantities of dredged material has occurred over the last several decades. The first part included repeated bathymetric surveys and sampling for distribution and seasonal variations of sediment texture and mineral composition throughout the study area, and especially at designated disposal sites. Near-bottom hydraulic conditions (waves, tides, currents, turbidity) were also measured at several sites and during all seasons. The second part was related to an experiment in which 459,000 m^3 of material dredged from the Columbia River estuary were discharged at a designated site, which was monitored before, during, and after disposal. Sedimentological aspects of the study were to identify and map all deposits of dredged material and to recognize seasonal and long-term changes. The objectives of the hydraulic aspects were to document ambient near-bottom conditions, and their effect on the deposit at the experimental site.

Deposits of dredge material can be identified relative to the surrounding sediments, because they tend to maintain their identity for many years. Sedimentological evidence suggests that the deposits disperse northward at approximately 0.5 km per year. At the experimental disposal site the volume of the bottom deposit was 61% of the total material dumped. Calculations of bedload transport rates, based on seasonal measurements of bottom currents, suggest that 635 m^3 of material (0.2% of the total deposit) spread northward from the site at about 0.4 km per year. This is similar to the rates determined by analysis of sediment mineralogy.

INTRODUCTION

This paper represents the results of a portion of a study of the hydraulic regime and physical nature of bottom sedimentation near the mouth of the Columbia River. Specifically, it recounts the history of approximately 459,000 m^3 (600,000 yds^3) of medium-grained sandy sediment that was dredged from the mouth of the Columbia River and deposited on an open water, shallow continental shelf site with ambient fine-grained sand. The study was supported by the U. S. Army Corps of Engineers, Waterways Experiment Station, Vicksburg; and administered through the U. S. Army Corps of Engineers, Portland District.

From 15 August 1974 to 31 August 1975, a regional study was conducted to:

1. evaluate available bathymetric charts and establish the bottom sediment characteristics of the region seaward of the mouth of the Columbia River, and
2. document current speed and direction, wave activity, meteorological and tidal fluctuations, and characterize the area according to expected levels of sediment movement.

Sampling was carried out during a range of environmental conditions to evaluate the sediment response to specific energy sources.

Following the first years' results, the Corps of Engineers carried out an experimental disposal operation. The disposal experiment began on 9 July 1975 and continued until 26 August 1975, during which time 458,632 m^3 (599,868 yds^3) of material dredged from the Columbia River were released around a marker buoy at the study site. The area chosen for disposal (Site G, Figure 1) had not previously received dredged materials. Hence, the experiment represented an opportunity to investigate the fate of a dredged material deposit in an open water disposal environment. Objectives during this time were:

1. to monitor and evaluate near-bottom hydraulic processes to determine their effect on the materials placed at Site G, and
2. to document spatial and temporal changes in the sediment characteristics (texture and mineralogy) and morphology of the disposal mound following termination of the disposal experiment.

This paper deals with identification of the dredged material deposit resulting from the experimental disposal project and the response of this deposit to the seasonal hydraulic conditions occurring on the inner continental shelf.

BACKGROUND

BATHYMETRY

The Oregon and Washington continental shelf (Figure 1) is relatively narrow (50 km) and steep (5m/km). The shelf gradient is very uniform with the shelf break occurring at 200 m. Several submarine canyons incise the shelf.

The Columbia River is the largest river along the Pacific coast of North America. It contributes 60% of the winter runoff between San Francisco Bay and the Strait of Juan de Fuca, and 90-99% of the spring, summer, and autumn runoff (White, 1976). Figure 2 shows the general bathymetry seaward of the Columbia River as well as Disposal Sites A, B, and G. The Columbia River channel can be seen to pass through the north side of the entrance adjacent to Cape Disappointment, and then turn sharply southwest. The outer tidal delta is skewed to the north, the seaward edge of the flat top of the delta

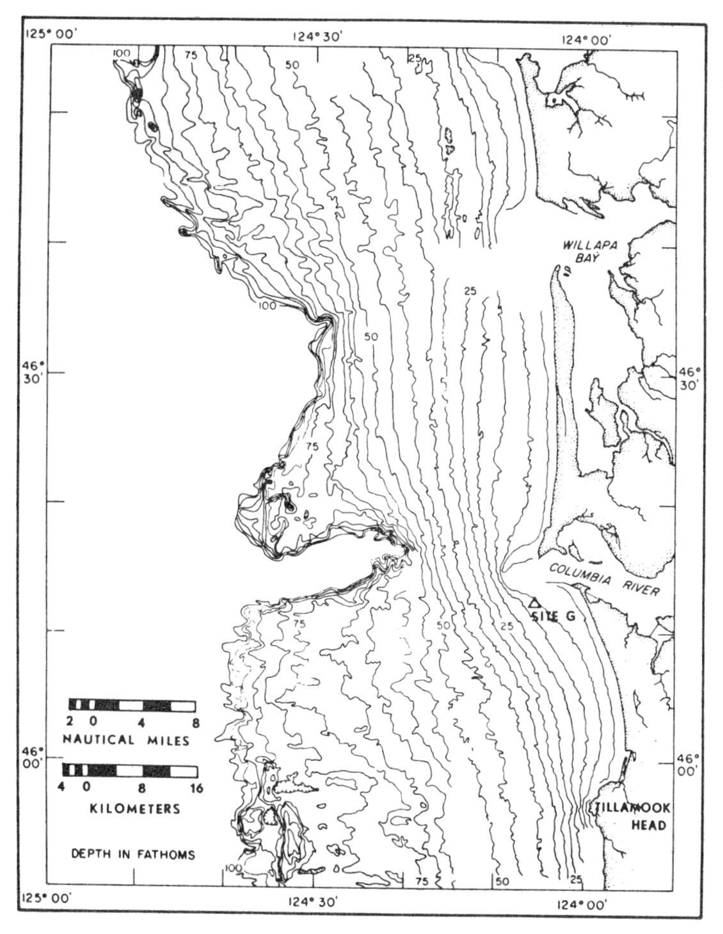

Figure 1
Bathymetry of Washington-Oregon continental shelf. Depths are in fathoms. Experimental disposal Site G is indicated by (▲).

being marked by the 54 ft (16.5 m) isobath. The position of this contour indicates a shoaling of over 90 ft (27.4 m) since 1902, most of which probably occurred since completion of the last jetty in 1917 (Lockett, 1959; Ballard, 1964). The steep north and west slopes of the tidal delta must mark areas of maximum growth and growth rate. Disposal of dredged material has been greater than the rate of removal due to natural processes, producing a secondary bathymetric feature between 66 ft (20 m) and 120 ft (36.6 m) at Site B (Sternberg et al., 1978).

HYDRAULIC REGIME

Circulation on the Washington shelf is dominantly controlled by large-scale weather systems. During the summer the Northeast Pacific Ocean is dominated by the North Pacific High. Winds are

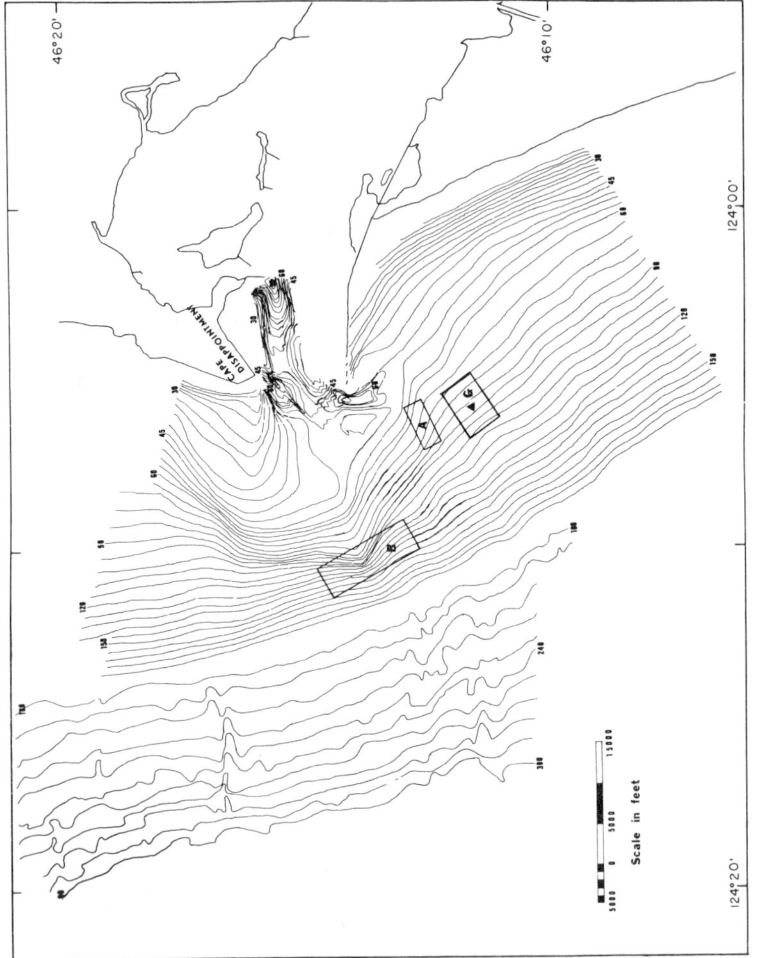

Figure 2
Bathymetry of the continental shelf off the mouth of the Columbia River. Depths are in feet. Disposal sites A, B, and G are shown for reference. The position of the marker buoy is shown by (▲).

characteristically from the north and northwest with speeds of about 5-15 knots (Barnes et al., 1972). At the surface, currents flow to the south and offshore; the bottom currents flow northward. Toward the end of the summer, the North Pacific High weakens and wind patterns are dominated by low pressure systems migrating west to east across the coast. Disturbances occur as individual storms of 3-7 days duration. Winds are from the south to southeast with average speeds of 10-20 knots and maximums of 50-55 knots (Barnes et al., 1972). Northward flowing currents develop at the surface, and coastal downwelling destroys stratification, allowing these currents to reach the bottom. This northward dominance of the bottom current patterns during all seasons has been documented by direct current measurements (summarized by Hopkins, 1971; Smith and Hopkins, 1972; Sternberg and McManus, 1972), the distribution of radionuclides in bottom sediments (Barnes and Gross, 1966), and the movement of seabed drifters (Barnes, Duxbury, and Morse, 1972).

Bottom currents seaward of the Columbia River are the result of numerous interacting phenomena. Sternberg et al. (1978), investigated the effects of tides, river hydraulics, winds, and waves. Each of these components was seen to have its own speed and direction characteristics, some of which also varied seasonally. Their results are summarized below.

The system of tidal currents was the basic velocity component upon which all others were superimposed. Tidal currents generally flowed parallel to the isobaths with flood currents flowing northward and ebb currents southward. The magnitude of the semi-diurnal component of $\overline{U}_{100}$ was estimated at 15-20 cm/sec (Sternberg et al, 1978).

The river flow in the vicinity of the river mouth was extremely complex in space and time. Bottom currents associated with the salt wedge intrusion tended to fluctuate in diurnal, semi-diurnal, and higher frequency (approximately 3-hr) modes, depending on the ratio of river discharge to tidal prism. During maximum river discharge, the bottom velocity record was most complex. During minimum discharges (August to September) the river estuary was more homogeneous, both vertically and horizontally, and river-induced bottom flows were less complex, showing semi-diurnal dominance (Sternberg et al., 1978).

An important correlation was observed between surface wind speed and the non-tidal component of $\overline{U}_{100}$ (Sternberg et al., 1978). It was observed that strong southerly winds, associated with individual storms sweeping the area, generated significant bottom currents which tended to flow parallel to the isobath trend. A net northward coastal flow was seen regardless of wind speed. The exact cause of this could not be determined from available data.

Comparisons of the maximum values of non-tidal $\overline{U}_{100}$ and wind speed during six major storms in the study area suggested that wind-generated bottom currents sufficient to erode sediment (approx. 25 cm/sec) require wind speeds in excess of 13 m/sec. Mean wind speeds of approximately 10 m/sec would generate $\overline{U}_{100}$ of 10 cm/sec which could cause sediment movement when reinforced with the local tidal flows. Analysis of 1975 wind data from the Columbia River Lightship indicated the annual frequency distribution of wind events greater than 10 m/sec (Figure 3). This indicates that a net northerly transport of water and sediment would be expected.

The results of a wave hindcast study for a deep-water site west of the study area have shown that the deep-water characteristics of

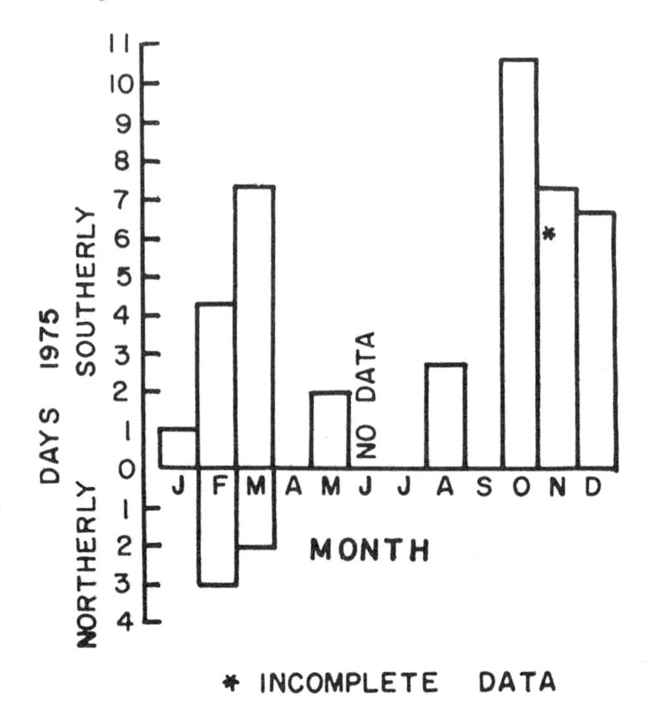

Figure 3
Frequency of time in days during each month of 1975 that mean winds exceeded 10 m/sec blowing from the north or the south. Winds were measured at the Columbia River Lightship and all data were averaged for 25 hr periods prior to analysis.

waves associated with the ten most severe storms occurring between 1950 and 1959 have a significant height (Hs) of 7-9.1 m and a significant period (Ts) of 11-14 sec propogating from the southwest or south-southwest (National Marine Consultants, 1961). The level of wave activity, generally agreed with wind conditions observed along the coast (Sternberg et al., 1978). Hindcast studies over a 3-year period (National Marine Consultants, 1961) estimated that significant waves 3.0 m and greater occur 48 days per year and waves 6.1 m and greater occur about 3 days per year.

SEDIMENTS

An estimate of the sediment load of the Columbia River, based on suspended sediment concentrations, is approximately 10^{10} kg/yr and bedload is assumed to be approximately 10% of the suspended load (Judson and Ritter, 1964; Whetten et al., 1969). The distribution of sediments on the Washington-Oregon shelf is the result of three factors: the relict sediments and topography from the recent rapid rise in sea level, the modern shelf as it would be without the influence of the Columbia River, and the sedimentary environment due to the presence of the river.

Sediment Texture

Sediments on the Washington-Oregon shelf near the Columbia River have dominant modes of 2.75 Ø and 3.25 Ø, except in areas of dredged material disposal where 2.0 Ø and 2.5 Ø mode sediments are prominent (Sternberg et al., 1978). South of the Columbia River, at least as far as Tillamook Head, Oregon, McManus (1972) described the sediments out to a depth of 70 m as nearshore modern sands. Beyond this, the topography was rough and the sediments appeared to be relict. Kulm et al. (1975) saw parallel zones of progressively finer sands offshore south of Tillamook Head. Generally, silt and clay increased offshore in classic fashion.

North of the Columbia River, modern sandy sediments derived from the river extended out to a depth of 50-60 m (Harman, 1972; McManus, 1972; Smith and Hopkins, 1972). Mixed silt and sand occurred between the 60 m and 100 m isobaths which Smith and Hopkins (1972) suggested was modern Columbia River silt overlying or mixed with relict fine sand (3.0 Ø). The silty sediments from the Columbia River are transported in suspension to the north-northwest during storms (Smith and Hopkins,1972). This same northward dispersal pattern has been documented by Barnes and Gross (1966) in a study of radionuclides from the Columbia River. During non-storm periods, temporary deposition of silty materials in the mid-shelf region between the Columbia River and Willapa Canyon may be significant. Significant quantities of silt and clay occur west of Disposal Site B and from there north-northwest along the 150 ft (45.7 m) to 180 ft (54.9 m) isobaths, which represent the natural position of deposition of silt and clay from the Columbia River (Nittrouer et al., in press). These sizes are not present in the disposed material.

Sediment Mineralogy

Columbia River bottom sediments are derived primarily from mechanical weathering of the andesitic volcanic material of the Cascade Mountains (Whetten et al., 1969 ; Sternberg et al., 1978). Residence time of these grains in the river appears to be relatively short.

Sediments found on the shelf, beyond the tidal delta and Site B, have a low abundance of fresh plagioclase. In comparison with Columbia River sediments, ambient shelf sediments have less orthopyroxene and andesitic lithic fragments, but are richer in altered lithic fragments, opaque minerals (especially magnetite), clinopyroxenes, altered plagioclase and basaltic lithic fragments.

METHODS

BATHYMETRY

Five bathymetric surveys of Site G were conducted by the Corps of Engineers during July, August, and September 1975 and in February 1976 using an automated hydrographic survey system, based on an Atlas precision echo-sounder and Del Norte Trisponder navigation. Sub-bottom reflection and side-scan sonar were also used on the 7-8 July and 25-26 August surveys. A summary of cruise data is presented in Table 1. The Del Norte navigation system is capable of high resolution positioning (±3 m), but our inability to resolve sea-level fluctuations during all surveys leads to a depth accuracy of ±0.5 ft.

Table 1
Summary of Bathymetric Cruise Data for Site G

Survey Dates	Trackline Orientation	Trackline Spacing
2-3 July 1975	N-S	250 ft (76 m)
7-8 July 1975	E-W	500 ft (152 m)
25-26 August 1975	E-W	250 ft (76 m)
2-3 September 1975	N-S	250 ft (76 m)
		200 ft (61 m) near buoy
2 February 1976	E-W	500 ft (152 m)

Bathymetric maps were constructed for each of the five available cruises at Site G using a 1 ft (0.3 m) interval. Side-scan sonar and subbottom reflection profiles were used to identify bathymetric irregularities due to sand waves. These irregularities were smoothed out of the records because of the difficulty their migration would cause in monitoring the bathymetric changes due to dredged material disposal. The sand waves occurred in the northeastern corner of Site G, well removed from the area affected by the test disposal.

HYDRAULIC REGIME

Instrumentation and Data Collection

Measurements of flow conditions in the bottom boundary layer (within 2 m of the seabed) were carried out during all seasons, using an instrumented tripod that freely descends from the sea surface. It can remain on the bottom for up to 30 days and:

1. continuously measure speed and direction 1 m off the bed with a Savonius rotor current meter and direction vane,
2. measure differential pressure 2 m off the seabed to estimate tides and pressure fluctuations from surface-wave motion,
3. activate a beam transmissometer located 1.4 m off the seabed each 30 min in order to estimate the concentration of suspended particulate matter, and
4. photograph the sea floor each 30 min.

All data are recorded internally and the tripod is retrieved by acoustic command or after a preset elapsed time. A complete description of the instrumented tripod is given by Sternberg, Morrison, and Trimble (1973) and Sternberg (unpublished manuscript). A station location chart is shown in Figure 4 and a summary of pertinent station information is given in Table 2. Complete data summaries can be found in Sternberg et al. (1978).

Analytical Procedure

Bottom currents and winds. Bottom current measurements of speed and direction were averaged over half-hour periods and are characterized by a mean velocity 1 m off the bed ($\overline{U}_{100}$). The direction vane

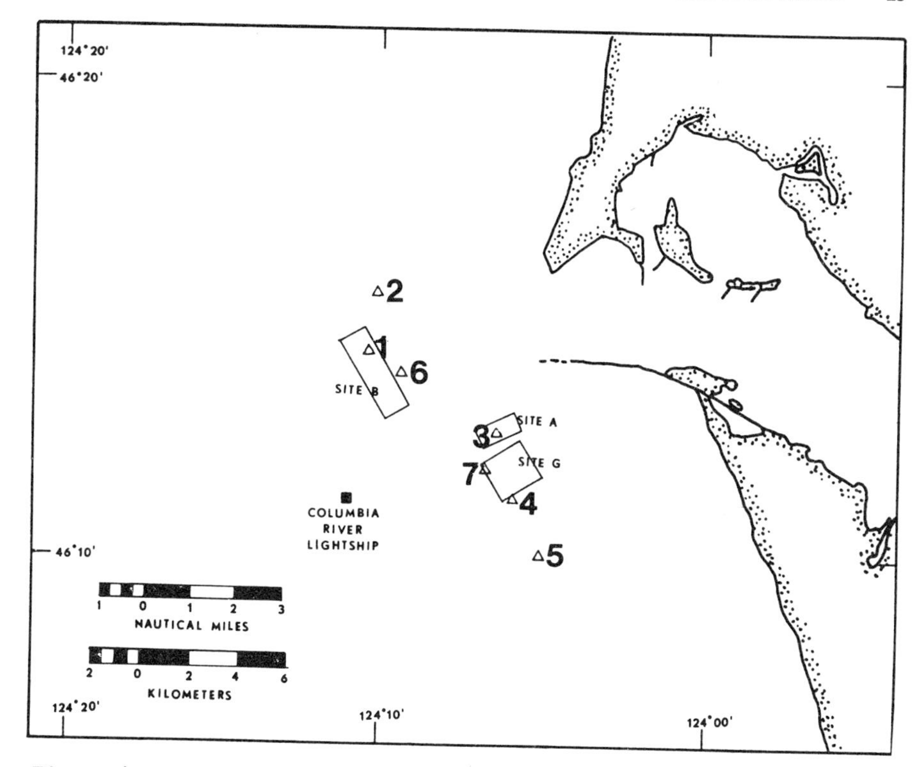

Figure 4
Location of the Columbia River Lightship (■) and instrumented tripod stations (△). Disposal Sites A, B, and G are shown for reference.

output was converted to true bearing. Current speed and direction data were plotted in four ways:

1. as individual plots of speed and direction versus time,
2. as eastward and northward components versus time,
3. as progressive vector diagrams to reveal net movements of bottom water over the total sampling period, and
4. as 25-hr time averages to remove most of the tidal component and higher frequency variations from the record, permitting the investigation of other types of currents (e.g., wind-generated currents), termed the "nontidal" component of flow. Wind records were subjected to similar analytical procedure.

Pressure. Pressure measurements were made with a potentiometer-type differential pressure transducer which senses up to ± 5 psi pressure changes while on the bottom. The output from the pressure was recorded without filtering and appeared as a composite of the tidal and higher frequency pressure fluctuations. The pressure record was visually averaged on half-hourly intervals to give the mean pressure due to tides. The high frequency fluctuations are directly related to surface wave activity. These fluctuations are obtained by measuring the width of the envelope containing the scatter of pressure data

Table 2
Instrumented Tripod Deployment Summary

Station	Deployment Times, Dates	Latitude/ Longitude	Site	Depth ft (m)	Total Operating Hours	Data Recovery (Hours)				
						Speed	Direction	Photo	Press	Trans.
1	1858 4-12-75 to 0810 5-6-75	46°12.28' 124°10.3'	B	98(30)	565	565	565	565	565	420
2	2020 4-12 75 to 0910 5-6-75	46°15.25' 124°10.56'	Site B Control	102(31)	565	565	565	565	565	No Instru-ment
3	1918 6-15-75 to 1015 7-8-75	46°12.5' 124°06.5'	A	79(24)	543	543	543	543	543	543
4	1241 8-19-75 to 0815 9-12-75	46°11.25' 124°06.5'	G	94(28)	570	570	570	570	570	570
5	1213 8-19-75 to 0902 9-12-75	46°10.0' 124°05.00'	Site G Control	103(31)	572	572	572	50	572	572
6	0837 12-12-75 to 1405 1-6-76	46°13.96' 124°09.97'	B	79(24)	605	605	605	351	605	333
7	0900 12-12-75 (instrument not recovered)	46°11.53' 124°06.1'	G							
			Total Hours			3166	3420	2644	3420	2438
			Total Days			131.9	142.5	110.2	142.5	101.6

points about the tidal signal (hence, the tidal signal is removed from the pressure fluctuations measurement).

Turbidity. The output of a Montedoro-Whitney Inc. transmissometer (0.3-m folded light path) was recorded on an arbitrary scale of 0-10 which represented a variation from low to high light attenuation. For calibration purposes, three General Oceanic Inc. suspended sediment sampling bags were mounted on each tripod. One bag was activated by the beam transmissometer output at specific values of light attenuation equivalent to each of low, medium, and high concentrations of suspended sediment. The water samples were filtered, dried, and weighed and the results used to construct an empirical calibration curve relating the transmissometer output to suspended sediment concentration (mg/l) 1.4 m above the seabed.

SEDIMENT ANALYSIS

Bottom sediment samples were collected by the University of Washington and Oregon State University (see Table 3). For samples collected using radar and sextant fixes, station locations are believed accurate to 0.9 km (0.5 n mi). Del Norte Trisponder system locations are estimated accurate to ± 250 ft (76.2 m).

Samples collected by the University of Washington, obtained using a Shipek grab sampler, were placed in vinyl bags, sealed, and stored cold until analysis. In cases where stratification was evident in the sample, an effort was made to collect and analyze material from each layer. Oregon State University used a Smith-MacIntyre bottom sampler. Several grabs were made at each station, one of the samples being preserved for the University of Washington.

Texture

Textural analyses consisted of homogenizing and then successively quartering the bottom sample until about 20 to 40 g of sediment remained. This subsample was wet sieved through a 4-Ø (0.625 mm) screen into a 1000 ml cylinder. The sediment remaining on the 4-Ø screen was dried and sieved (10 min) to 0.25Ø fractions. The pan fraction was added to the cylinder and a pipette analysis was used to determine the size distribution of the silts (at 0.5-Ø intervals) and clays (at 1.0-Ø intervals) (Krumbein and Pettijohn, 1938).

The data were subjected to a Q-mode factor analysis (Imbrie and Van Andel, 1964) to assist in determining the relationships among samples. From this, the important size classes become quite obvious. The relationships among variables (size classes) could have been determined using R-mode factor analysis, but because the general nature of the important size classes was already known from previous work (McManus, 1972; Smith and Hopkins, 1972; Kulm et al., 1975) this technique was not applied. The Q-mode factor analysis with final oblique rotation was run using the 29 grain-size classes (in Ø notation) as variables. Seven factors, which explained 99% of the variability among samples, were selected. The large number of factors was retained in order to differentiate the apparently important relationships among the fine to very fine sand fractions. Trial runs distinguished extremal samples for each of the chosen factors. These are the samples with grain size distributions most similar to those

Table 3
Bottom Sediment Sample Summary

Dates	Number of Stations	Number of Grabs UW	OSU	Navigational Control*
28-30 Sept 74	15	15		RB
30 Sept-30 Oct 74	153	153		RB
16-17 Nov 74	88			DN
4-8 Dec 74	58		58	RL
11-12 Dec 74	86	84		DN
20-25 Jan 75	49		49	DN, RL
19-21 Apr 75	26		26	DN
23-27 July 75	37		37	DN
20 Aug 75	27		27	DN
12-15 Sept 75	72		72	DN
21-22 Oct 75	29		29	DN
8-9 July 75	51	51		DN
11-12- Dec 75	97	95		DN
5-9 Jan 76	37		37	DN
19-30 Apr 76	12		12	DN
7-8 June 76	12		12	DN

*RB-Range and Bearing; DN - Del Norte Trisponder system; RL - Range and Loran A.

of the factors. From coarse to fine, the factors and their representative extremal samples can be characterized by their modes: 1.75 Ø to 2.25 Ø (Factor 1), 2.5 Ø (Factor 2), 2.75 Ø to 3.0 Ø (Factor 3), 3.25 Ø (Factor 4), 3.75 Ø (Factor 5), 4.5 Ø (coarse silt, Factor 6), and what is herein termed 12 Ø (fine silt and clay, Factor 7).

Mineralogy

A split of the <4 Ø (>0.0625 mm) sedimentary material recovered during size analysis was impregnated with Fiberlay-epoxy base resin and used to make a thin section (Sternberg et al., 1978). The section was etched using hydrofluoric acid and then stained with sodium cobaltinitrate solution to aid in determining potassium feldspar content. A glass cover slip was mounted to the slide using Canada balsam as the mounting medium. A minimum of 300 points per slide were counted and a Mineral Index (MI) for each sample was calculated using the following formula:

MI = (% Fresh Plagioclase + % Potassium Feldspar)/ (% Altered Lithic Fragments + % Altered Plagioclase + % Opaques).

A split of the same <4 Ø (>0.0625 mm) sedimentary matrial that was used in the point count analysis was also used in magnetic separations. A Frantz magnetic separator was used at a back tilt angle of 25° and a current of 1.7 amps. The front face of the magnet was covered with paper on which the magnetic fraction was retained. A 10 to 20 gram sample was used in this analysis. The magnetic fraction and the non-magnetic fraction were weighed and the ratio of the weights was used to obtain the Magnetic Ratio (MR), defined as:

MR = (weight of non-magnetic fraction)/ (weight of magnetic fraction).

Electron microscopy, x-ray fluorescence, and x-ray diffraction analyses were also performed on a few samples during the early stages of mineralogic characterization of the sediments.

RESULTS

The site for the experimental disposal was chosen to be south of the tidal delta. Sharp contrasts, both textural and mineralogical, between Columbia River and ambient shelf sediments facilitated identification and monitoring of the disposed dredged material (Sternberg et al., 1978). Between 9 July and 27 August 1975, approximately 459,000 m^3 (600,000 yds^3) of material were released in the vicinity of a marker buoy at Site G. Dredges usually circled the buoy at a distance of about 213 m (700 ft) releasing the dredged material on the southern side. Sea and weather conditions resulted in dumping north of the buoy about 20% of the time (Charles Galloway, personal communication).

BATHYMETRY

Prior to the disposal experiment, the isobaths at Site G were regular and shoaled to the northeast (Figure 5a). Post-disposal bathymetry (Fig. 5b) indicates that the measurable effect of the disposal mound was confined to a radius of 460 m about the marker buoy with accumulation dominantly to the south and west of the buoy. The measurable sediment deposit immediately following the experiment had a volume of 278,298 m^3 (364,000 yds^3) (Fig. 5c) which represents 61% of the total quantity of material released over this site. A later survey in February 1976 indicated that prominent highs at the disposal site had been smoothed out with 71% of the volume of the original disposal mound still remaining as a measurable bathymetric feature. The resolution of this method ($\pm$ 0.5 ft) suggests that as the deposit continues to spread, covering a greater area, further bathymetric monitoring of this site will give only qualitative results.

SEDIMENT TEXTURE

The textural composition of the extremal samples chosen by factor analysis is given in Table 4. Only Factor 1, 2, 3 and 4 sediments (1.75 ϕ to 2.25 ϕ, 2.5 ϕ, 2.75 ϕ to 3.0 ϕ, and 3.25 ϕ modes, respectively) occurred with loadings >0.4 before, during, or after the experimental disposal at Site G. Distributions of these factors prior to 9 July 1975 are shown in Figure 6. Before the disposal experiment, Site G was characterized by a dominance of Factor 3 and 4 sediments. Loadings of Factor 3 sediments were high in the entire region around the marker buoy, especially to the west and southwest. High loadings of Factor 4 sediment occurred where Factor 3 sediment was less

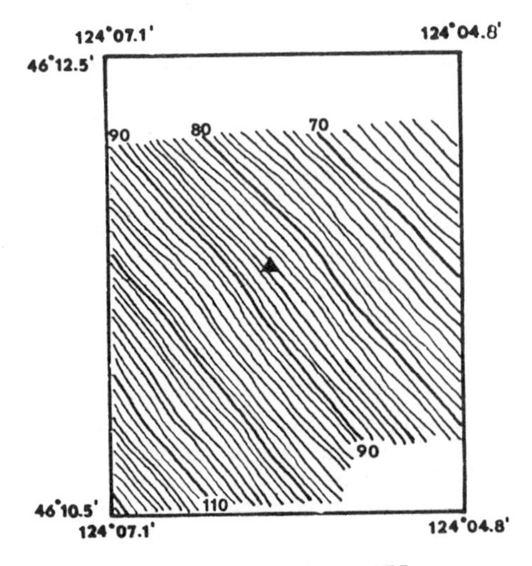

a. 2-3 JULY 1975

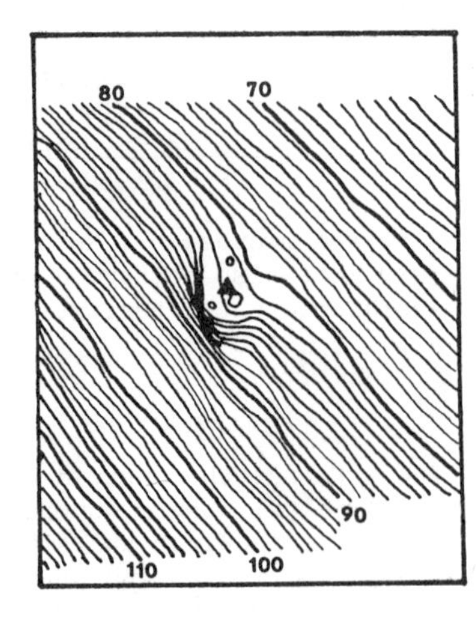

b. 2-3 SEPTEMBER 1975

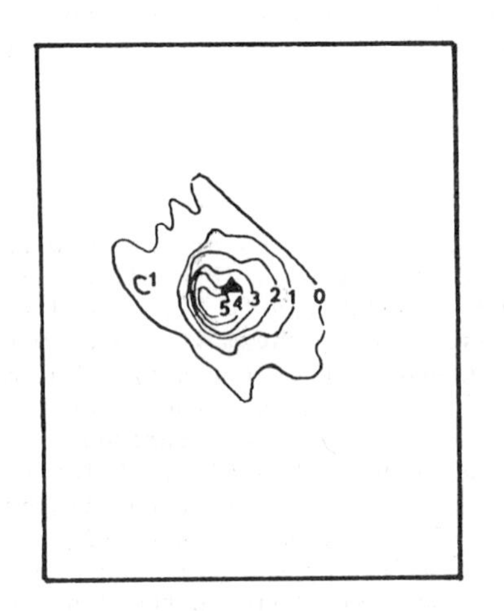

c. ISOPACH

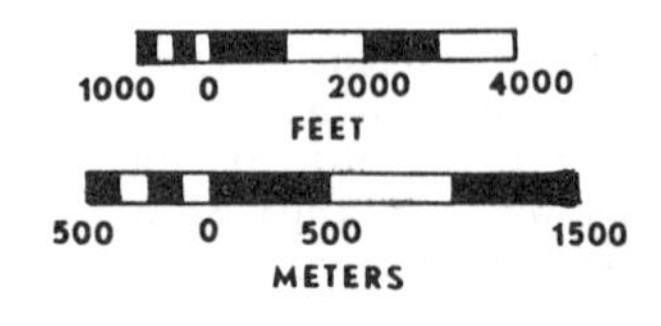

Figure 5
Bathymetry at Disposal Site G; (a) prior to the disposal experiment (2-3 July 75) and (b) immediately after completion of the disposal operations (2-3 September 75). An isopach map of the thickness of the material above the 2-3 July contours is shown in (c). The marker buoy is shown by (▲).

prominent, notably in the east and southeast. Factor 2 sediment occurred as small patches of moderate loading throughout the area. No Factor 1 sediment was present prior to the disposal.

Figures 7 to 10 indicate the distributions of Factor 1 to 4 sediments during the six post-disposal surveys. In September, immediately after conclusion of the disposal of dredged material, Factor 1

Table 4
Textural Data for Extremal Samples of Each Factor (Rounded to whole percent for each class)

Factor	1.*00	1.25	1.50	1.75	2.00	2.25	2.50	2.75	3.00	3.25	3.50	3.75	4.00	4.50	5.00	5.50	6.00	6.50	7.00	7.50	8.00	9.00	10.00	11.00	≥12.00
1	1	1	6	17	32	19	11	6	2	1															1
2			1	4	13	16	47	9	3	2	1	1													1
3					1	5	15	32	24	11	2	1		1	1										3
4					1	1	2	8	15	32	18	10	4	5	1							1		1	1
5			1	3	8	8	7	6	5	8	5	28	14	3											1
6								1	1	3	5	5	4	42	13	7	3	2	2	1	1	1	2	1	4
7			1	1	1	1	1	2	2	3	2	1	1	2	9	8	5	8	8	5	6	7	7	6	12

* Size in phi units

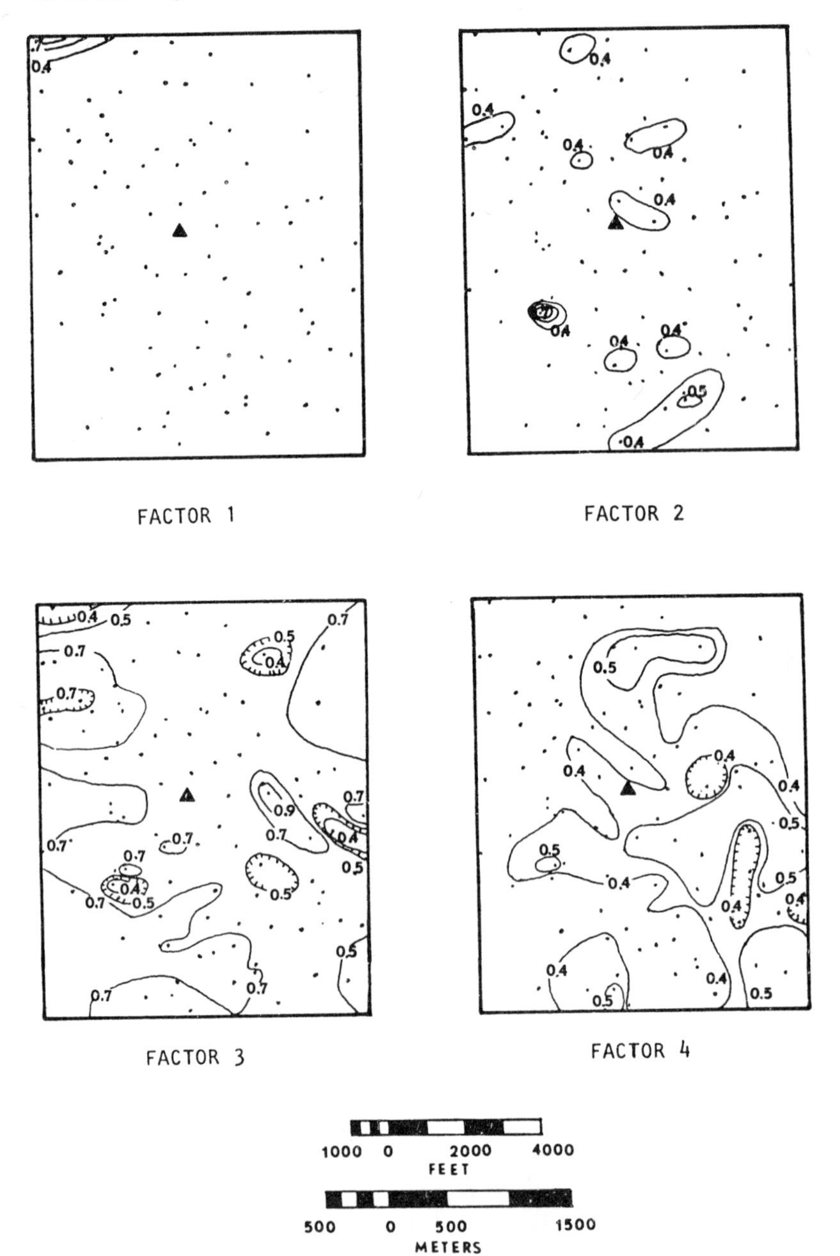

Figure 6
Distribution of Factor 1, 2, 3 and 4 sediments at Site G prior to the experimental disposal operations. Isolines are drawn for factor loadings of 0.4, 0.5, 0.7, and 0.9. Data are from samples collected in September, October, November, and December 1974, and January and April 1975. Sample locations are shown by (•) and the marker buoy is shown by (▲). The area shown is the same as in Figure 5.

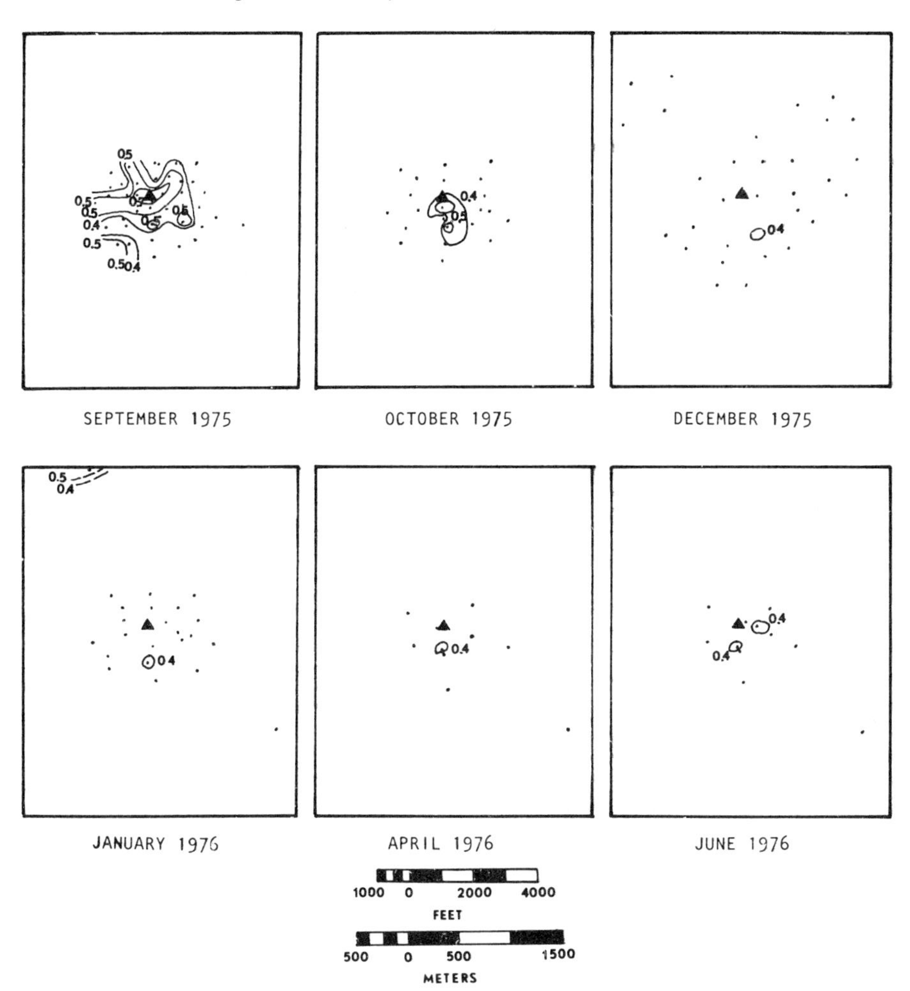

Figure 7
Distribution of Factor 1 (2.0 Ø mode) sediments at Site G following the disposal operations. Isolines are drawn for factor loadings of 0.4, 0.5, 0.7 and 0.9. Sample locations are shown by (•) and the marker buoy is shown by (▲). The area is the same as in Figure 5.

sediments dominated near the marker buoy. They were generally confined to distances less than 1400 ft (430 m) from the buoy, the largest concentrations occurring along the curved disposal arc south of the buoy. Factor 2 sediments occurred in large concentration as a "halo" around Factor 1 sediments, presumably reflecting a grain size sorting associated with the currents produced by impact of the disposal sediment with the bottom. Factor 3 sediments occurred as a narrow "halo" around Factor 2 sediments, and Factor 4 sediments appeared only in the southeasternmost corner.

October results indicated that the areal extent of Factor 1 sediment had diminished to a narrow north-south band located south of the marker buoy. Factor 2 sediment still occurred as a halo around

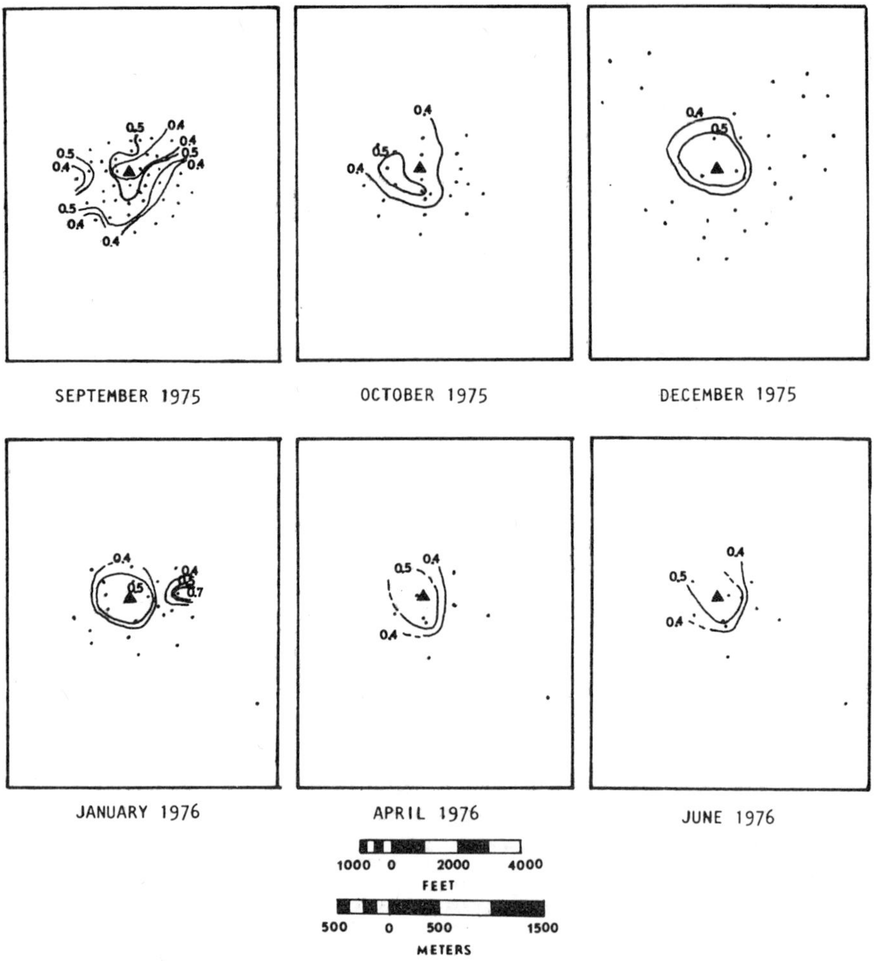

Figure 8
Distribution of Factor 2 (2.5 Ø mode) sediments at Site G following the disposal operations. Isolines are drawn for factor loadings of 0.4, 0.5, 0.7, and 0.9. Sample locations are shown by (•) and the marker buoy is shown by (▲). The area shown is the same as in Figure 5.

Factor 1 sediment, but loadings were smaller and its extent was more restricted. Factor 3 sediment occurred over a larger area, and much closer to the marker buoy, than in September, and Factor 4 sediment had reappeared in the southwest corner.

Areal restriction of Factor 1 sediment continued through December, when it was confined to a small area south of the marker buoy. Factor 2 and 3 sediments occurred in moderate loadings around the buoy. The limited distribution of samples made it difficult to determine that these factors were returning to the patchy distribution that had characterized them prior to the experiment. Factor 4 sedi-

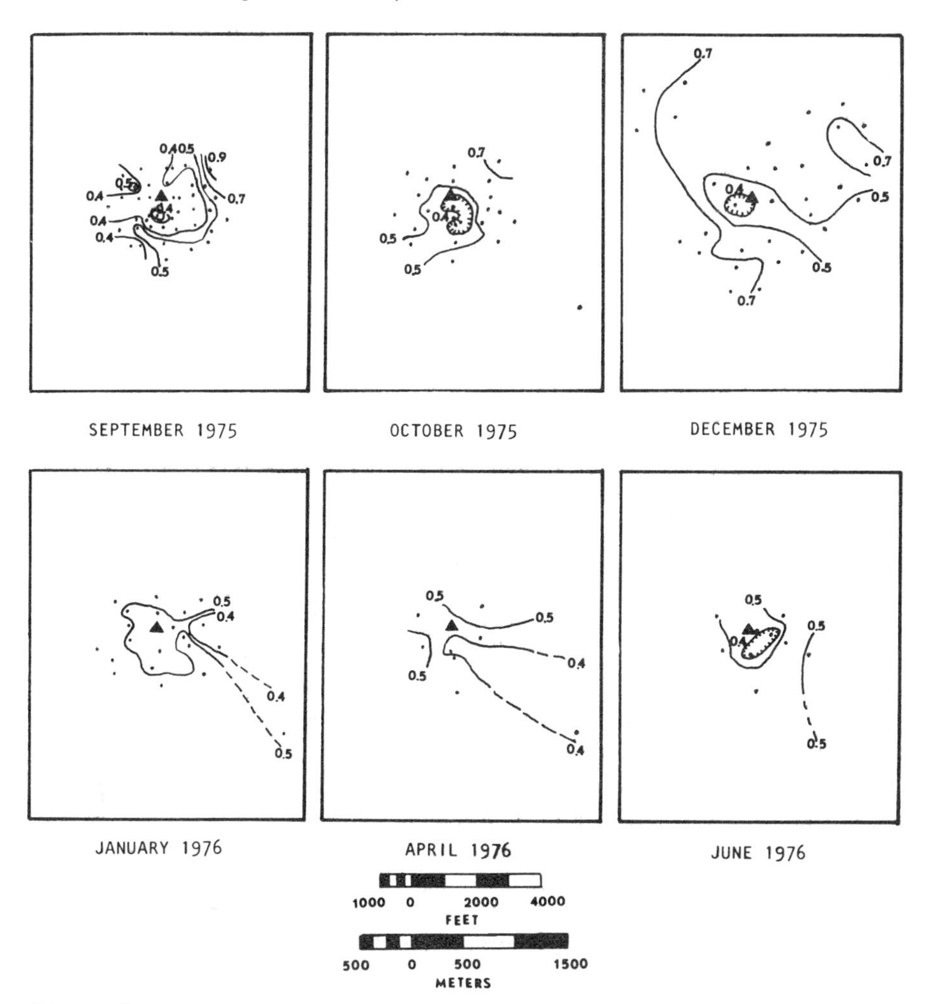

Figure 9
Distribution of Factor 3 (2.75 Ø mode) sediments at Site G following the disposal operations. Isolines are drawn for factor loadings of 0.4, 0.5, 0.7, and 0.9. Sample locations are shown by (•) and the marker buoy is shown by (▲). The area shown is the same as in Figure 5.

ment showed negligible recovery. These same trends were seen in the January, April, and July 1976 surveys. Factor 1 sediment persisted in a small area south of the buoy. Factor 2 and 3 sediments appeared to have recovered their predisposal distributions and Factor 4 sediment showed very little recovery.

The small number of samples collected, especially in later surveys, made isolining of factor loadings very subjective, therefore, it is useful to look at temporal changes in sediment texture by comparison of samples collected at approximately the same positions at different times. Figure 11 shows the location of groups of samples collected within 0.1 n mi of each other. Definite temporal changes

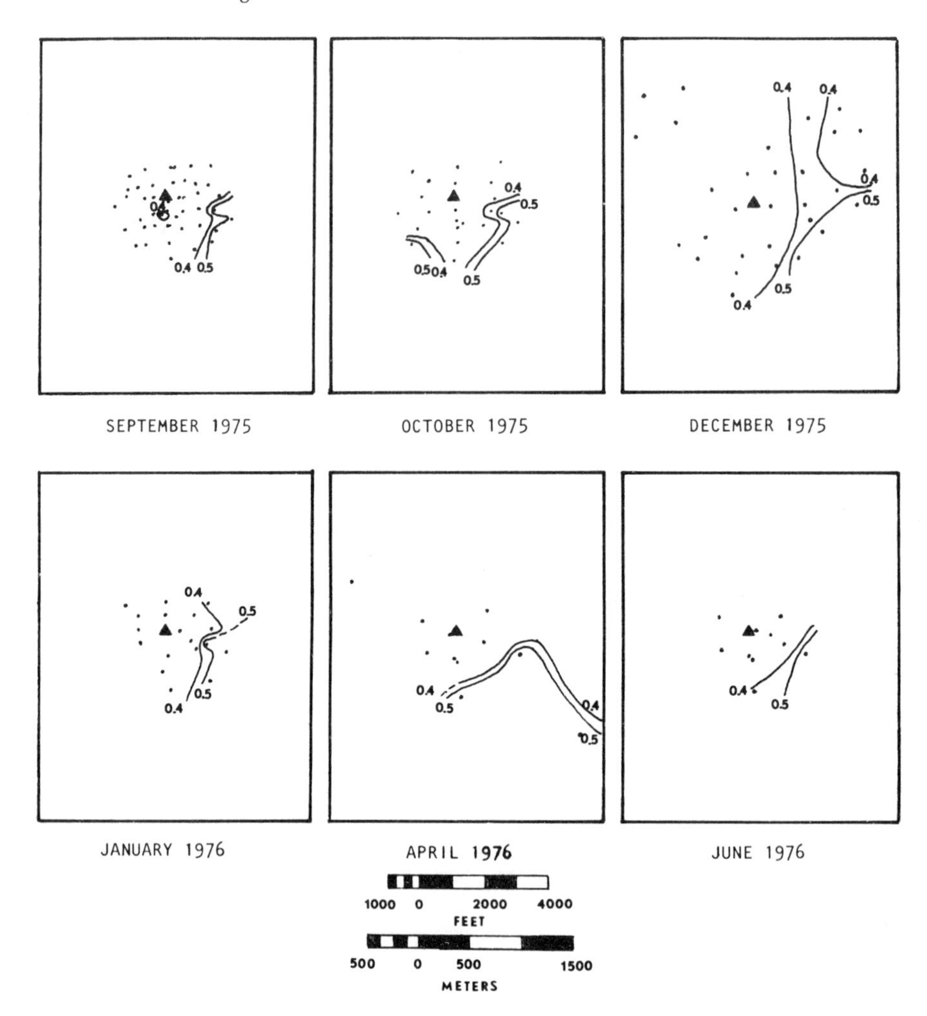

Figure 10
Distribution of Factor 4 (3.25 Ø mode) sediments at Site G following the disposal operations. Isolines are drawn for factor loadings of 0.4, 0.5, 0.7, and 0.9. Sample locations are shown by (•) and the marker buoy is shown by (▲). The area shown is the same as in Figure 5.

in the sediment size and factor loadings can be discerned at these locations.

Sample groups immediately surrounding the marker buoy (72, 75, 76, 78, 79, 80 and 84) showed little to no recovery to ambient fine-grained sands at the end of 9 months (June 1976). Groups 76, 77, 78, and to some extent 72, even showed some coarsening of sediments during the winter months. Location of these samples on and near the central bathymetric high of the disposal mound probably facilitated removal of any existing fine-grained sediments and prevented surrounding sediments from covering the disposed materials.

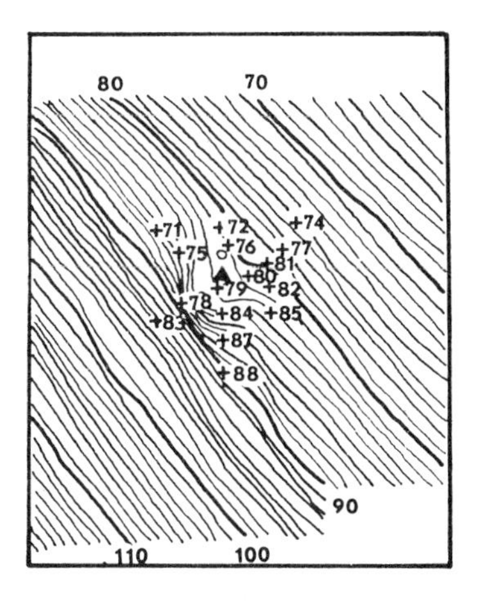

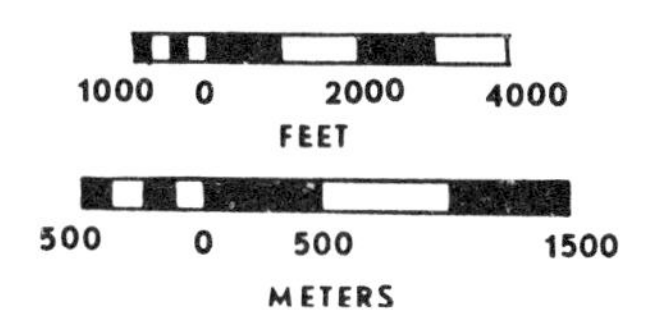

Figure 11
Location of groups of samples collected within 0.1 nautical miles of each other. Bathymetry (in feet) is shown for reference. The marker buoy is shown by (▲). The area shown is the same as in Figure 5.

Groups of samples further from the marker buoy showed more recovery to predisposal conditions. On the west and southwest sides (71, 83 and 87) slight to moderate recovery was indicated by June 1976. Samples on the east side (81, 82 and 85) showed more definite trends toward recovery of predisposal conditions. Migration of the surrounding shelf sediments northwest along the isobath trend would account for the fastest recovery having occurred on the southeastern side. Samples at 74 in the northeast corner, and at 88 in the south showed complete recovery to pre-disposal conditions.

SEDIMENT MINERALOGY

Generally, an initial offshore increase in Magnetic Index (MI) of the sediments is seen as the fraction of magnetite decreases, but beyond the mid-shelf MI decreases with the increasing concentration of altered lithic fragments and altered plagioclase. Values of MI immediately seaward of the river mouth and at Site B are high, reflecting the mineralogy of the Columbia River sediment placed there by dredging activities. Isolines of magnetic ratio (MR) indicate an offshore increase in MR due to a rapid offshore decrease of magnetite and increase in altered lithic fragments and altered plagioclase.

A rough estimate of the migration rate of the dredged materials at Site B can be made based on the MI isolines. The distance from Site B to the most westerly corner of the 1.0 MI plume (approximate

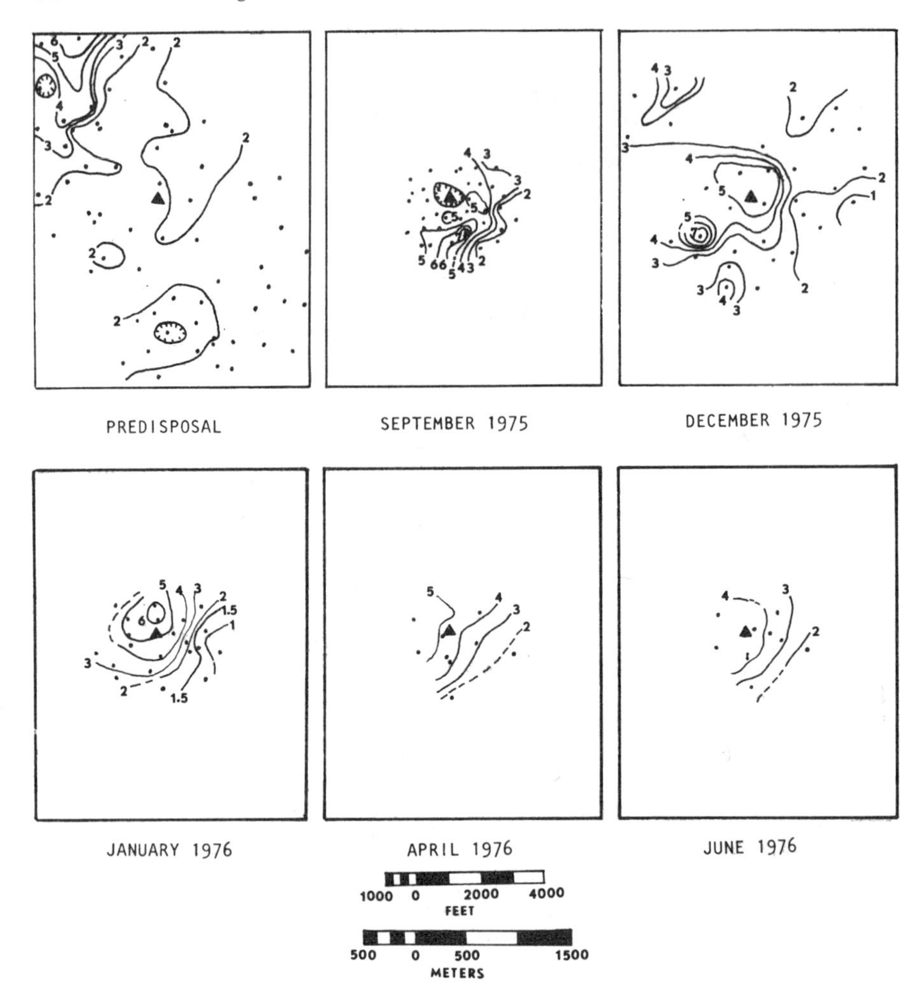

Figure 12
Isolines of Magnetic Ratio (MR) at Site G. Sample locations are shown by (•) and the marker buoy is shown by (▲). The area shown is the same as in Figure 5.

background level) which extends north-northwest away from Site B is 10.2 km. Records of the Portland District Corps of Engineers show that the major disposal of dredged material at Site B was carried out during the past 20 years (Sternberg et al., 1978). Assuming that the observed MI plume results from the selective migration of dredged material away from Site B, the annual migration rate for sediments at the head of the plume is 500 m/yr.

Isolines of magnetic ratios (MR) at Site G are shown in Figure 12 for the surveys in December 1974, September and December 1975, and January, April, and June 1976. It can be seen that prior to the experimental disposal, values of MR generally varied from 1 to 2, especially in the vicinity of the marker buoy. Higher values, from 3 to 5, occurred in the far northwestern corner of the area, close to the region probably influenced by former Disposal Area A (Fig. 2). In September, after conclusion of the disposal operation, MR values ranged generally from 4 to 5, with highs of 6 occurring south and east of the buoy. These high MR values reflect the Columbia River origin of the sediments. The far eastern side of the area shows MR values of 2 to 3, delimiting the extent of disposal-related perturbation in that direction.

By December the high MR values of 5 to 6 were smoothed out and peripheral values had decreased to 1.5 to 3. January isolines showed a trend of decreasing MR, in all directions, away from the marker buoy. April and June MR isolines indicate a continuance of this trend, with a very slight decrease in the areal extent of the high MR contours.

Examination of MR values at the sample groups shown in Figure 11 gives more specific information than the MR isolines. Temporal changes in MR corroborate the sediment textural changes. Samples close to the marker buoy (72, 75, 76, 78, 79 and 84) experienced little to no recovery; in fact, the MR increased from a value of 5 in December to 6 in January. This is either due to transport of coarse non-magnetic fractions into the area, or to transport of finer magnetic fractions away from the area. Size analysis results at these stations show no significant changes in sediment size, so both processes may have occurred. Sample groups 71, 77, 80, 82, and 87 show slight to moderate recovery. Good to complete recovery is seen at groups 74, 81, 85, and 88.

SEDIMENT TRANSPORT

The sediment characteristics and bottom current measurements were analyzed to reveal the bottom sediment transport at Site G and to predict the future stability of the disposal deposit. The primary considerations were the mode of transport, frequency of grain movement, and mass transport of sediment away from the disposal deposit both as suspended load and as bedload.

Mode of Transport

Sedimentary particles may move by rolling or sliding along the bottom within several grain diameters of the bed (bedload transport), or may be carried up into the flow and maintained by turbulent forces and be transported as suspended load. These two modes of transport are endmember conditions and are separated by a transition region. The mobility of sedimentary particles is strongly dependent on the mode of transport.

Competency curves used to predict the threshold conditions for grain movement are published in the geological literature and to a certain degree have been investigated in the shallow marine environment (Sternberg, 1971). These curves relate the threshold velocity ($\overline{U}_{100\,t}$) or the threshold drag velocity (U_{*T}) to the mean grain size of the bottom sediment. Using the curves evaluated by Sternberg (1971), the threshold velocity for erosion of each of the modal grain sizes characterizing the bottom sediments in the study area has been determined and is shown in Table 5.

As discussed by Smith and Hopkins (1972), the criterion for suspended load transport once the critical value for erosion has been exceeded is:

$$\frac{Ws}{kU_{*t}} \leq .8 \qquad (1)$$

where Ws is the settling velocity for sediment, k is von Karman's constant (=0.4), and U_{*t} is the friction velocity as defined as $(\tau_0/\rho)^{\frac{1}{2}}$ with τ_0 the boundary shear stress and ρ the fluid density. The relation between U_{*t} and $\overline{U}_{100_t}$ as given by Sternberg (1972) is:

$$U_{*t}^2 = 3 \times 10^{-3} \overline{U}_{100_t} \qquad (2)$$

Table 5
Required Bottom Currents ($\overline{U}_{100_t}$) for Bedload and Suspended Load Transport and the Percent of Time that these Conditions were Exceeded at the Tripod Stations

Factor	1		2	3	4	5	6	7
Modal Size (ϕ)	2.00 - 2.25		2.5	2.75	3.25	3.75	4.5	6.0
Modal Size (mm)	0.25 - 0.21		0.18	0.15	0.11	0.071	0.044	0.016
Bedload $\overline{U}_{100_t}$*	30 - 29		29	29	29	29	29	29
Suspended $\overline{U}_{100_t}$	156	118	90	65	34	29	29	29
Station 1								
Threshold Exceeded	8		8	8	8	8	8	8
Suspended Load	0		0	0.2	5	8	8	8
Station 2								
Threshold Exceeded	11		11	11	11	11	11	11
Suspended Load	0		0	0.1	6	11	11	11
Station 3								
Threshold Exceeded	0		0	0	0	0	0	0
Suspended Load	0		0	0	0	0	0	0
Station 4								
Threshold Exceeded	3		3	3	3	3	3	3
Suspended Load	0		0	0	2	3	3	3
Station 5								
Threshold Exceeded	3		3	3	3	3	3	3
Suspended Load	0		0	0	2	3	3	3
Station 6								
Threshold Exceeded	66		66	66	66	66	66	66
Suspended Load	0		0	4	43	66	66	66

*Assumes a low degree of consolidation for the fine sedimentation.

Using equations (1) and (2), the critical values of $\overline{U}_{100_t}$ that must be exceeded to produce suspended load transport for the factors have been calculated and are given in Table 5. The percentage of time that $\overline{U}_{100}$ exceeded the threshold of erosion and suspended load values is given in Table 5.

This analysis shows that the coarser sediments in the study area do not undergo the suspended mode of transport. For example, even during the strongest currents observed in the area (80.1 cm/sec), Factor 1 and 2 sediments would not be suspended. The size limit for full suspension appears to be Factor 3 (2.75 Ø) which was suspended during 4% of the winter experiment (Station 6) and only for a small percentage of the time during the spring experiment (Stations 1 and 2). The finer size sediments (Factors 5, 6, 7) will normally be carried in suspension once threshold conditions are surpassed, hence the percentages under "suspended load" are the same as for the "threshold exceeded" values. Factor 4 is transported more frequently in suspension than as bedload and hence is also quite mobile.

Frequency of Movement

The frequency of motion in the study area is related to the magnitude of the bottom currents over time and the threshold conditions for the bottom sediment. Wave motions in the study area also tend to suspend bottom sediment and are discussed in a later section of this paper. Table 5 shows that bottom-current induced grain motion did not occur during some summer months (i.e., Station 3 in the June-July period), but occurred as frequently as 20 days/mo (66%) in the December-January period. Sediment movement during the transition months (April-May and August-Sept) may be expected to vary greatly from year to year depending on the storm conditions. During 1975 the frequency of time that threshold conditions were exceeded averaged 3 days/mo (10%) for the April-May period (Stations 1 and 2) and 1 day/mo (3%) for the August-September period (Stations 4 and 5).

In general it would be expected that the frequency of motion estimates from bottom currents would compare with the severity and frequency of storms as shown in Figure 3. Sediment movement would occur on almost a daily basis during the winter season whereas during the summer, weeks could pass without threshold conditions being exceeded.

Bedload Transport

The results of Table 5 suggest that the dredged material deposited at Site G (Factor 1 and 2 sediments) is transported as bedload. Due to the complexities of the bottom flows measured at the tripod stations it was difficult to find flow conditions where computations of bedload transport on a reasonably steady basis were possible. Three storm episodes occurred during measurement periods which resulted in bottom currents that both exceeded the threshold of grain motion and were relatively isolated in time (not immediately preceeded or followed by another storm). The instrumented tripod at Station 6 became damaged during the third storm, hence data were cut off about midway through the disturbance.

The data collected during the three storm periods have been used to estimate the bedload mass transport of sand. Calculations are based on a procedure described by Sternberg (1972). Estimates of ripple migration distances associated with a given storm have also been made using an empirical relationship described by Chang (1939) and modified according to data collected from the marine environment (Kachel and Sternberg, 1971; Sternberg, 1971). The storms analyzed represent different levels of intensity and although calculations of this sort are very crude, some indication is obtained regarding the quantities of sand transported by a given storm and the distance of sand migration. The results of these calculations are given in Table 6. The partial measurement of storm conditions from the 24-26 December period is also included; the calculations for this period have been proportionally increased to account for the instrument failure after recording about 55% of the storm.

Table 6
Estimates of Mass Transport as Bedload and Travel Distance of Sand During Severe Storms

Station	4	5	6	6
Date	26-30 Aug	26-30 Aug	21-24 Dec	24-26 Dec*
Mean wind speed (m/sec)	13.5	13.5	14.8	20.0
Sediment modal size at station	2.75 Ø	2.75 Ø	2.5 Ø	2.5 Ø
Bedload transport				
gm/cm/storm	9.2	7.7	68.0	4.3×10^3
gm/deposit/storm	4.8×10^5	4×10^5	3.5×10^6	2.2×10^8
cm^3/deposit/storm	3.0×10^5	2.5×10^5	2.2×10^6	14.0×10
m^3/deposit/storm	0.3	0.25	2.2	140
Transport distance m/storm	2.1	1.8	11.0	53

*Data collection was interrupted about midway through the passage of the storm, hence these values represent about 55% of the total.

The bedload values shown in Table 6 are calculated in terms of mass of sediment transported per unit width of the bed per second (gm/cm/sec). These numbers have been calculated for every half-hour mean of $\overline{U}_{100}$ that exceeded the threshold and summed over the duration of each storm and are tabulated as:

1. gm/cm/storm, and then multiplied by the east-west width of the Site G disposal deposit (1700 ft or 520 m) to show

2. the mass of sand transported northward across the total deposit per storm (gm/deposit/storm).

Additionally, the mass transport has been converted to volume transport:

$$Q\rho_s \ (0.6) = j \tag{3}$$

where Q is the volume transport in cm^3/deposit/storm, ρ_s is the sediment density, 0.6 converts sediment density to bulk density (after Kachel and Sternberg, 1971), j is the mass transport in gm/deposit/storm.

Applying these results to the study area as a whole suggests that individual storms transport on the order of 0.25-250 m^3 of Factor 1 and 2 sediments per storm. The distance travelled by these sediments is on the order of 2 to 100 m, respectively. In an effort to estimate the annual mass transport as bedload, these results are plotted in

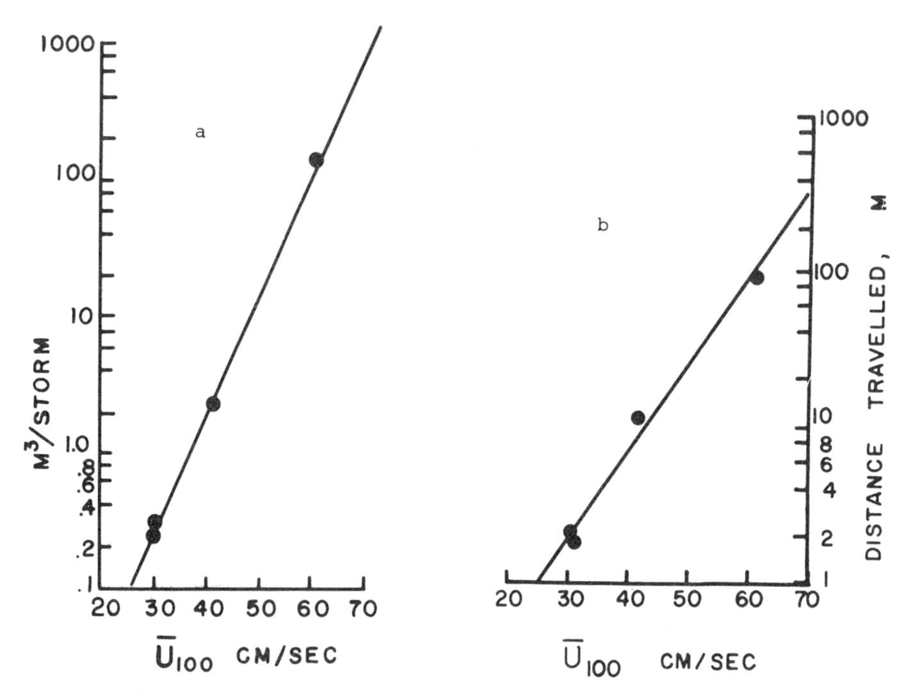

Figure 13
Estimates of sediment transport as bedload (a) and sediment migration distances (b) vs. the maximum wind-generated bottom velocity ($\overline{U}_{100}$) associated with individual storms occurring in the study area. The volume transport is computed as m^3/storm travelling across the total width of the disposal deposit.

Figure 13 which shows an empirical relationship between the volume transport and distance transport per storm vs. the maximum value of the nontidal or wind-generated component of $\overline{U}_{100}$ associated with each storm. Combining Figure 13 with the wind-bottom and mean bottom current relationship (discussed on page 5 in this manuscript), and reviewing the annual wind pattern obtained from the Columbia River Lightship for 1975, provides a means of estimating the annual transport as bedload in the study area. This comparison has been carried out and the results are shown in Table 7. This analysis suggests that storms occurring in the study area transported on the order of 635 m^3 (830 yds^3) of sediment northward as bedload from Site G during 1975. The travel distance for this material was on the order of 440 m for the year.

It should be emphasized that these estimates should only be considered as order of magnitude approximations. They do suggest, however, that the characteristic sediments placed at Site G by the disposal experiment will tend to be dispersed very slowly if at all and would be displaced in a north-northwest direction at a very slow rate.*

*Due to the conical shape of the deposit which rises above the surrounding sea floor, it is expected that the bottom currents would tend to spread and flatten the deposit due to increased local shear stresses generated over the bathymetric feature.

Table 7
Calculations of Mass Transport and Displacement of Sediments due to Storm Activity in 1975

Month	Events with mean speed greater than 13m/sec m/sec	Estimated non-tidal $\overline{U}_{100}$ cm/sec	Estimated bedload transport m^3/deposit/storm	Estimated Sediment Displacement m/storm
January	0.0	60	160	100
February	20.0	35	1	4
March	15.3	41	2	8
	16.4	(45)	(5)	(13)
	(17.3)*			
April	0.0	41	2	8
May	16.4			
June	No Data			
July, Aug, Sept	0.0			
October	16.4	41	2	8
	19.6	58	115	75
	20.0	60	160	100
November	13.0	26	<1	1
	15.0	34	1	2
	18.6	52	30	33
December	17.0	44	5	11
	14.8	33	1	3
	20.0	60	160	100
	13.4	26	<1	<1
TOTAL 1975			635 m^3 northward across Site G	441 m^3 displaced north of Site G

*Winds from the north; the calculated values have been subtracted from the total.

This assumes that the deposit remains uncovered by seasonal deposits of silty sediments which would tend to protect it from further erosion. The annual mass transport is estimated at only 0.2% of the total deposit at Site G and migration is only about 440 m per year. This migration rate is in agreement with the rate determined independently using the distribution of mineralogy for the dredged materials deposited at Site B over the past several decades (see page 22).

Suspended Load Transport

The analysis of sediment texture and mean bottom currents indicates that the primary transport mechanism of the dredged materials is by bedload transport. Observations of pressure fluctuations, bottom photographs, and near bottom turbidity suggest, however, that strong wave activity does cause sediment movement in the suspended transport mode.

All hydraulic and turbidity data and bottom photographs were analyzed during periods of high waves and relatively low bottom currents to reveal the level of wave activity (pressure fluctuation) that causes bed deformation and sediment suspension. All data were then analyzed to determine the percent of time that the threshold level of wave activity was exceeded and the results are summarized in Table 8. The percent of time that wind-generated currents exceeded the threshold velocity (from Table 5) is also included for the purpose of comparison.

This analysis suggests that sediment movement due to wave activity occurs frequently at the disposal site. Calculations were not made of the mass transport as suspended load** hence it is not possible to compare the wave-induced mass or volume transport to the current-induced transport; however, it is suspected that mean current-induced transport of Factor 1 and 2 sediments is the major component. Bedload mass transport is related to the third power of velocity (Sternberg, 1972), whereas, wave action tends to suspend particles but does not generate strong net bottom currents. Therefore, the wave activity would be instrumental primarily in helping to spread and smooth the disposal deposit and would suspend particles to be moved by ambient mean currents. Even if the total mass transport were to double because of wave action (i.e., 0.2%/yr to 0.4% of the total deposit/yr), the general conclusion suggesting relative stability of the dredged material deposit would not change.

Table 8
Percent of Time that Threshold of Grain Motion was Exceeded by Waves and Currents at the Tripod Stations

Station	Month	Wind-generated Currents	Waves
1,2	Apr/May	11%	21%
3	June	0	21
4,5	Aug/Sept	3	6
6	Dec/Jan	66	93

**Transmissometer data were incomplete during some of the major storm episodes.

SUMMARY

Sedimentological Characteristics and Bathymetric Results

1. Materials dredged from the mouth of the Columbia River can be distinguished, using texture and mineral composition, from other shelf sediments even after decades of residence on the shelf. Factor analysis of sediment texture shows that the dominant size modes on the Columbia River tidal delta and surrounding shelf are 2.75 Ø (Factor 3) and 3.25 Ø (Factor 4). Materials dredged from the mouth of the Columbia River are characterized by 2.0 Ø (Factor 1) and 2.5 Ø (Factor 2) modes. Columbia River sediments are richer in orthopyroxene and andesitic lithic fragments, but poorer in altered lithic fragments, opaque minerals (mainly magnetite), clinopyroxenes, altered plagioclase and basaltic lithic fragments, than are the ambient shelf sediments. The Columbia River sediments are thus characterized by high Mineral Index (MI) and high Magnetic Ratio (MR), relative to shelf sediments at the disposal site.
2. The deposit resulting from the disposal of 459,000 m^3 (600,000 yds^3) of material dredged from the Columbia River mouth was a recognizable mound. Immediately following the disposal experiment, the deposit had a generally conical shape 760 m (2500 ft) in radius and 1.5 m (5 ft) in elevation, and contained 278,300 m^3 (364,000 yds^3) of sediment. This represented 61% of the material released over the site. Five months later, after onset of the winter storm period, this deposit remained observable.

HYDRAULIC REGIME AND SEDIMENT TRANSPORT CALCULATIONS

1. Wind-generated bottom currents are the dominant cause of motion of the coarser sediments in the study area. They are best developed with strong southerly winds and may add a non-tidal component to existing tidal currents that exceeds 60 cm/sec flowing to the north-northwest. The maximum observed bottom current ($\bar{U}_{100}$) was 80.1 cm/sec. Strong southerly winds occur primarily in the months of October through March as individual storms that continue for 3 to 7 days. During 1975, winds strong enough to generate bottom currents sufficient to move bottom sediments occurred approximately 35 days from the south and 5 days from the north.
2. Sedimentary materials coarser than 0.18 mm (2.5 Ø) can only move as bedload in response to bottom currents measured in the area. These coarse materials are the dominent sizes associated with the Columbia River dredging at the river mouth; hence the dredged sediment deposits, although agitated frequently by winter storms (thus spreading them laterally), would be relatively stable with time. Estimates of mass transport of sediment as bedload suggest that 635 m^3/yr (830 yds^3/yr) or 0.2% of the original deposit may be transported northward from the disposal site about 0.4 km/yr (0.25 n mi/yr).
3. The 2.75 Ø sediments represent the transition size between total bedload and total suspended load transport. Sedimentary materials finer than 0.15 mm (2.75 Ø) are frequently moved as suspended load by waves and currents. Because suspended material moves

with the water mass, these finer sediments are very mobile, are frequently deposited between storms and then resuspended, and show significant variations as a result of seasonal variations in river input and winter storm activity.

POSTDEPOSITIONAL SEDIMENTOLOGICAL RESPONSE

1. During the 9-month period following the disposal experiment, systematic changes were observed in sediment texture and mineral composition at the disposal site. Immediately around the center of the disposal deposit, sediment texture and mineral composition have remained unchanged (significant amounts of Factor 1 and 2; MR = 5), thus showing little or no recovery toward predisposal conditions. Surrounding this central area, the sediment characteristics exhibit slight to moderate recovery with modal size decreasing (significant amounts of Factor 2 and 3) and magnetic minerals increasing (MR = 3). The peripheral region on the south and southeastern sides of the deposit has undergone almost complete recovery to ambient textural and mineralogical conditions (Factors 3 and 4 predominant, and significant amounts of Factor 2 remain; MR = 1-2).
2. Dredged materials will remain more or less in place with a slight tendency to migrate northward parallel to the isobaths. The deposit will spread laterally as a result of waves and currents, hence its bathymetric expression will be reduced. The sediment texture and composition will continue to be a recognizable feature although finer grained sediments (<0.15 mm) will move seasonally across the area depending on river input and hydraulic conditions.

REFERENCES

1. Ballard, R. L. 1964. Distribution of Beach Sediment near the Columbia River. Univ. Washington, Dept. Oceanography Tech. Report No. 98.
2. Barnes, C.A., A. C. Duxbury, and B. A. Morse. 1972. The circulation and selected properties of the Columbia River effluent at sea. In: A. T. Pruter and D. L. Alverson (Editors), *The Columbia River Estuary and Adjacent Ocean Regions*. Univ. Washington Press, Seattle.
3. Barnes, C. A., and M. G. Gross. 1966. Distribution at sea of Columbia River water and its load of radionuclides. In: *Disposal of Radioactive Wastes into Seas*. International Atomic Energy Agency, Vienna.
4. Chang, Y. L. 1939. Laboratory investigations of flume traction and transportation. *Trans. Am. Soc. Civil Engineers*, 104: 1246-1313.
5. Harman, R. A. 1972. The distribution of microbiogenic sediment near the mouth of the Columbia River. In: A. T. Pruter and D. L. Alverson (Editors), *The Columbia River Estuary and Adjacent Ocean Regions*, Univ. Washington Press, Seattle.
6. Hopkins, T. S. 1971. On the circulation over the continental shelf off Washington. Ph.D. thesis, Univ. Washington, Seattle, 204p. (unpublished).

7. Imbrie, J., and Tj. van Andel. 1954. Vector analysis of heavy-mineral data. *Bulletin Geol. Soc. America,* v. 75, p. 1131-1156.
8. Judson, Sheldon and Dale F. Ritter. 1964. Rates of regional denudation in the United States. *Journal Geophys. Research,* v. 69, p. 3395-3401.
9. Kachel, N. B., and R. W. Sternberg. 1971. Transport of bedload as ripples during an ebb current. *Marine Geol.,* v. 19, p. 229-244.
10. Krumbein, W. C., and F. J. Pettijohn, 1938. *Manual of Sedimentary Petrography.* Appleton-Century-Crofts, New York.
11. Kulm, L. D., R. C. Roush, J. C. Harlett, R. H. Neudeck, D. M. Chambers, and E. J. Runge. 1975. Oregon continental shelf sedimentation: Inter-relationships of facies distribution and sedimentary processes. *J. Geology,* v. 83, p. 145-175.
12. Lockett, J. B. 1959. Interim consideration of the Columbia River entrance. *Proc. Am. Soc. Civil Engrs., J. Hydraulics Div.,* v. 85, p. 17-40.
13. McManus, D. A. 1972. Bottom topography and sediment texture near the Columbia River. In: A. T. Pruter and D. L. Alverson (Editors), *The Columbia River Estuary and Adjacent Ocean Waters.* Univ. Washington Press, Seattle.
14. National Marine Consultants. 1961. Wave statistics for three deep-water stations along the Oregon-Washington coast. Santa Barbara, California, 16p.
15. Nittrouer, C. A., R. W. Sternberg, and D. A. McManus. In press. Sedimentation on the Washington Continental Shelf. In: W. E. Pequegnat and R. Darnell (Editors), *The Ecology and Management of the Continental Shelf.* Gulf Publishing Co.
16. Smith, J. D. and T. S. Hopkins 1972. Sediment transport on the continental shelf off Washington and Oregon in light of recent current measurements. In: D. J. P. Swift, D. B. Duane, and O. H. Pilkey (Editors), *Shelf Sediment Transport: Process and Pattern.* Dowden, Hutchinson, and Ross, Stroudsburg, Pa., p. 143-180.
17. Sternberg, R. W. 1971. Measurements of incipient motion of sediment particles in the marine environment. *Marine Geology,* v. 10, p. 113-119.
18. Sternberg, R. W. 1972. Predicting initial motion and bedload transport of sediment particles in the shallow marine environment. In: D. J. P. Swift, D. B. Duane, and O. H. Pilkey (Editors), *Shelf Sediment Transport: Process and Pattern.* Dowden, Hutchinson, and Ross, Stroudsburg, Pa., p. 61-82.
19. Sternberg, R. W., and D. A. McManus. 1972. Implications of sediment dispersal from long-term bottom-current measurements on the continental shelf off Washington. In: D. J. P. Swift, D. B. Duane, and O. H. Pilkey (Editors), *Shelf Sediment Transport: Process and Pattern.* Dowden, Hutchinson, and Ross, Stroudsburg, Pa., p. 181-194.
20. Sternberg, R. W., D. R. Morrison, and J. A. Trimble, 1973. An instrumentation system to measure near-bottom conditions on the continental shelf. *Marine Geol.,* v. 5, p. 181-189.
21. Sternberg, R. W., J. S. Creager, W. Glassley, and J. Johnson. 1978. Aquatic Disposal Field Investigations, Columbia River Disposal Site, Oregon, Appendix A: Investigation of the Hydraulic Regime and Physical Nature of Bottom Sedimentation. U.S. Army Corps of Engineers , 332p.

22. Whetten, J. T., J. C. Kelley, and L. G. Hanson. 1969. Characteristics of Columbia River sediment and sediment transport. *J. Sediment. Petrol.*, v. 39, p. 1149-1166.

23 White, S. M. 1967. The mineralogy and geochemistry of the sediments on the continental shelf off the Washington-Oregon coast. Ph.D. thesis, Univ. Washington, Seattle, 213p.

Geologic Effects of Ocean Dumping on the New York Bight Inner Shelf

S. Jeffress Williams

ABSTRACT

High resolution seismic reflection records, sediment cores and deep borings, and comparison of bathymetric charts from 1845 to 1973 provide evidence that ocean dumping of assorted solid materials has significantly filled parts of the Hudson shelf channel, and is an important geologic process. Ocean disposal of natural and man-made wastes was officially initiated seaward of New York Harbor in 1888 to relieve health problems, congestion and accelerated shoaling of navigation channels long associated with uncontrolled disposal within the city and adjacent waterways. Records show that about 850 million m^3 of liquid and solid wastes have been dumped in the past 85 years. This has resulted in creation of several mounds with relief of about 15 m covering an area of about 9 thousand hectares. The calculated volume of anthropogenic solids filling the Hudson channel is 318 million m^3. Much of the material is similar in character to indigenous sediment. The results indicate most materials except sewage sludge are fairly stable and remain in the original dump sites. In spite of large volumes of sludge dumped at the same site since 1924, no evidence of significant accumulation on the seafloor has been found.

INTRODUCTION

Mankind with its many activities has had an impact on modifying the morphology and environment of subaerial parts of the earth for at least the past ten thousand years. Until recently most impacts were primarily related to physical effects such as excavations and accumulations of debris from natural catastrophy (e.g., floods, earthquakes, volcanoes, fires) and warfare. In ancient times when towns were destroyed, the inhabitants would simply smooth and pack the debris, cover it with a thin layer of soil and then rebuild. Gunnerson (1973) provides information on this debris accumulation process where in one example he shows that from 3000 BC to 1300 BC, Troy went through seven stages of occupation and was built three times at the same site. Total accumulation of solid waste debris is about

14.5 m (Gunnerson, 1973).

Such examples of Man as a significant geologic agent are fairly numerous in historic time for continental areas, but Man's impact on the oceans, estuaries and the sea floor are poorly documented at best. Since the late 1960's, impacts by humanity on the environment and degradation of air, water, and soil have received widespread attention by the public and increased levels of funding to perform scientific studies. New York City with its practice of disposing of millions of tons of waste materials annually in the ocean seaward of the harbor has received much of the attention, as well as federal agencies (e.g., Corps of Engineers, Coast Guard, Environmental Protection Agency, National Oceanic and Atmospheric Administration) charged with administering policies of ocean dumping. In spite of recent publicity and concern, the practice of transporting large volumes of wastes by barge to officially designated ocean dumpsites has been going on since 1888, and unofficial dumping even earlier. Historical records show that officials and segments of the public were aware of at least the more obvious effects of dumping from the beginning. However, Bettmann (1974) shows that prior to initiation of ocean dumping, New York was being smothered in debris, garbage, animal remains and wastes; and, that such accumulations fouled the environment and were a serious threat to general health of the city. Sanitation systems were primitive, garbage pickup was not available or inept, and many streets were so crowded with garbage and debris that travel was nearly impossible. Waterways and harbors adjacent to the city were commonly used for disposal of wastes, and dumping platforms were used until 1872 so trash wagons could more easily empty their contents into the Hudson and East Rivers. Much of the lighter material was frequently carried by coastal currents to recreation beaches on New York and New Jersey, and led one official in 1898 to state that the city's wastes "defiled the beaches to an intolerable degree" (Bettmann, 1974). Consequently, federal and local authorities recognized the severity of the problem and chose ocean dumping as a more favorable alternative. Historical detail is lacking, but apparently removing wastes from the city did improve health and general living conditions, and shoaling of navigation waterways was less of a problem though regular patrolling of the harbors and dump sites was necessary to ensure compliance with the regulations. The objective of this paper is to examine the geologic and physical character of the New York Bight presently and during the past almost nine decades to decipher the geologic and long-lasting effects that dumping has had on the inner continental shelf area seaward of New York City.

DATA, EQUIPMENT AND PROCEDURES

Most seismic and core sediment data reported here were collected in separate surveys in 1964 and 1969 under contract with Alpine Geophysical Associates, Inc., as part of the U.S. Army Coastal Engineering Research Center (CERC) Inner Continental Shelf Program (ICONS). Results of the studies for this region are contained in CERC reports by Williams and Duane (1974), and Williams (1976). For the shelf region shown in Figure 1, 825 km of high resolution continuous seismic reflection profiles and 61 vibratory cores were taken, and logs from several borings in the vicinity of Ambrose Light were used.

The CERC cores are 10 cm in diameter with a mean length of 3.3 m, and the borings are a maximum of 76 m long.

The seismic records were obtained with a 50 to 200 J Sparker and were examined to determine the geologic nature of the sea-floor morphology, and to characterize the principal and secondary acoustic reflectors in the shelf subbottom. The more persistent and continuous reflectors were mapped, and many seismic profiles were interpreted and reduced to line drawings.

The cores were visually described and logged, and representative sediment samples were removed for detailed microscopic examination and textural analysis. Grain size analyses were done on samples at regular depth intervals from each core. Several of the cores containing artifacts or interesting lithologies were split longitudinally to allow detailed study of the sediment stratigraphy and composition. For the deeper borings, engineering logs (McClelland Engineers, 1963) were used to identify the deep stratigraphy evident

Figure 1
CERC data for inner New York Bight consist of 825 km of high resolution sparker seismic records and 61 vibratory sediment cores. Also shown are locations of eight deep engineering borings. Heavy lines are sections of seismic profiles shown in Figure 6 over a. Diamond Hill mound, b. cellar dirt site, c. dredge spoil site, d. sewage sludge site, and e. acid site.

on the seismic profiles but deeper than the cores penetrated, and to determine the composition of Diamond Hill mound.

Whenever possible, stratigraphic boundaries in the cores and borings were correlated with acoustic reflectors on the seismic records to yield information on the regional extent of natural sediments and ocean dumped anthropogenic solid materials. These data were used with historic bathymetric information to construct isopach maps and seafloor profiles.

The last primary source of information is the sequence of bathymetric charts made of New York Bight in 1845 (Fig. 2), 1888 (Fig. 3), and 1936 (Fig. 4) by the U.S. Coast and Geodetic Survey, and the latest (1973) hydrographic chart (Fig. 5) surveyed by the Corps of Engineers for the National Oceanic and Atmospheric Administration (NOAA). These charts enabled careful comparison of contour positions to be made which identified areas where long term deposition or erosion occurred. Isopach maps, line profiles, and rates of elevation change were derived.

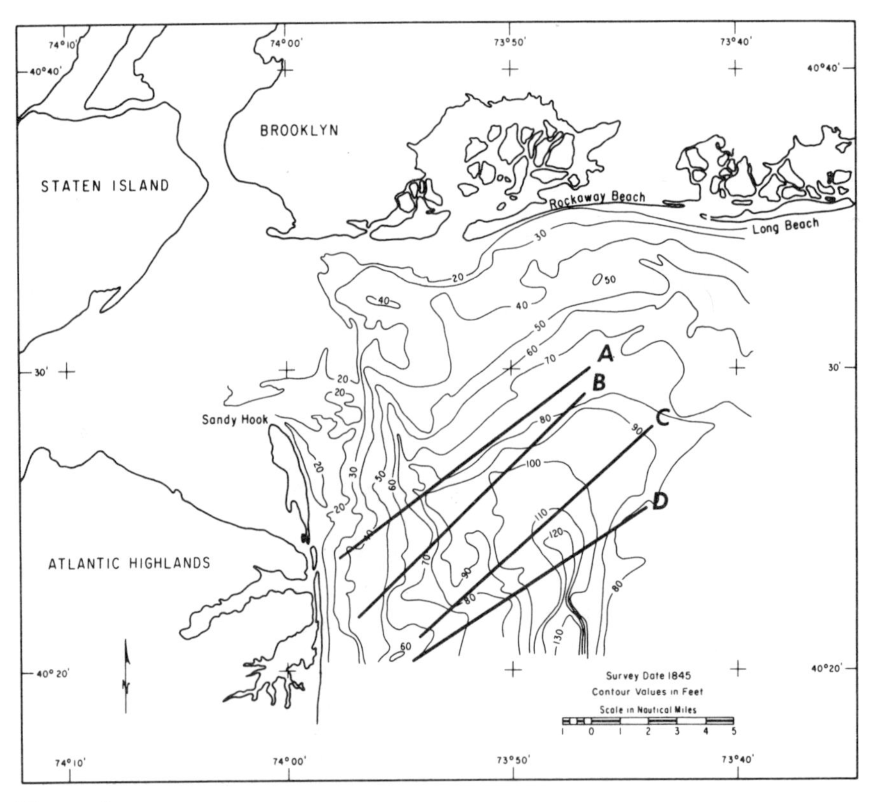

Figure 2
Earliest detailed bathymetric map of the New York Bight made in 1845 prior to initiation of ocean waste disposal shows the Hudson Channel and adjacent shelf had characteristic natural morphology in contrast to physiographic evidence of dumping on later maps. Lettered sea floor profiles are in Figure 8. Contours in feet. Source: Williams, 1975.

INNER SHELF GEOLOGY AND GEOMORPHOLOGY

Inner New York Bight (Fig. 5) covers an area of about 650 km^2, which includes the inner shelf off the Sandy Hook spit region of New Jersey, and off Rockaway and Long Beach on the Long Island south shore. The bight is immediately south of the Harbor Hill moraine marking the most southerly advance of Pleistocene continental glaciers in the southern New England region. Consequently, much of the present geologic character and geomorphology of the mainland and shelf result from complex Pleistocene glaciofluvial processes which modified the land when sea level was considerably lower than at present. As sea level has risen during the past 10 to 15 millenna effects of Holocene marine processes have been superimposed. The

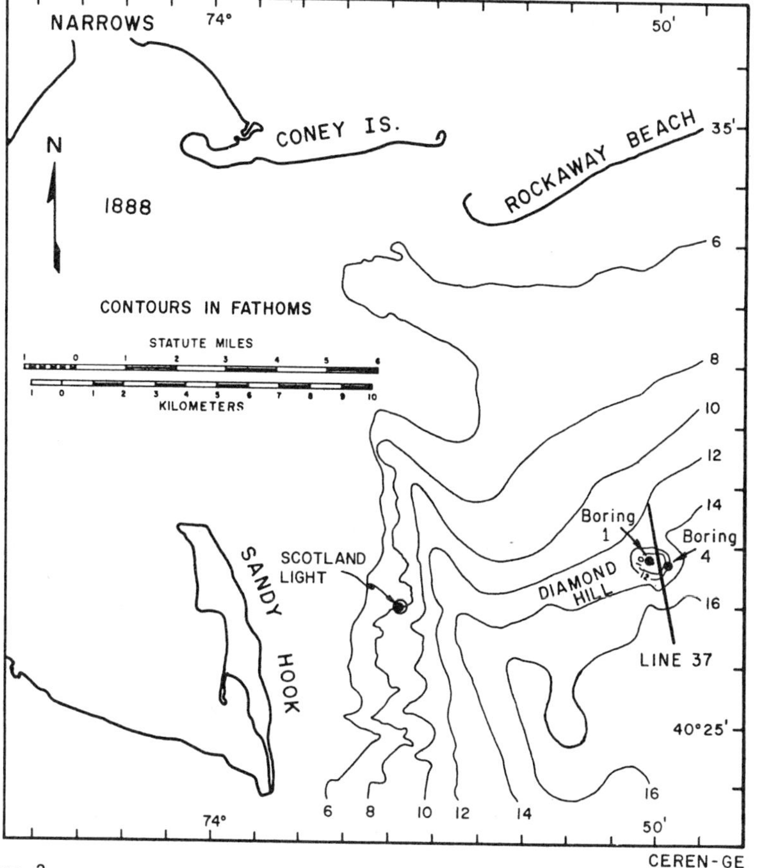

Figure 3
Bathymetric map of the New York Bight based on the 1885-1888 survey. The Hudson Channel in comparison with 1845 data (Figure 2) appears to have the same morphology. However, the Diamond Hill mound is an anomolous feature shown on seismic line 37 (Figure 6a) that originated from very controlled dumping of cellar dirt solid materials. Composition of the mound is revealed in Ambrose Light borings 1 and 4 (Table 2). Source: Modified from Veatch and Smith, 1939.

Long Island coast is a low-relief relict sand plain where offshore barrier islands and elongate spits predominate. In contrast, Atlantic Highlands of northern New Jersey is a nonglaciated headland region exhibiting over 80 m of relief. The New Jersey coastline is straight and regular except for Sandy Hook spit which has grown and recurved during the Holocene epoch into Lower New York Bay as a result of a large littoral drift to the north. The Raritan and Hudson Rivers drain the Piedmont Province to the west and discharge into Raritan Bay and Lower Bay, respectively. Both were major agents in transporting large volumes of sediment into, through, and

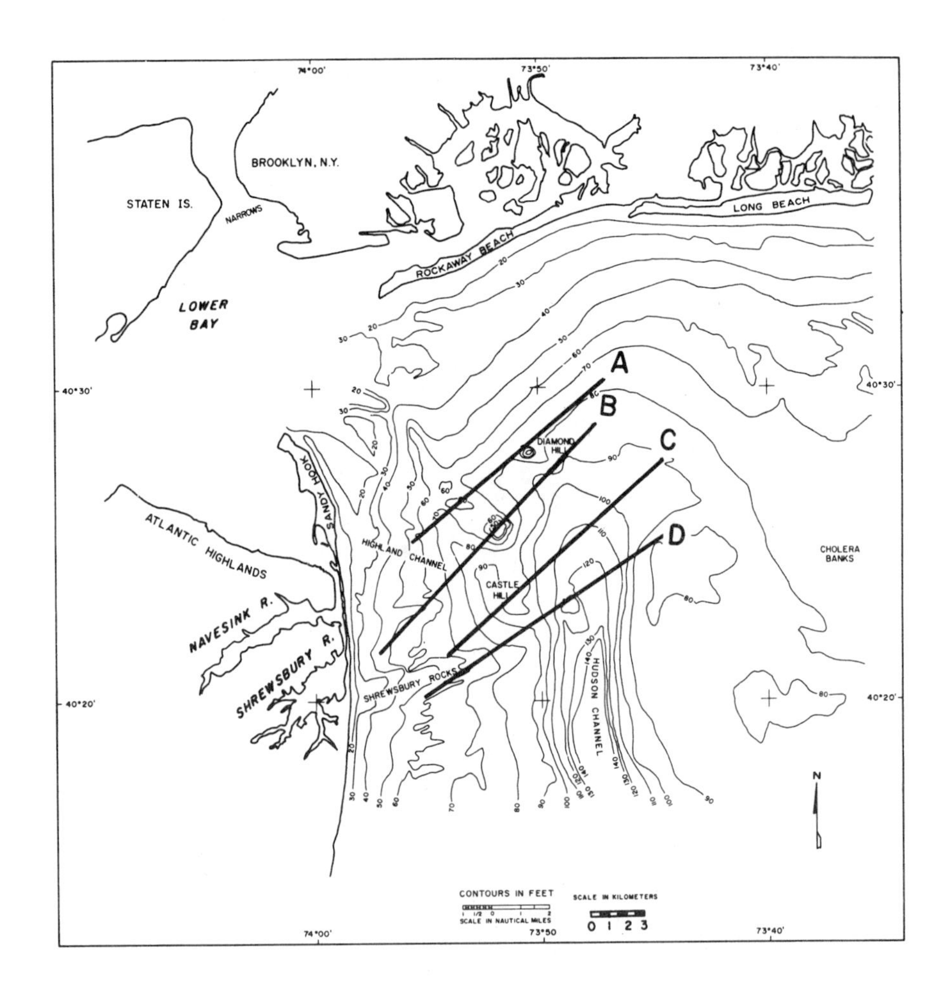

Figure 4
Bathymetric map of the New York Bight from National Ocean Survey chart 1215, based on the 1936 survey. Presence of the four topographically positive features at Castle and Diamond Hills in the Hudson Channel is attributed to accumulation of ocean dumped waste materials. Lettered profiles are shown in Figure 8. Contours in feet. Source: Williams, 1975.

out of the bight when sea level was lower and they flowed across the exposed shelf. The Hudson has a well-defined channel cut into the shelf which can be traced from Verrazano Narrows, where it breaches the Harbor Hill moraine, through Lower Bay and across 200 km of inner and outer shelf to the shelf edge. It is the best developed shelf channel and canyon of the many off the U.S. Atlantic ccast and has remained unfilled. However, Williams (1976) has shown that as many as eight other ancestral Hudson channels (Fig. 6b) are buried in the shelf of western Long Island. The Raritan shelf channel exhibits little bathymetric expression, but a buried ancestral

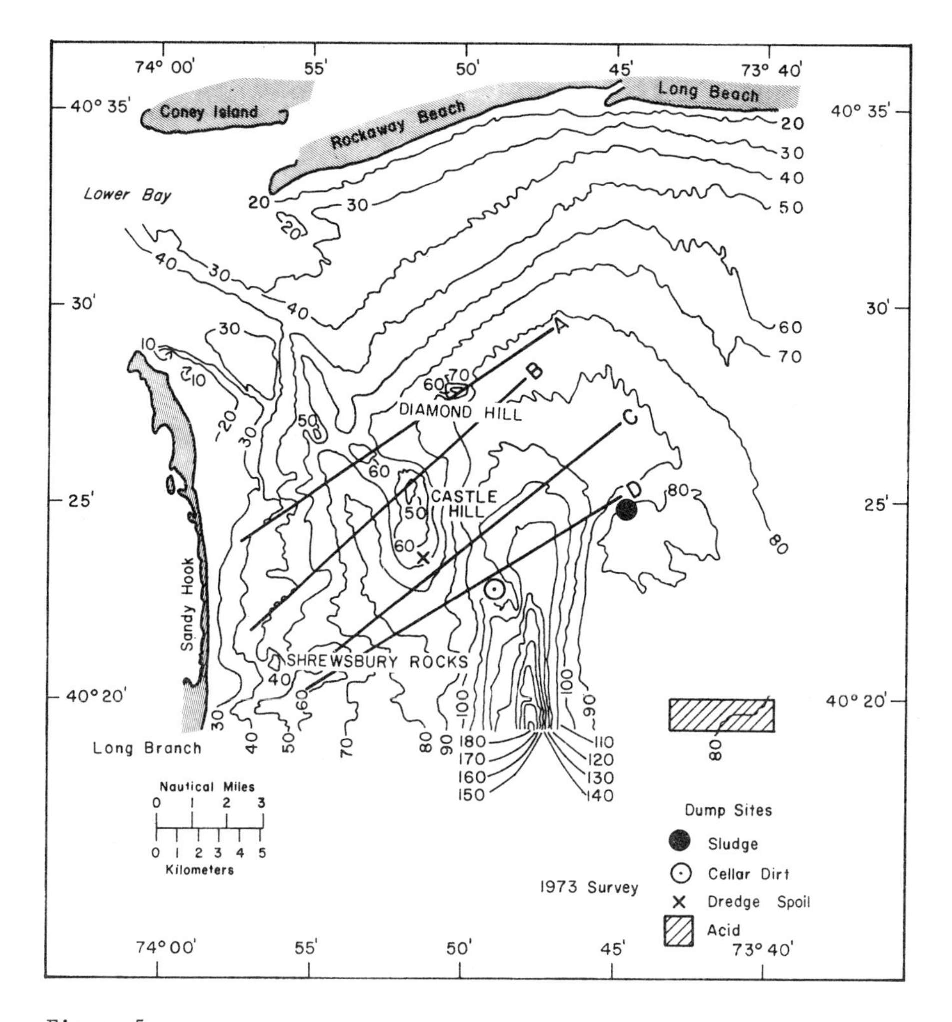

Figure 5
Bathymetric map based on 1973 survey showing detailed seafloor morphology. Four contemporary disposal sites adjacent to the Hudson Channel are shown while older sites at Castle and Diamond Hills are manifested only by their topographic expression. Lettered seafloor profiles are in Figure 8. Contours in feet. Source: Williams, 1975.

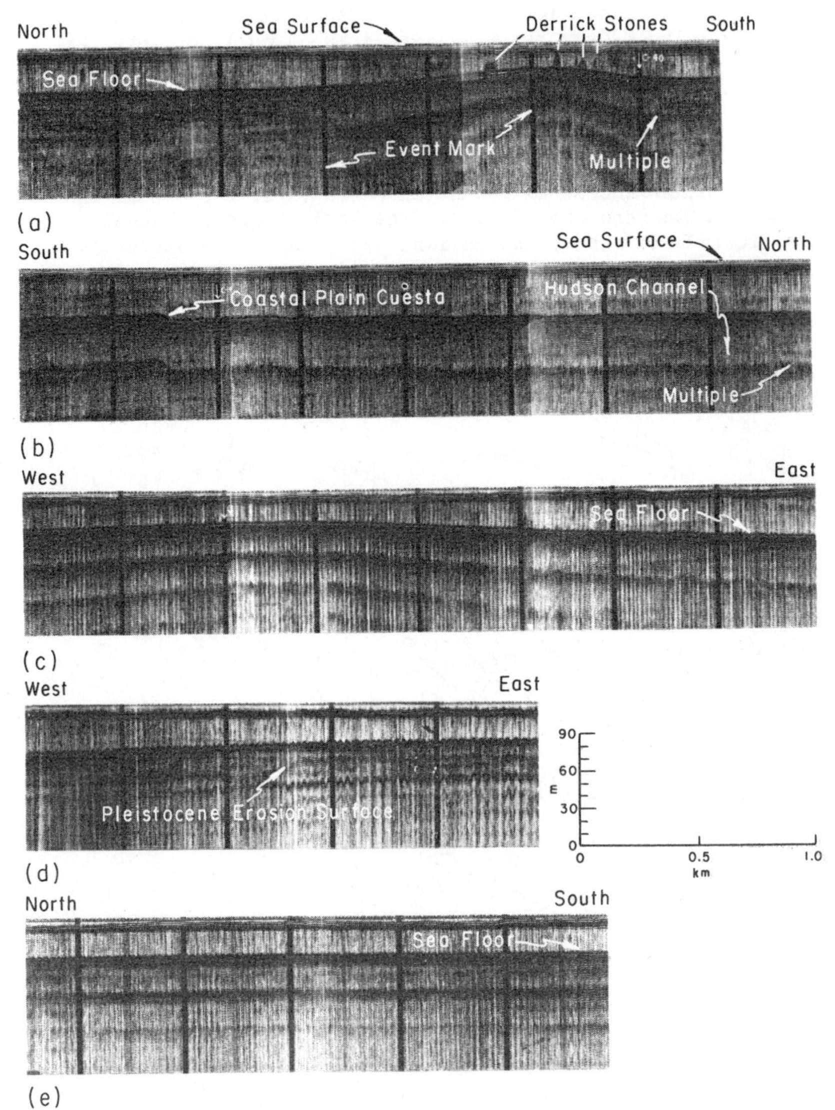

Figure 6
Sections of seismic-reflection profiles at several of the New York Bight disposal sites are: a. line 37 at Diamond Hill shows 9 m of accretion at the center; b. continuation of line 37 at the cellar dirt site shows the sea floor is relatively flat in spite of evidence that about 3 m of accretion has occurred since 1936; c. line NN at the dredge spoil site shows a broad mound resulting from 10 m of accretion of accorted dredge materials dumped between 1936 and 1973; d. line 44 at the sewage sludge site exhibits an apparent natural sea floor with no significant accumulation of sludge since 1925 to mask underlying acoustic reflectors; e. line 38 at the acid site east of the Hudson Channel shows no apparent effects of acid disposal since 1948. Profile positions are shown in Figure 1.

channel in Raritan Bay is present which underlies Sandy Hook spit at a depth of -40 m, and intersects a buried Hudson channel about 20 km east of Sandy Hook, New Jersey (Williams, 1976).

The sea floor of the inner bight has a gentle seaward slope with fairly uniformly spaced contours from the shoreface to the limits of coverage at -30 m (Fig. 5). Shrewsbury Rocks are a dominate feature as elongate shoals projecting northeast from the New Jersey coast. The shoals are truncated at the -30m contour by the Hudson Channel, and then continues to the Long Island shelf as Cholera Banks on the eastern side of the channel. The cuesta forming Shrewsbury Rocks comprises the eroded edges of Coastal Plain sedimentary strata dipping to the southeast and underlying the entire New York Bight. These are primarily clastic sands and gravels with glauconitic silty sands and clays of late Cretaceous or early Tertiary age. Several CERC seismic records in this region (Fig. 6b) contain numerous prominent acoustic reflectors which correspond in attitude to Coastal Plain strata and correspond in outcrop position with the shoals. South of Shrewsbury Rocks, Tertiary strata have been truncated close to the sea floor and are thinly mantled($\approx$1 m) by Quaternary sediments. Cores in the region contain either fine-grained cohesive sediments or silty medium to very coarse sand overlying reddish-brown or glauconitic sandy gravels. In contrast, the shelf north of Shrewsbury Rocks contains thick ($\approx$35 m) accumulations of Quaternary age sand and gravel overlying the deeply eroded Upper Cretaceous strata.

Figure 2 is based on an 1845 bathymetric survey which is the earliest complete survey for the bight region. It shows morphology of the coastal landforms and seafloor features in a "natural" state prior to major effects from man-made modifications such as construction of coastal engineering works (e.g., jetties, groins, channels, land fills) or ocean dumping of solid wastes. Offshore, the Hudson Channel was clearly the dominant feature, and was open and continuous from the 40 m depth contour for about 15 km to the northwest into Lower New York Bay. The contours show that the channel bottom and side slopes were fairly even and regular, and the only irregularities were places where the cuesta crosses the channel north of the 40 m contour, and the region east of Sandy Hook at the confluence of the ancestral Raritan and Hudson Channels.

HISTORY OF OCEAN DUMPING ACTIVITIES

The New York Harbor U.S. Congressional Act of 1888, established that the Supervisor of New York Harbor granted permits for transport and disposal of various wastes in the ocean seaward of New York Harbor. The first dump site for all solid debris (Table 1) was 4 km south of Coney Island; however, immediate shoaling and fouling of adjacent beaches caused the site to be relocated in 1900 to an open shelf area seaward of Lower Bay. The seafloor at the first site off Coney Island apparently still contains remnant debris and is known for excellent fishing by the name "Tin Can Grounds" (Jensen, 1975). Also, during dredging operations for a recent beach nourishment project at Rockaway Beach an offshore source of sand fill at East Bank shoal off Coney Island was used which contained fragments of red bricks. The new site southeast of Sandy Hook was in the thalweg at the head of the Hudson Channel which mariners knew about since the bathymetric surveys of Lindenkohl in 1842-44, and the geologic work by James Dana in 1863.

Table 1
Representative Chronology of Official New York Bight Dumping Grounds*

Date	Location
1888	Mud buoy, 2.5 miles south of Coney Island, for the deposit of all refuse, including garbage and city refuse.
1900	A point one-half mile southward and eastward of Sandy Hook Lightship.
1903	A point 1.5 miles east of Scotland Lightship in 12 fathoms of water.
1906	A point 2 miles southeast of Scotland Lightship in 14 fathoms of water.
1908	For cellar dirt, mud, "one-man stone", steam ashes and floatable material, a point 3 miles southeast of Scotland Lightship.
1913	Limit of water for deposits, 15 fathoms.
1914	For material containing floatable matter, (street sweepings) not less than 4 nautical miles east-southeast of Scotland Lightship, in not less than 17 fathoms of water. For larger "derrick stones", a spot 4 miles southeast of Scotland Lightship is not less than 19 fathoms of water.
1920	Garbage dumped 15 miles southeast of Scotland Lightship as temporary measure.
1924	Essentially same site locations as in 1908 and 1914, except that the site for "derrick stones" was moved because of shoaling problems to 6 miles from Scotland Lightship in $\geq$ 19 fathoms of water. Dumping sewage sludge was initiated not less than 8 miles southeast of Scotland Lightship. Garbage dump site moved 19 miles from Scotland Lightship.
1934	Dumping of garbage and floatable material halted.
1950	For cellar dirt materials, 4 miles southeast of Scotland Lightship in 15 fathoms of water. For "derrick stone", 6.25 miles southeast of Scotland Lightship in 17 fathoms of water. Site for sewage sludge same as in 1924.

*Compiled from U.S. Army, Corps of Engineers Annual Reports, 1885 to 1968.

Composition of solid wastes dumped is vague in the early official reports, but the general classes were: mud, one-man stone, steam ashes, derrick stone, sheet sweepings, and cellar dirt. Garbage was also dumped at sea until a reduction plant was put into operation in 1897. Dumping of garbage resumed in 1920 (Table 1) when the plant burned, and the practice continued, much to the objection of coastal communities on New Jersey and Long Island, until 1934. Cellar dirt is a general term for excavation and construction materials such as masonary material, brick, tile, wood, and natural soil and rock. Barging of solid wastes to the ocean dump sites became more popular as previously used disposal areas on the mainland were closed or used for urban expansion. Also, as New York and surrounding communities expanded in area and population, old buildings and structures were demolished, and new sites were excavated. Also, several vehicle tunnels under the Hudson and East Rivers and a network of subways on Manhattan were built which required massive excavation efforts. At this same time, New York City grew as a commercial shipping center and large public works projects were undertaken to deepen and widen ship channels leading from the Lower Bay to the Upper Bay-Manhattan region. Wigmore (1909) provides insight to a dredging project from 1900 and 1907, when the Ambrose ship channel was constructed from the open sea into New York Harbor. Records show a total of 38 million m^3 of mostly sand and gravel were removed from the sea floor, transported by barge, and dumped in the Hudson Channel at the Castle Hill site east of Scotland Light. As Table 1 and Fig. 7 show, the cellar dirt and dredge spoil ocean dumping sites were frequently moved seaward in line with the Hudson Channel thalweg as old sites were filled and a positive topographic relief became evident. The Corps of Engineers, Annual Reports dating to 1885, report that lead-line surveys were regularly made at the dump sites to monitor accretion, and that patrol boats in the harbors tried to prevent dumping of debris in unauthorized areas.

CONTEMPORARY DUMP SITES

SEWAGE SLUDGE

The site for dumping processed and raw sewage sludge was established in 1924, considerably later than for cellar dirt and dredge spoil. The site remains in the original location in about 27 m of water (Fig. 5) on the eastern flank of the Hudson Channel.

ACID WASTE

The site for discarding industrial waste acid is about 10 km southeast of the sewage sludge site (Fig. 5), and consists of two adjacent areas; one used in summer and the other in winter. The site has been used since 1948 and is probably the most obvious from the water surface of all the sites because of the striking reddish-brown color of the sea during a dump due to the large amount of iron compounds mixed with the acid.

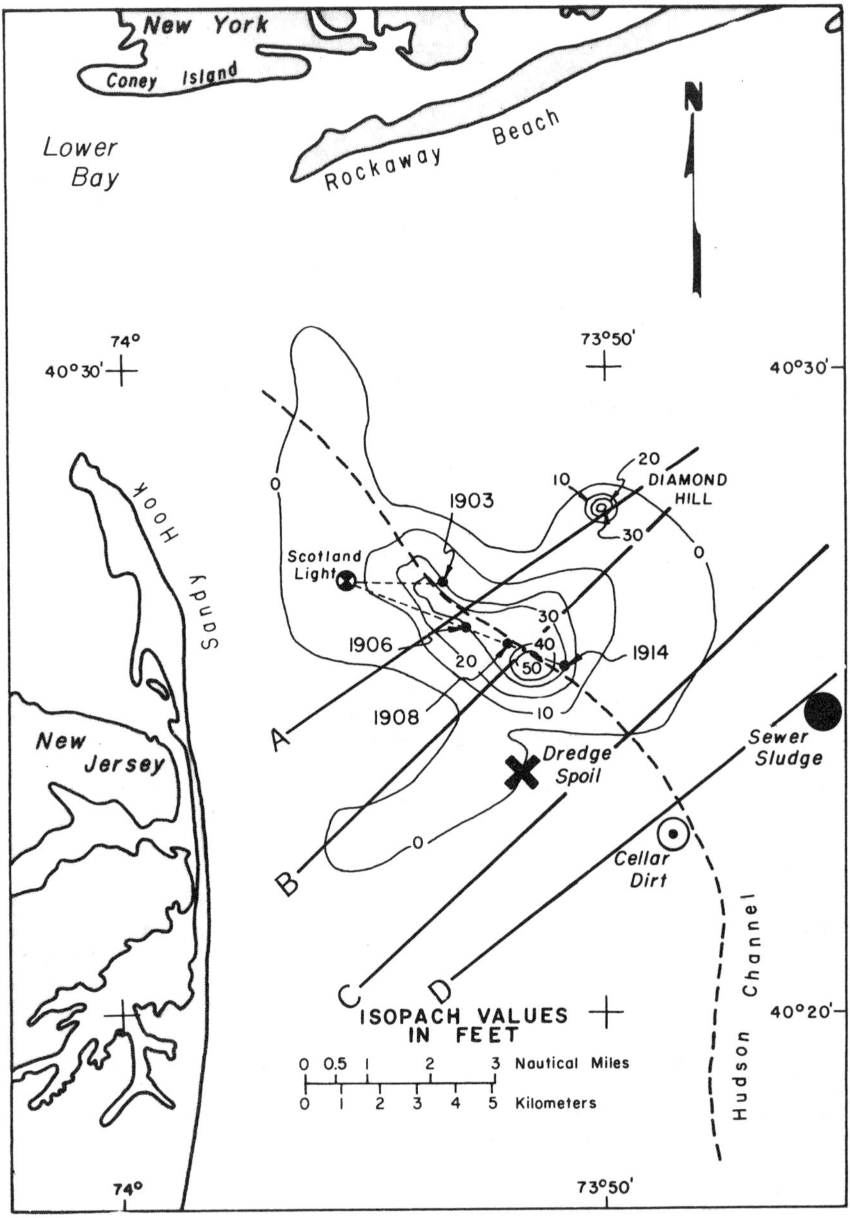

Figure 7
Isopach map showing lateral and vertical extend of solids that have accumulated in the Hudson Channel region as a result of ocean dumping. Data result from comparison of bathymetric information from the 1845 (Figure 2) and 1936 (Figure 4) hydrographic surveys. Official dump sites from 1903 to 1914 from the Scotland Light, shown as presented in Table 1 are landward of contemporary sites. Source: Williams, 1975.

WRECK SHIPS

The dump site for wrecks is 22 km south of Ambrose Light in the middle of the Hudson Channel at a water depth of about 55 m. Unsalvagable hulks are transported by private individuals or disposed by the Corps of Engineers when they are a hazard to navigation under authority of the River and Harbor Act of 1899 (Pararas-Carayannis, 1973).

CHEMICAL WASTE

The official dump site for toxic chemical wastes and explosives is located about 190 km southeast from New York, but because of the high transportation costs alternatives to ocean dumping have been used; hence this site has been little used and exact descriptions and volumes of wastes dumped are not readily available.

EFFECTS OF OCEAN DUMPING

DIAMOND AND CASTLE HILLS REGION

Figure 3 is a portion of a map from Veach and Smith (1939), based on 1885-1888 lead-line surveys. Comparison of the contour positions with Fig. 2 shows that the Hudson Channel morphology is essentially unchanged; however, presence of the Diamond Hill mound 6 km east of the channel is striking. It is not shown on the 1845 map, and a feature so large and anomolous (1.3 km diameter, 9 m relief) would have certainly been known to the early surveyors. Inclusion of Diamond Hill on the 1888 map proves that it originated prior to that survey, but no reference was found in any of the official Corps of Engineers Annual Reports dating to 1885 to indicate more precisely when dumping occurred; and, in fact, that area was apparently never designated as an approved ocean dumping site. Composition and possible origin of Diamond Hill is suggested by the name "Subway Rocks" applied by local fishermen who find the fishing excellent because large "derrick stones" are present on the sea floor which act as a rock and rubble reef and attract marine life. Jensen (1975) surmises that Diamond Hill is composed of bedrock material possibly excavated from Manhattan during construction of the subway system and disposed of at the Diamond Hill Site. However, he cites no official record or chronology of such dumping and has no additional information on its origins (Jensen, per. comm., 1976). A second possible explanation for the origin of the cellar dirt debris comprising Diamond Hill is that the region was used for dumping excavation and demolition material resulting from construction of the Brooklyn Bridge from 1867 to 1883. McCullough (1972) gives an interesting account of the history of construction and states that the bridge was a major engineering achievement. Considerable demolition of existing houses, roads, and warehouses and excavation of bedrock and soil for the foundation and access roads was required and possibly a buoyed site at Diamond Hill was chosen by Roebling and his engineering colleagues as the answer to their disposal problem. Seismic line 37 in Fig. 6a transects the area and provides additional proof that large rocks or similar solid masses are present at Diamond Hill. Parabolic-shaped reflectors above the seafloor are evidence suggesting that large coherent blocks of material are either exposed on the seafloor or thinly buried. But while the record shows that large objects are present it also suggests that the mound is composed primarily of much smaller size materials deposited in such a way that the mound is acoustically opaque. There is little evidence of internal stratification and the acoustic energy penetrating the mound was scattered and absorbed such that deeper underlying Pleistocene reflectors are barely visible. This contrasts with the clear subbottom stratigraphy on line 37 north and south of the hill and on other records in the area.

In 1963, the U.S. Coast Guard planned to replace the Ambrose lightship with a permanent pile structure, and they contracted with McClelland Engineers, Inc. to gather and develop engineering information necessary for design of the pile foundation. Because of the relatively shallow water, the Diamond Hill area was chosen as the general site for investigation and a detailed fathometer survey was made to precisely map sea floor morphology. Following this survey a boring located on the crest of the mound (40°27'48"N, 73°50'05"W) in 14.6 m of water was made. This first boring was terminated at 15 m below the seafloor because of the presence of 6 to 9 m of cellar dirt material and rubble fill as shown in Table 2. Consequently, the site was abandoned because of the impracticality of driving piles through the rubble fill, and a second site in the next most shallow water depth was chosen off the northwest slope of Diamond Hill in about 23 m of water. Two additional borings made at the second site revealed a natural stratigraphic sequence to 76 m below seafloor with only small amounts of cobbles and brick in the top 30 cm (McClelland, 1963). Subsequently, a third site was chosen on the southeast slope of the hill in 22.6 m of water and boring 4 was made. Table 2 shows that the top 5 m of the sea floor is composed of cellar dirt fill material and visual inspections of the site by divers revealed that the seafloor consists primarily of coarse sand and scattered rubble, with the largest piece being about 0.15 m^3 (McClelland, 1963). This site was finally selected for construction of the Ambrose tower.

Comparing bathymetry in Fig. 4 from the 1936 survey with that of 1845 (Fig. 2) and 1888 (Fig. 3) reveals some remarkable changes. Diamond Hill seems to have retained its general size, shape, and relief during the 48 year period (1888-1936) and no evidence is present to suggest that sea floor currents removed significant quantities of material; probably owing to the general coarseness of the component fill. However, obvious changes did occur at Castle Hill within the Hudson Channel. Figure 4 shows that major accretion took place and three circular mounds were created which did not exist in 1888. The magnitude of accretion is shown by the isopach map in Fig. 7, and by comparison of seafloor profiles in Fig. 8. Figure 7 shows that maximum aggradation corresponds in position with Castle Hill in the thalweg of the Hudson Channel and with Diamond Hill. Maximum fill thickness is 15 m at Castle Hill and 9 m at Diamond Hill; however, these areas superimpose a much larger mound of fill with maximum dimensions of 14.5 km long 11.5 km wide covering an area of about 9 thousand hectares. The elongate protrusion of the zero base line toward the New Jersey coast in Fig. 7 appears anomolous, and probably is due to lack of survey accuracy or insufficient sounding data, rather than actual accumulation of fill.

TABLE 2

Geologic logs of deep borings taken on Diamond Hill mound for engineering studies for the Ambrose Light Tower confirm the anthopogenic origin of the sea floor feature. (Fm. McClelland, 1963). Units in feet. See Fig 3 for locations.

Ambrose Light - Boring No. 1

Location: 40 27'48"N, 73 50'05"W Water Depth: 48 feet (MLW)

Top to 21.5	Fine to coarse grained gravel and cobbles with some gray clayey silt and sand (fill) mixed with pieces of brick and concrete. Sand and coarse gravel below 9 ft mixed with some rubble. Boulder or concrete from 19 to 21.5 feet
21.5 to 29	Gravelly with seams of coarse brown sand and clay with traces of wood and pieces of glazed tile and brick in dark gray clayey silt at 29 feet.
29 to 49	Gray silty fine sand with shell fragments and traces of clay.

Ambrose Light - Boring No. 4

Location: 40 27'32"N, 73 49'51"W Water Depth: 72 feet

0 to 16	Fill. Coarse brown sand with pea gravel. Black organic clayey silt 1.5 to 2 feet. Silt, sand and gravel below 2 feet with pieces of brick, roofing paper, wood and miscellaneous materials.
16 to 26	Gray silty fine sand with shell fragments.
26 to 52	Soft gray silty clay with shell. Firm by 30 feet with flecks of wood. Micaceous. Less silty at 48 feet.
52 to 62	Gray fine to medium sand.
62 to 121	Brown fine to coarse sand with pea gravel below 62 feet. Some 1.5 inch gravel below 70 feet. Coarse gravel 94 to 97 feet. Coarse sand with pea gravel below 97 feet.
121 to 196	Lignite layer, 121 to 122 feet, then light gray silty fine sand with traces of lignite and wood. Much decomposed wood at 169 feet.
196 to 250	Hard gray shaly clay with sand seams to 200 feet grades into soft shale by 218 feet, slightly lignitic. Some thin limestone seams below 238 ft.

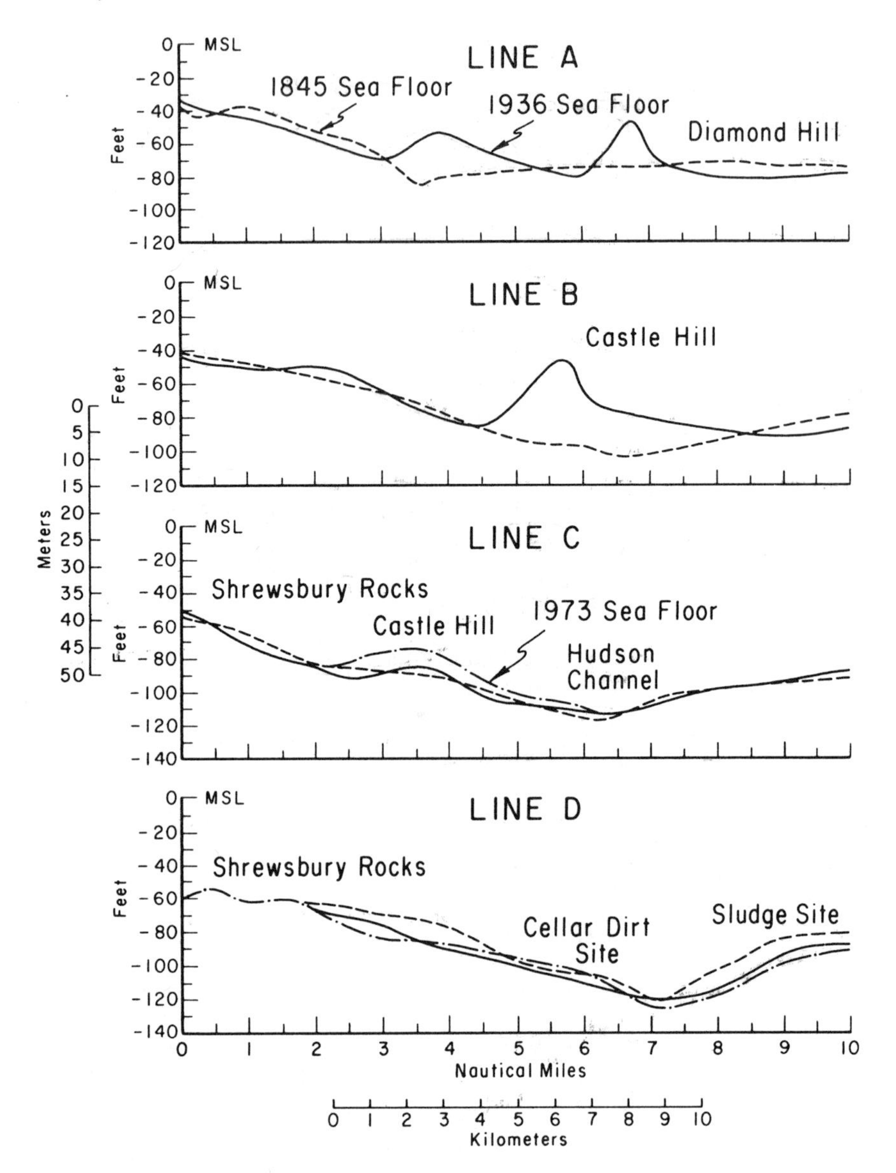

Figure 8
Sea floor profiles across the inner shelf based on data from 1845 (Figure 2, 1936 (Figure 4) and 1973 (Figure 5) bathymetric surveys. Major accretion and erosion on lines A and B occurred prior to 1936, and no significant changes in sea floor elevation are evident between 1936 and 1973. Line C shows evidence of continued net accretion at Castle Hill due to the long history of ocean dumping. Line D exhibits apparent long term net erosion except for 3 m of accretion since 1936 at the contemporary cellar dirt site. No accumulation of sewage sludge is apparent at the site east of the Hudson Channel. Source: Williams, 1975.

Dashlines in Fig. 7 from the old Scotland Lightship correspond to ship tracks to the official dump sites for some dates listed in Table 1. The disposal sites used from 1903 to 1914 and later for some materials clearly correspond with the location of maximum accretion in the Castle Hill region. Ocean dumping of cellar dirt and dredge spoil continued uninterrupted from 1914 in the same general vicinity of Castle Hill, but as shoaling became evident the dump sites were moved progressively seaward into deeper water to their present positions shown on Fig. 5.

As part of a long term study undertaken by the NOAA to determine the consequences of sewage sludge dumping in the bight, a detailed bathymetric map (Fig. 5) was made in 1973. Seafloor profiles in Fig. 8 A and B show no significant change in seafloor elevation on the mounds at Diamond and Castle Hill between 1936 and 1973. However, line C in Fig. 8 shows 5 m of accretion occurred on the southeast flank of the dredge spoil (mud dump) site immediately south of Castle Hill during that period. Greater detail is provided in Fig. 9 which clearly shows that the spoil pile has been elevated 10 m and extended south in the form of broad ridge as shown in Fig. 6C. The isopach map in Fig. 9C shows a nearly circular mound about 4.8 km in diameter has resulted from ocean dumping of dredge spoil during the

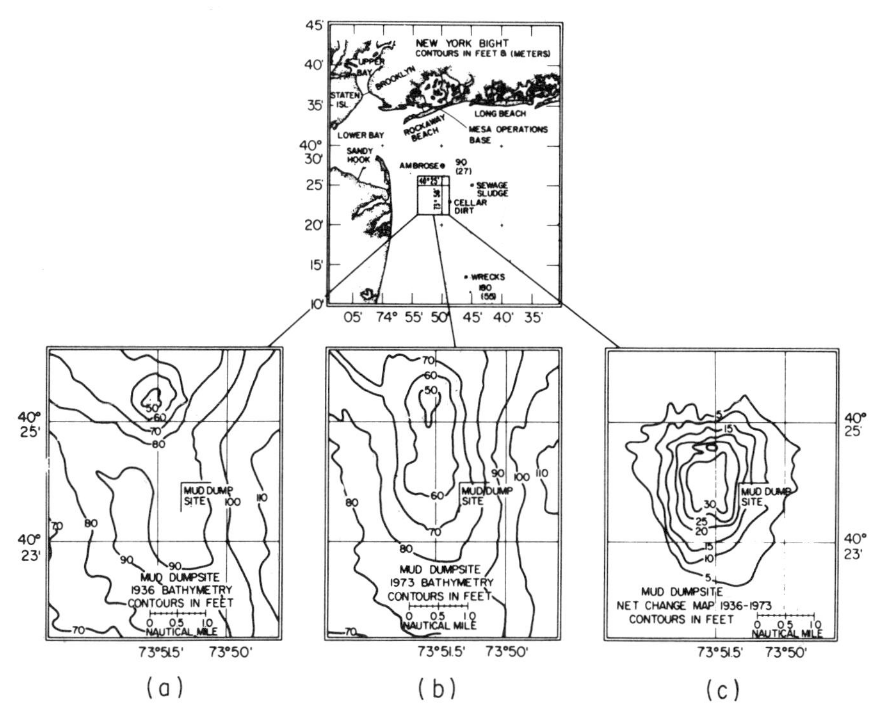

Figure 9
Changes in sea floor topography at the dredge spoil (mud) site: a. 1936 bathymetry shows an isolated mound at -50 feet resulting from earlier ocean dumping; b. 1973 bathymetry shows an elongate ridge; and c. shows net accretion to be 10 m. Source: G.L. Freeland and G.R. Merrill's studies (NOAA, 1975).

37 year period. Geometry of the mound suggests that the dumping was well controlled and restricted to the official site, and that little of the material has been transported to adjacent seafloor areas by slumping or current action. Profile D in Fig. 8 transects the contemporary dump sites for cellar dirt and sewage sludge, and shows that the cellar dirt site was apparently eroded about 1 m from 1845 to 1936, and has since aggraded by about 3 m due to recent ocean dumping. The sludge site has apparently undergone fairly continuous erosion between 1845 and 1973, in spite of the large volumes of sludge (3.7 million m^3 in 1974; Dewling and Anderson, 1976) dumped in the area since 1924. This apparent lack of accumulation of sludge at the official site is also supported by evidence from 11 CERC cores and several CERC seismic lines (Fig. 6d) in that region and by recent results from the NOAA studies. The seafloor appears to be a hard substrate with very little absorption of acoustic energy to mask subbottom geology. The 11 cores are fairly evenly distributed and only three (Cores 90, 94, 105) contain small thicknesses ($\leq$ 15 cm) of sludgelike material on top. Some of the other cores contain minor quantities of artifacts (brick, glass, wood, coal, ash clinkers) similar to those in Fig. 10, but the major component is gray silt and silty fine to medium sand. An explanation is that sludge is almost neutrally buoyant and consequently, when it is discharged from the barges most of the material remains suspended in the water column and may be carried with the prevailing currents. Final disposition and ultimate fate of the sludge is still unknown.

VOLUMES OF WASTE MATERIALS

Deriving accurate figures on the total volume of waste materials dumped at the various sites in the bight since initiation of granting permits in 1888 until the present day is difficult and full of uncertainties. Records kept by the Supervisor of New York are lacking in detail; specific composition of much of the material dumped is excluded, and the volumes given include not only solids but liquids, degradable organic matter and light floatable material too. During the dumping operation, all of the liquids and floatables would be removed, and the less dense and finer grain-size solids would remain suspended in the water column for relatively long time periods while settling to the sea floor. During the settling period some portion of the fine material would be removed by currents and carried to shelf areas adjacent to the dump sites. Consequently, close agreement between volumes of material dumped and calculations of actual in-place anthropogenic solids is understandably poor.

Ratios of solids to liquids and relative stabilities of the materials dumped varies greatly with the nature of the waste material. Cellar dirt consisting of construction and excavation debris, rock and soil, and dredged sand and gravel contains high percents of solids and are stable. However, dredge spoils composed of mud contain only about 50% solids, while industrial acids, chemicals, and sewage sludge contain 5 to 10% solids (Gross, 1972). Also, organic and water soluable materials degrade quickly and are not geologically significant.

According to the Annual Report of 1915 (U.S. Army, Corps of Engineers), 225 million m^3 of waste materials were dumped in the bight between 1890 and 1915; yielding a mean annual volume of 9 million m^3. The Annual Report of 1960, reports that from 1890 to 1960, 709 million m^3 of material were deposited at the sites in the Hudson Channel; yielding a mean annual volume of 10 million m^3. Dewling and Anderson (1976) show an average dumping volume of 10 million m^3 for 1960, and 10 million m^3 for 1965. Pararas-Carayannis (1973) shows average annual volume dumped from 1965 to 1970 to be 10 million m^3, and Dewling and Anderson (1976) show a significant increase in dumping activity since 1970 to a volume of 13.5 million m^3 for 1974. Therefore, using an annual volume of 10 million m^3, the calculated volume of all waste materials deposited at the ocean sites

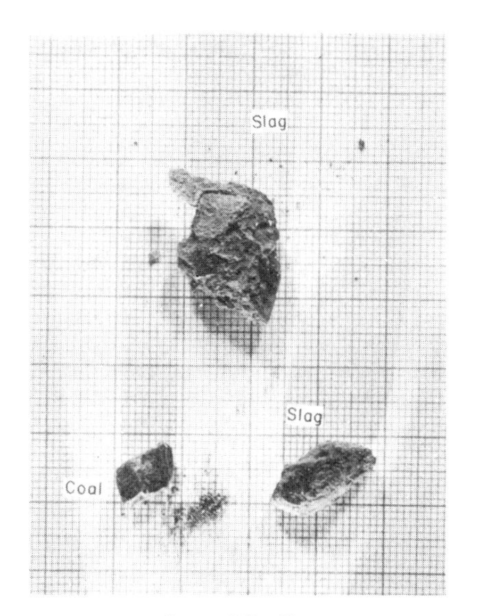

Core 99 Top

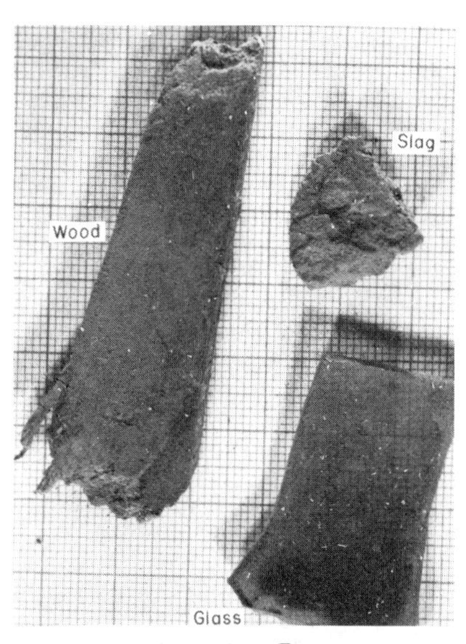

Core 99 Top

Core 52 Top

Figure 10
Photographs of typical solid waste materials found in two CERC cores. Core 99 is located at the present cellar dirt site, while core 52, south of Rockaway, is several kilometers from any known past or contemporary dump site. Grid in mm.
Source: Williams & Duane, 1974.

from 1888 to 1973 is 850 million m^3. Gross (1976) states that about 250 million m^3 of waste solids have been deposited from 1888 to 1973, with 88 million m^3 of the total having been dumped, primarily as dredge materials, between 1963 and 1973. These figures (850 and 250 million m^3) suggest the overall average of solids in the disposed material to be about 29%, which is judged to be a conservative figure.

Calculations on the isopach map in Fig. 7 yield a total in-place volume of 216 million m^3, excluding Diamond Hill. Separate measurements of the Diamond Hill mound yield a volume of 4 million m^3. Freeland and Merrill (1976) calculated the volume of accretion at the dredge spoil site (Fig. 9C) to be 93 million m^3, and 5 million m^3 at the cellar dirt site between 1936 and 1973. The addition of these four volumes of in-place solids is 318 million m^3 which represents total accretion in the immediate dump area of the bight from 1845 to 1973. This is 37% of the volume of material dumped. These figures include effects of natural sedimentation and erosion in the area which may account for a small portion of the discrepancy with Gross' figure of 250 million m^3. However, the greatest sources of error are judged to be in planimetering the isopach areas from maps used, and in determining volumes of actual solids dumped from figures in reports on the total volumes dumped. Another source of discrepancy between reported volumes of materials dumped and volumes of solids in-place on the sea floor is the degree of compaction and settlement of the solids after dumping through time. The reduction in volume would vary depending upon composition, but it could be significant.

SUMMARY

Like ancient Troy and many other cities through history, New York City has had problems of how to effectively and economically dispose of waste materials generated by both inhabitants and industries. As a Dutch, English, and finally American settlement in the 17th and 18th centuries a common solution was to incinerate or bury the materials at the periphery of the city or to dump them into adjacent waterways and wetlands. However, by the late 19th century the volume and diversity of wastes were so great that the old solutions were no longer viable, and living conditions became intolerable. The solution chosen was to barge materials to designated ocean dumpsites at the head of the Hudson Channel. Information in this paper shows that during the almost nine decades of ocean dumping, an estimated 318 million m^3 of assorted solids have accumulated in the topographic depression of the channel resulting in a topographic inversion. Several large mounds composed of anthropogenic materials have relief of 15 m above the sea floor and cover an area of about 9 million hectares. While ocean dumping activities have been shown to be a significant geologic agent in changing the sea floor morphology and sediment character, the long term effect on ecological balance of organisms and possible permanent degradation of the shelf environment is still under scientific study. Results from these studies should be used to provide intelligent guidance to decide the future of ocean dumping versus other alternatives to process and recycle waste materials.

ACKNOWLEDGEMENTS

Much of the information contained in this paper result from a CERC sponsored research effort supervised and aided by D.B. Duane (NOAA) while at CERC. Additional study was completed with help from M.G. Gross (Chesapeake Bay Institute), and H.E. Malde (U.S.G.S.). C. Everts (CERC) provided helpful review and P. Davis typed the paper.

Data presented were collected as part of the civil works function of the U.S. Army, Coastal Engineering Research Center. Conclusions presented are unofficial unless designated by other authorized documents.

REFERENCES

1. Bettmann, O.L., 1974. *The Good Old Days - They Were Terrible*, Random House, New York, 207p.
2. Dewling, R.T., and Anderson, P.W., 1976. New York Bight I: Ocean Dumping Policies, Oceanus, summer issue, 2-10.
3. Freeland, G.E., and Merrill, G.F., 1976. Deposition and Erosion in Dredge Spoil and other New York Bight Dumping Areas, ASCE Proceedings of the Specialty Conference on Dredging and its Environmental Effects, P.A. Krenkel, J. Harrison and J.C. Burdick, 3rd, Editors 936-946.
4. Gross, M.G., 1972. Geologic Aspects of Waste Solids and Marine Waste Deposits, New York Metropolitan Region, Geol. Soc. Am. Bull., 83: 3163-3176.
5. Gross, M.G., 1976. New York Bight II: Problems of Research, Oceanus, summer issue, 11-18.
6. Gunnerson, C.G., 1973. Debris Accumulation in Ancient and Modern Cities, ASCE Journal of Environmental Engineering Division, 99 EE3: 229-243.
7. Jensen, A.C., 1975, Artificial Fishing Reefs, MESA New York Bight Atlas Monograph 18, New York Sea Grant Institute, Albany, NY, 23 pp.
8. McClelland Engineers, Inc., 1963. Fathometer Survey and Foundation Investigation Ambrose Lt. Station New York Harbor Entrance, (Report No. 63-162-1, unpublished).
9. McCullough, D., 1972. *The Great Bridge*, Simon & Schuster, New York, 636p.
10. National Oceanic & Atmospheric Administration, 1975. Ocean Dumping in the New York Bight, Technical Report ERL 321-MESA-2: 78p.
11. Pararas-Carayannis, G., 1973. Ocean Dumping in the New York Bight, An Assessment of Environmental Studies, U.S. Army Coastal Engineering Research Center, Technical Memorandum 39: 159p.
12. U.S. Army Corps of Engineers, 1885-1968. Annual Reports of the Chief of Engineers, Supervision of New York Harbor.
13. Veatch, A.C., and Smith, P.A., 1939. Atlantic Submarine Valleys of the United States and the Congo Submarine Valley, Special Paper No. 7, Geol. Soc. Am.
14. Wigmore, H.L., 1909. Memorandum on Dredging Work in Ambrose Channel, U.S. Army Prof. Memo., Engineer Bur., 1:57-62.

15. Williams, S.J., and Duane, D.B., 1974. Geomorphology and Sediments of the Inner New York Bight Continental Shelf, U.S. Army Coastal Engineering Research Center Technical Memorandum 45: 81p.
16. Williams, S.J., 1975. Anthropogenic Filling of the Hudson River (Shelf) Channel, Geology, 3(10): 597-600.
17. Williams, S.J., 1976. Geomorphology, Shallow Subbottom Structure and Sediments of the Atlantic Inner Continental Shelf Off Long Island, NY, U.S. Army Coastal Engineering Research Center, Technical Report 76-2: 123pp.

Mud Deposits near the New York Bight Dumpsites: Origin and Behavior

George L. Freeland
Donald J. P. Swift
Robert A. Young

ABSTRACT

A detailed investigation of the floor of the New York Bight apex reveals that muddy sediment occurs in the Christiaensen Basin, in the fringing lagoons and marshes near the mouths of tidal inlets, and in ephemeral patches on the Long Island shoreface, where mud settles into the troughs of sandwave-like bedforms. The appearance and disappearance of these nearshore mud patches has led to an erroneous concept of a sewage-sludge "front" advancing toward the Long Island beaches from the sewage-sludge dumpsite. While Bight apex mud deposits do contain sewage sludge, this contamination is usually not apparent in gross physical properties or composition; sophisticated chemical techniques are required to detect the contamination. Fine-grained deposits near the New York Bight sewage-sludge dumpsite are primarily of natural origin and should, therefore, be called mud, not sludge. Fine-grained sediment at the dredge-spoil dumpsite is also mostly natural sediment, but it is anthropogenic in that it is dredged harbor and estuary sediment.

The impact of sewage-sludge dumping in the New York Bight apex depends to a large extent on the pathways and rates of the natural sediment transport system. The rate of sewage-sludge input is higher than the estimated rate of fine sediment input from land runoff. However, the input of suspended fine sediment into the Bight apex by coastwise shelf currents, though not precisely known, may be much higher than the input from either dumping or land runoff. The rate at which organic matter is cycled through the suspended-sediment system via biological production and consumption of organic suspended sediment is estimated to be almost an order of magnitude higher than the rate of suspended-sediment input from either land runoff or dumping. Such limited information as is available suggests that the rate of sewage-sludge dispersal is high relative to the rate of input. Our model leads us to infer that 1) sewage sludge is rapidly diluted in the Bight apex by natural sediment, and 2) the more stable sewage-sludge components are widely dispersed through the natural mud deposits of the New York Bight system.

INTRODUCTION

Intense public interest has been generated in recent years concerning the nature and distribution of mud deposits on the Long Island inner shelf (Figs. 1, 2). The interest stems from public concern over the high rate of sewage-sludge dumping at the dumpsite 25 km seaward of New York Harbor entrance (4 x 10^6 m^3, containing 5% solids, in 1973), and confusion concerning the difference between anthropogenic sewage-sludge deposits and naturally occurring ocean floor muds (Soucie, 1974). There has been a tendency, in public discussions, to refer to any fine-grained sea floor deposit as

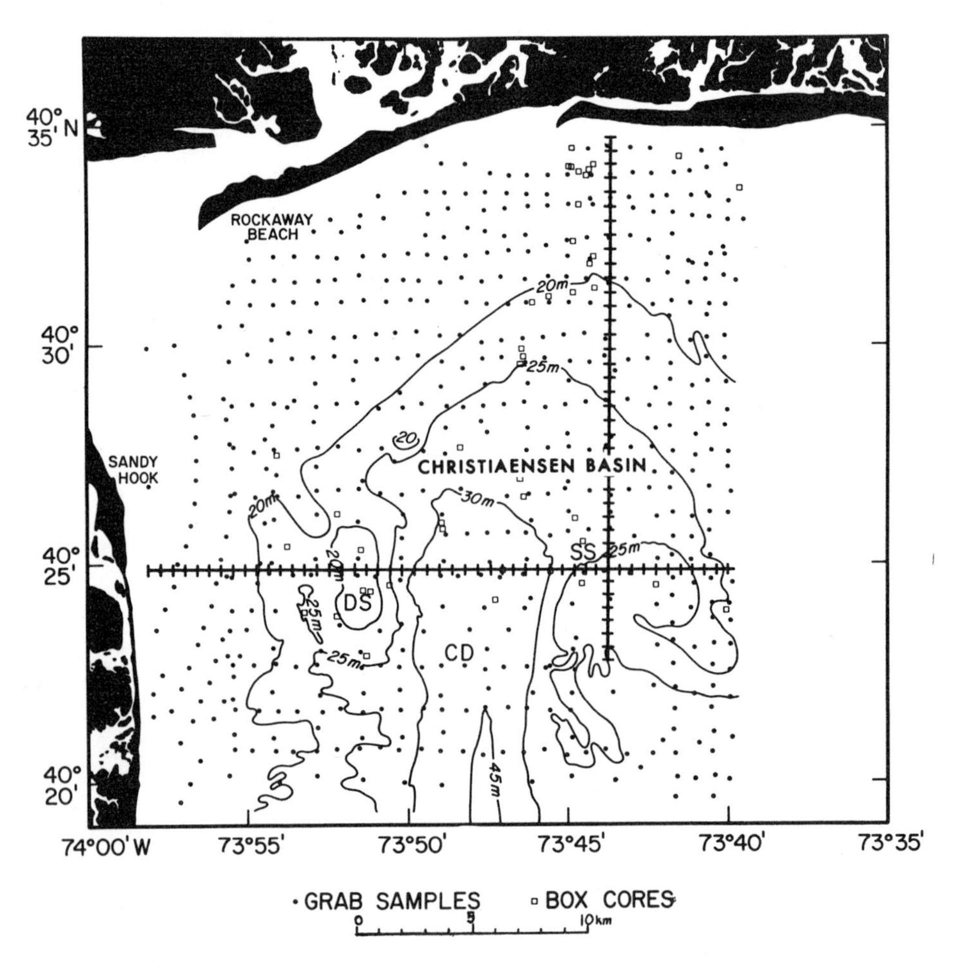

Figure 1
Bathymetry of the New York Bight apex mud grab-sample net on which map of bottom sediment distribution is based. CD is cellar-dirt dumpsite, DS is dredge-spoil dumpsite, SS is sewage-sludge dumpsite, HSV is the upper Hudson Shelf Valley, CV is Cholera Bank, and ticked lines are quarterly monitoring transects.

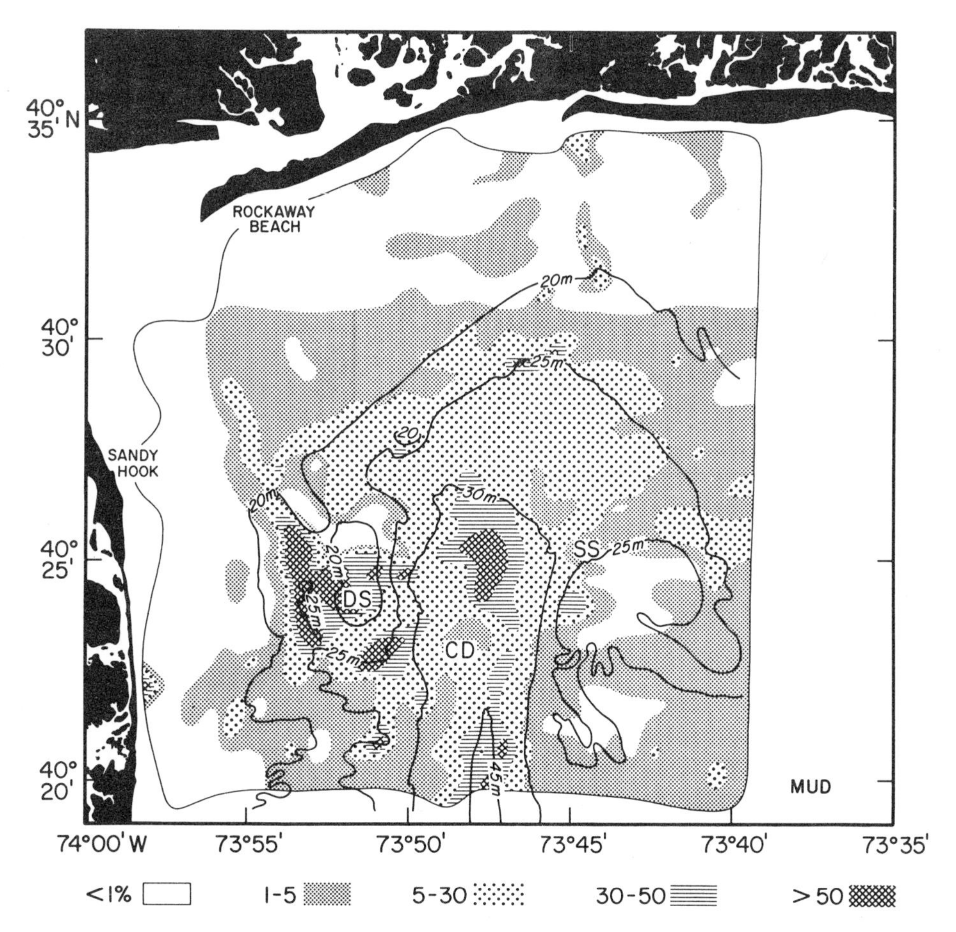

Figure 2
Map of percent mud (sediment finer than 62μ) on the floor of the New York Bight apex. From Freeland and others, in press.

"sewage sludge", and newspaper articles have referred to "sludge beds" and a sludge "front" creeping from the dumpsite toward the Long Island beaches (Pearson, 1974a, b, c). In addition to sewage-sludge particles, approximately 0.5 x 10^6 m^3/yr. of dumped dredge-spoil fine particles settle slowly to the bottom and are widely dispersed away from the dredge-spoil dumpsite proper (Freeland and Merrill, 1976). They also become mixed with natural and sewage-sludge muds.

Recent studies indicate that the mud deposits of the New York Bight apex (the inner shelf in the vicinity of New York Harbor mouth; Fig. 1) and upper Hudson Shelf Valley generally contain no more organic carbon than is found in continental shelf muds from other areas (Hatcher and Keister, 1975; Duedall, 1976). The muds of the New York Bight apex may well contain significant amounts of anthropogenic organic matter in addition to naturally occurring organic matter, but the detection of the differing chemical signatures of the "natural" organic component of shelf-floor mud versus that of sewage sludge requires sophisticated analytical techniques (see Segar and Cantillo, 1976; Harris, 1976; Hatcher and Keister, 1976; Hatcher and others, 1977; Schaeffer, 1977 for discussion of such methods). Gross (1972) and Mueller and others (1976) have pointed out that the input of anthropogenic sediment into the New York Bight apex exceeds the input of natural land-derived sediment, but this ignores the contribution of the alongcoast flux of natural sediment, which may be much greater than either the contribution from land or from sludge dumping. Drake (1977, p. 222), calculated that the resuspension of bottom sediment in the New York Bight apex during a two-day period of strong wind-driven currents was equivalent to between 10 and 20 days of sewage-sludge dumping. Dilution of sewage sludge with other sediment must be rapid during the stormy winter months, when strong wind-forcing of water flow occurs every seven to ten days.

The purpose of this paper is to assess the origin and behavior of fine particles in the apex in order to gain insight to the ultimate fate of fine dumped materials. We interpret data on fine-grained bottom deposits in the apex in light of existing models of fine sediment transport on the inner continental shelf.

DISTRIBUTION OF DEPOSITS

Mud deposits in the New York Bight apex occur primarily in the amphitheater-like basin at the head of the Hudson Shelf Valley referred to by Veatch and Smith (1939) as the Christiaensen Basin (Figs. 1, 2). The pattern is compatible with the thesis of McCave (1972, Fig. 94) that mud deposition on the continental shelf does not simply occur in quiet-water settling basins, but is controlled by near-bottom suspended-sediment concentration, as well as by the near-bottom hydraulic activity. In Figure 2, the 5% mud isopleth thus defines an area within which mean near-bottom hydraulic activity is offset by high prevailing near-bottom suspended-sediment concentration so that deposition may occur. In the winter, concentrations are commonly 1 to 3 mg/l 2 m above the sea floor (Drake, 1977)

Repeated occupation of two sampling and sidescan sonar transects indicates the extent of grain-size variation and the patchy distribution of bottom sediment types in the apex (Figs. 3, 4). Such changes in grain size suggest primarily small-scale spatial variations, with the major changes due to temporal variation. Some of the small-scale changes are, no doubt, due to the ship's maneuvering error over small sampling areas. The positions of grab samples are known to $\pm$ 3 m, but successive samples from a single station may be tens of meters apart. Furthermore, Figures 3 and 4 indicate that the Christiaensen Basin deposits are primarily muddy sands rather than sandy muds; over-50% mud occurs only near the dredge-spoil dumpsite and in a relatively small area in the central basin.

Of special concern are the temporal and spatial patterns of mud distribution on the Long Island south shore, where beaches are

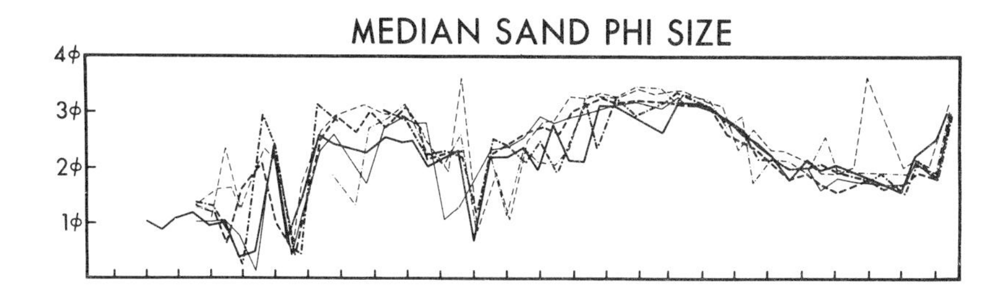

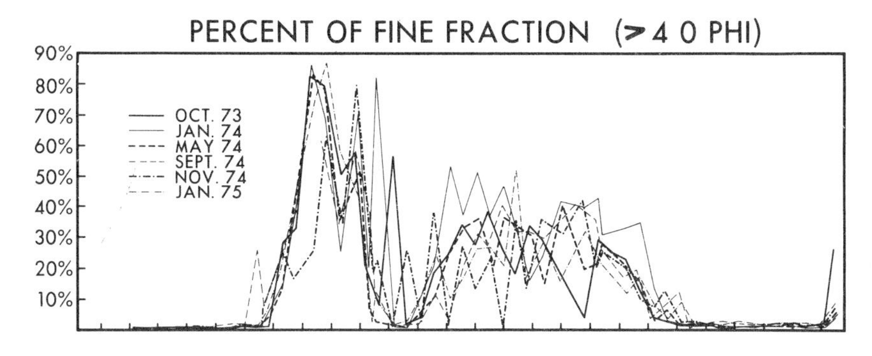

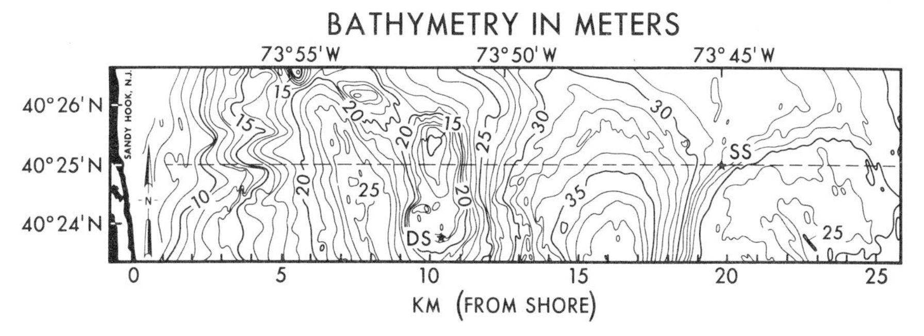

Figure 3
Median diameter of sand and percent mud on an east-west transect through the New York Bight apex for five consecutive reoccupations. From Stubblefield and others, 1977.

imputed to be threatened by the advance of "sludge beds" (Pearson, 1974a, b, c). The incomplete sample net shown on Figure 5 indicates that aureoles of greater than 5% mud in bottom sediment occur around the mouths of Fire Island Inlet and Jones Inlet; further sampling at the mouth of East Rockaway Inlet might show a similar aureole. Inside these inlets the lagoons of the western Long Island south shore are floored primarily by mud and fine muddy sand (Smith and Ali, 1973).

Highly localized mud deposits south of Atlantic Beach (Figs. 2, 5) have lead to reports of onshore movements of "sludge beds". Harris (1976) first called attention to the facts that these mud deposits tend to be found in the troughs of sandwave-like features,

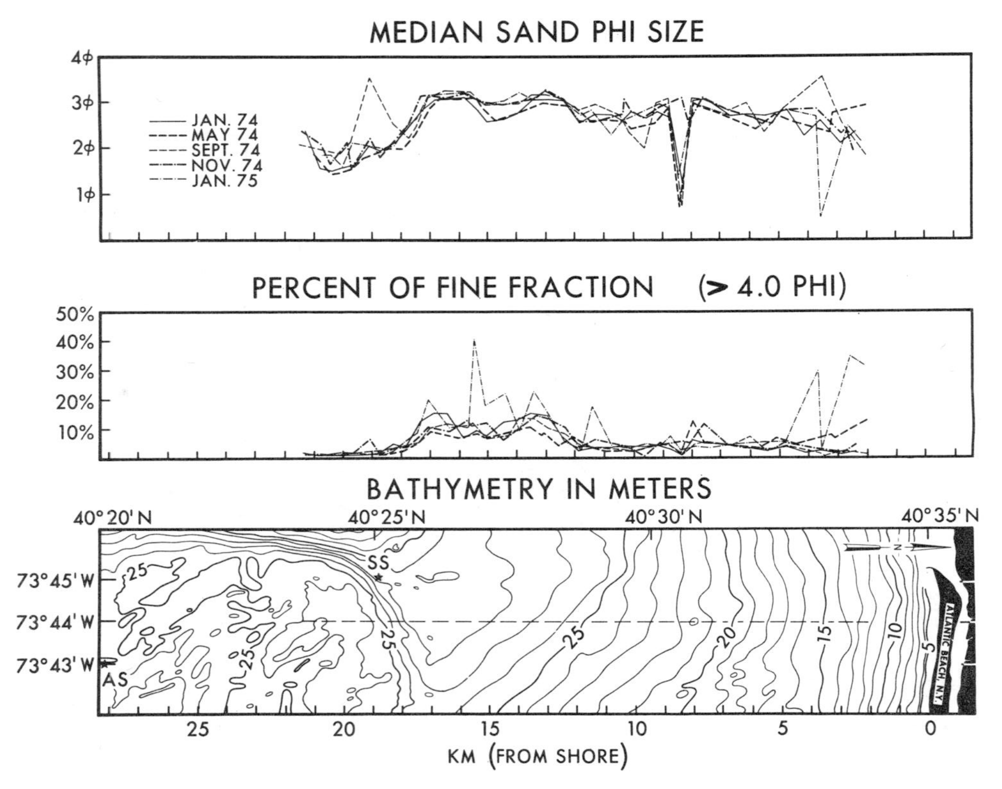

Figure 4 Median diameter of sand and percent mud on a north-south transect through the New York Bight apex for five consecutive reoccupations. From Stubblefield and others, 1977.

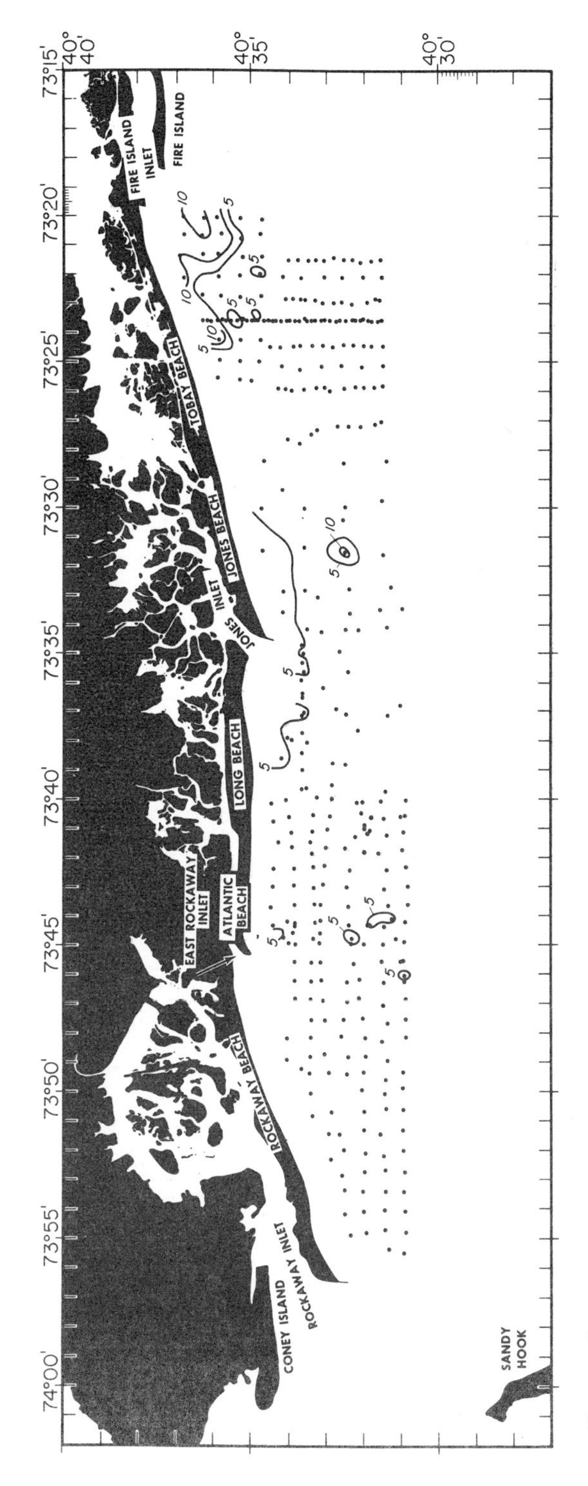

Figure 5
Percent mud on the western Long Island south shore. From Freeland and Swift, in press a)

and that their distribution changed markedly on a time scale of months. However, his results were hampered by inadequate precision in navigation positioning. Some of the same mud patches were encountered during the 1972-1974 grab sampling program of Freeland and others (in press), and in the 1974-1975 transects of Stubblefield and others (1977), especially their January 1975 survey (Fig. 4). However, the 0.5 km sample spacing was too wide for detailed resolution necessary to delineate the true extent of these deposits. Sidescan sonar records and grab samples taken in January 1975 did confirm that the mud patches are associated with the troughs of a sandwave-like pattern that slowly expanded during 1974 (Stubblefield and others, Fig. 14).

On the basis of fathometer records, Harris (1976) reported spacings of 3 to 10 m and heights of approximately 1 m for the sandwave-like bedforms of Atlantic Beach. The features observed on the transect of Stubblefield and others (1977, Fig. 14) had spacings of about 50 m. A detailed sidescan sonar study of the area off Atlantic Beach under calm conditions (Freeland and others, in press; Swift and Freeland, in press) indicates that the bedforms are primarily slight (less than 1 m) depressions, 2 to 200 m wide, floored by coarse, rippled, sand, gravel, and occasionally mud (Figs. 6, 7). The depressions are asymmetrical, with sharply defined northeast sides and gradational southwest sides (Fig. 7).

DISCUSSION

COASTAL TRANSPORT MODEL

The problem of the time and space distribution of coastal mud deposits must be reviewed in the light of existing models of coastal sedimentation. Several decades of studies, summarized by Postma (1967) and Drake (1976) indicate that estuaries and lagoons are generally more turbid than either the rivers that serve them or the adjacent shelf water mass. Hydrodynamic processes operating within these tidal water bodies (two-layer estuarine circulation; settling lag; scour lag, time-velocity asymmetry of the tidal cycle; see Postma, 1967) cause suspended-sediment particles to migrate into them from both rivers and the adjacent shelf. Particles which move into the estuary from both shelf and river sources are entrained and trapped in the turbidity maximum (Schubel, 1972; Schubel and Okubo, 1972). The sediment concentration increases until as much escapes as is added, and equilibrium is attained at a concentration that is many times that of the sources.

Outside the estuaries and lagoons, the zone of high suspended-sediment concentration decreases exponentially across the shelf. Near-bottom concentrations are high on the inner shelf as a consequence of more intense wave surge in the shallow water; near-surface values there are high because of the turbid plumes of brackish surface water that extend from inlets and estuary mouths during ebb tides and because of high primary biological productivity during spring and summer (Figs. 8, 9). Thus, there is a feedback loop on the inner shelf, especially in the nearshore turbid zone in the New York Bight apex analogous to the estuarine feedback loop causing the

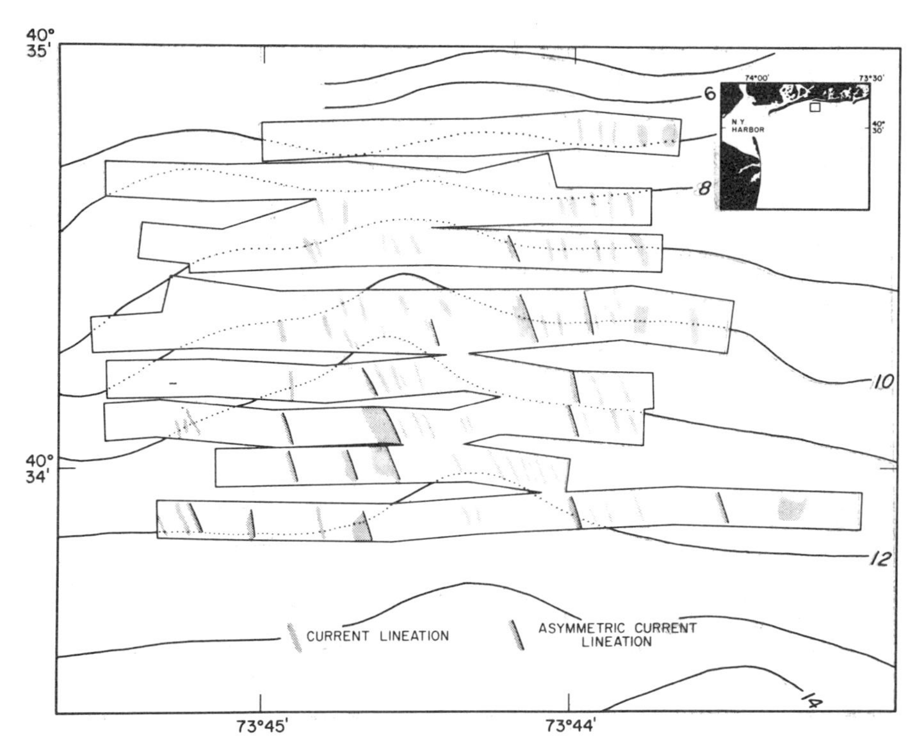

Figure 6
Array of sandwave-like bedforms at Atlantic Beach, Long Island. Asymmetric current lineations have sharply defined up-current sides and gradational down-current sides. Depths in meters. From Freeland and Swift, in press a.

estuarine turbidity maximum. Both the two-layer estuarine circulation system and the typical fairweather conditions of weak coastal upwelling (Lavelle and others, 1976; Baylor, 1973; Scott and Csanady, in press) serve as hydrodynamic traps to concentrate the suspended sediment thus introduced to the coastal water column. In the apex, another highly concentrated source is dredge-spoil and sewage-sludge dumping. While the bulk of the dredge spoils goes directly to the sea floor, most of the sewage sludge enters the inner-shelf and shelf-estuary feedback loops. An effective nearshore trapping mechanism is landward mass transport of more turbid bottom water, then seaward transport, by the wave-driven system of turbid rip-current

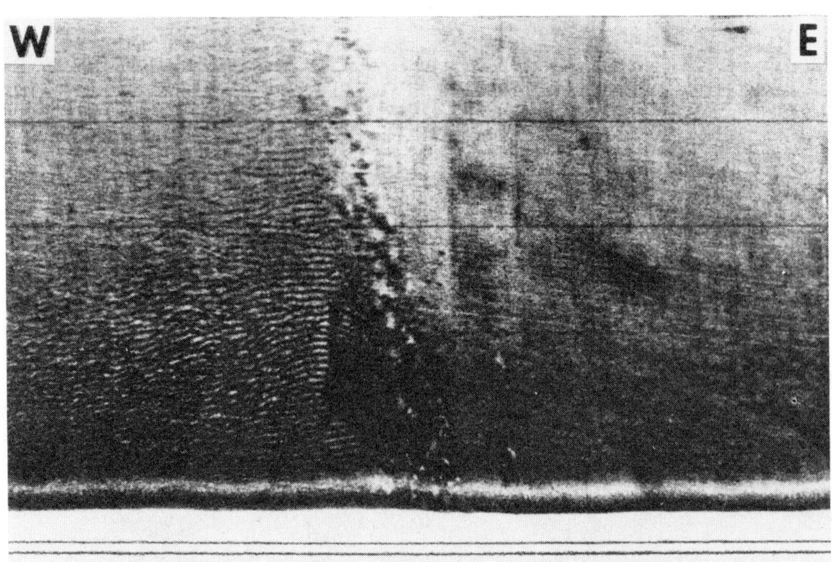

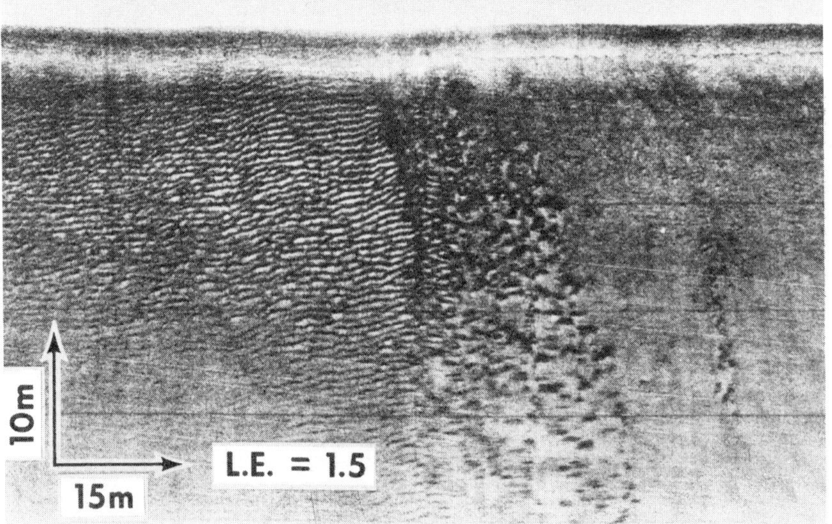

Figure 7
Sidescan sonar record of sandwave-like bedform at Atlantic Beach, Long Island. Fine sand to east, muddy gravel in center, coarse sand west, becoming finer toward west.

plumes (Fig. 10).

The suspended-sediment transport system in the New York Bight, then, may be viewed as consisting of two basic pathways (Fig. 11); 1) fluvial effluent, which joins 2) a coastwise transport path created by the net southerly drift of turbid inner-shelf water (Beardsley and others, 1976). In between are the feedback loops described above, and short to long term (days to centuries) storage

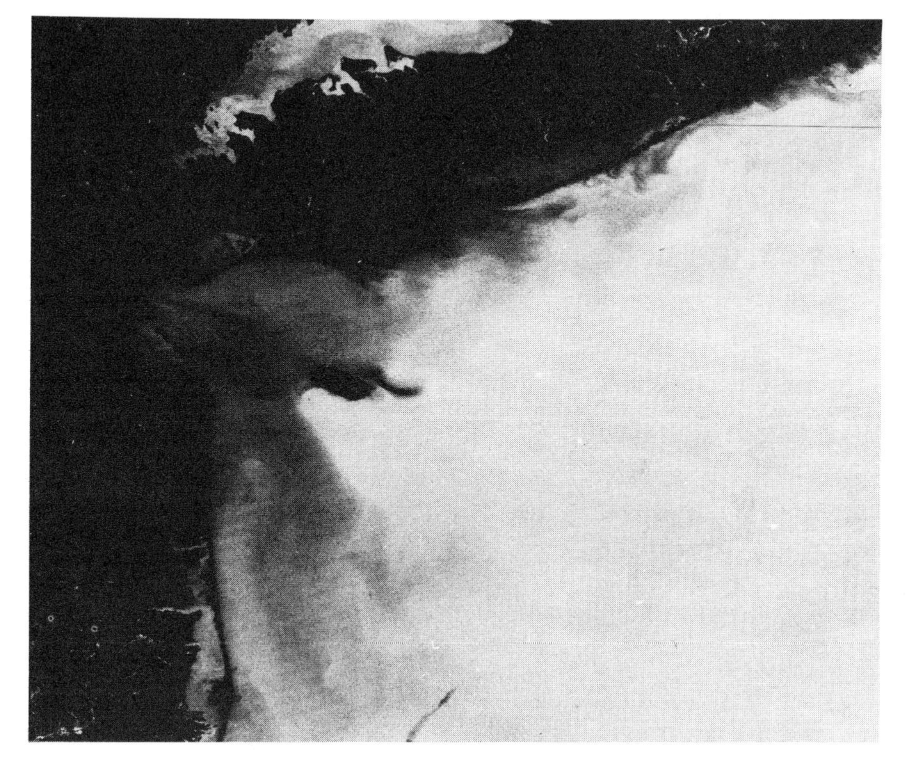

Figure 8
LANDSAT satellite image of the New York Bight apex revealing nearshore turbid water. Note Hudson-Raritan estuary plume, dark highly turbid water areas in the central apex, probably from acid-waste dumping, and nearshore Long Island and New Jersey turbid waters. Wind had been blowing an average of 15 to 22 mph from 240° to 300°T for the previous 58 hours. Scene 1258-15082, MSS Channel 5 (0.6 to 0.7 μm, orange-red), April 7, 1973, 1008 hrs EST. Courtesy of NASA and Dr. George Maul, AOML/PhOL.

on the estuary, lagoonal, or shelf floors, or on fringing salt marshes (Fig. 12). Sediment recycling on the estuarine feedback

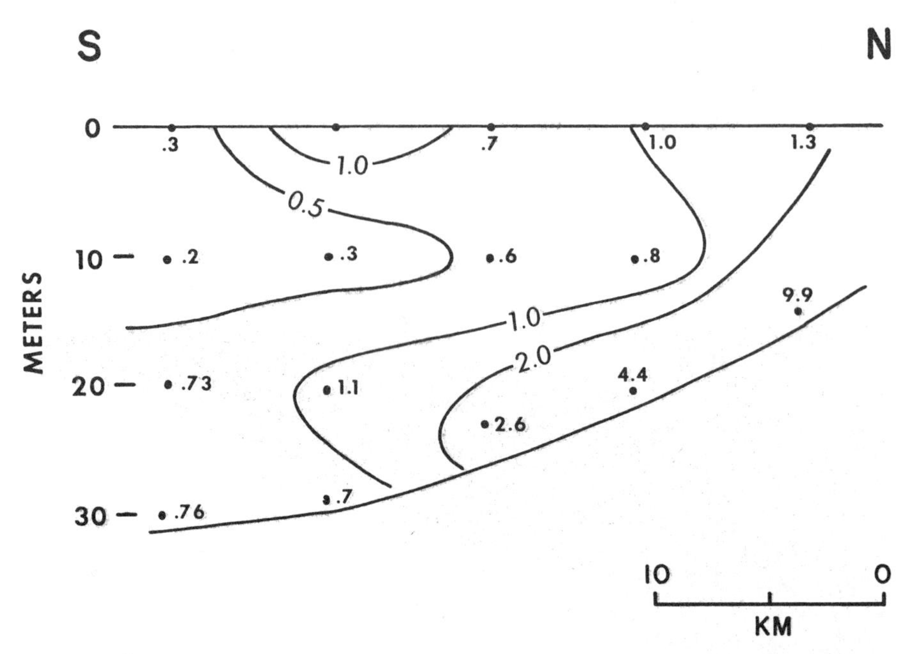

Figure 9
Vertical distribution of total suspended load on a north-south transect south of Atlantic Beach on September 16-20, 1973. From Swift and others, 1975; Drake, 1977.

loop is many times greater than is the throughput, as indicated by higher suspended-sediment concentrations in the Bight apex than in the adjacent coastal waters of New Jersey and Long Island (Drake, 1977; Manheim and others, 1970).

Application of this model to the mud distribution maps (Figs. 2-5) suggests that the Christiaensen Basin mud deposit is situated so as to constitute a reservoir in the cycle of suspended-sediment exchange between the Bight apex and the Hudson estuary. Suspended sediment can escape either downcoast to the south in the strong flows associated with the Hudson Shelf Valley (Gadd and others, in press) or north towards the long Island beaches when the circulation

Figure 10 (at right)
Schematic diagram of coastal circulation pattern believed to control nearshore suspended-sediment concentration. Above: wind and density-driven flow causes onshore bottom flow, upwelling, and offshore surface flow. Suspended sediment tends to settle out of this upper layer, so that sediment movement closes the "open loop" of fluid flux. In the very nearshore (below) this two-layer circulation is reinforced by landward mass transport of bottom water by waves, and seaward flowing rip currents on the surface. From Freeland and Swift, in press b.

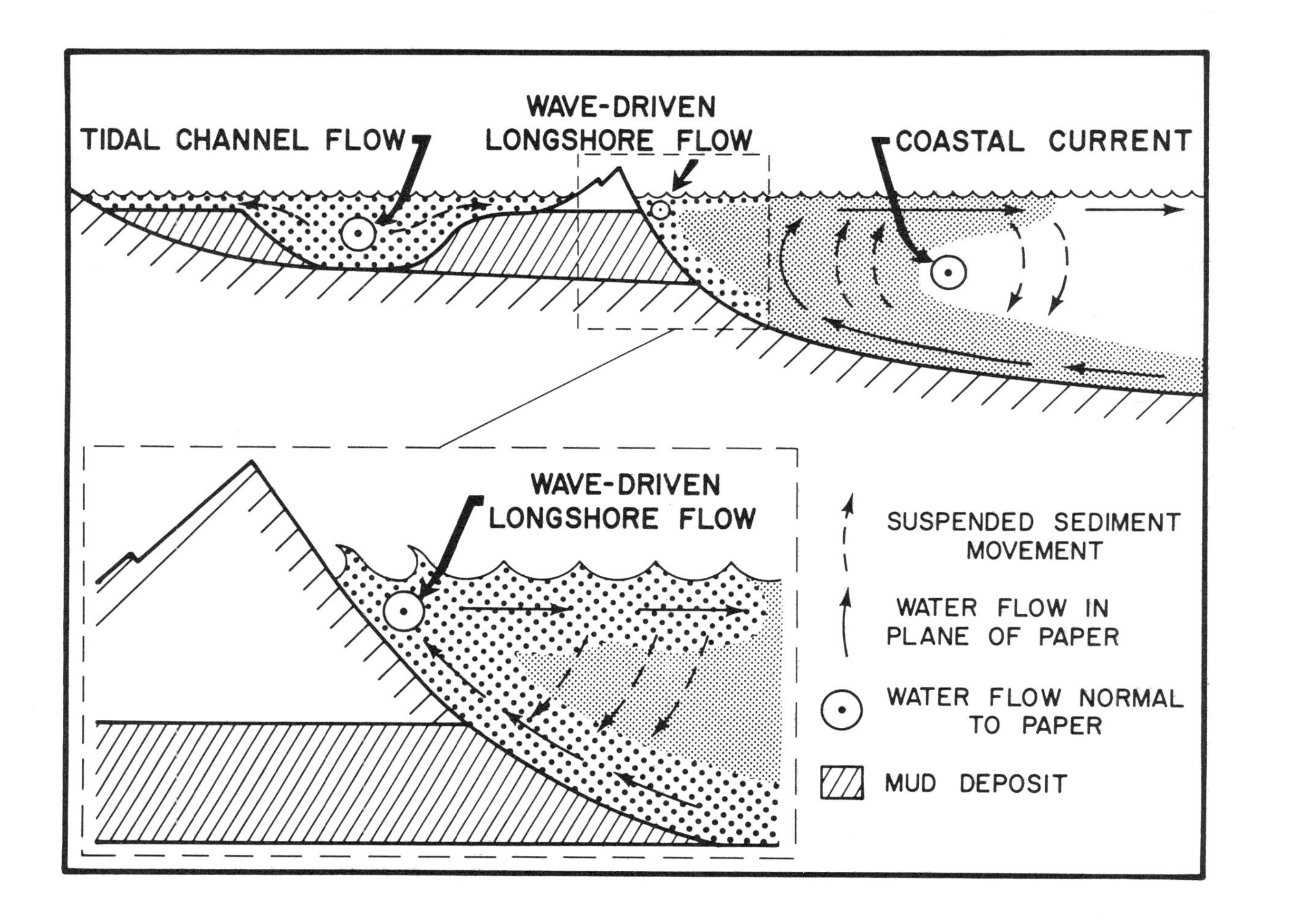

TIDAL CHANNEL FLOW
WAVE-DRIVEN LONGSHORE FLOW
COASTAL CURRENT
WAVE-DRIVEN LONGSHORE FLOW
SUSPENDED SEDIMENT MOVEMENT
WATER FLOW IN PLANE OF PAPER
WATER FLOW NORMAL TO PAPER
MUD DEPOSIT

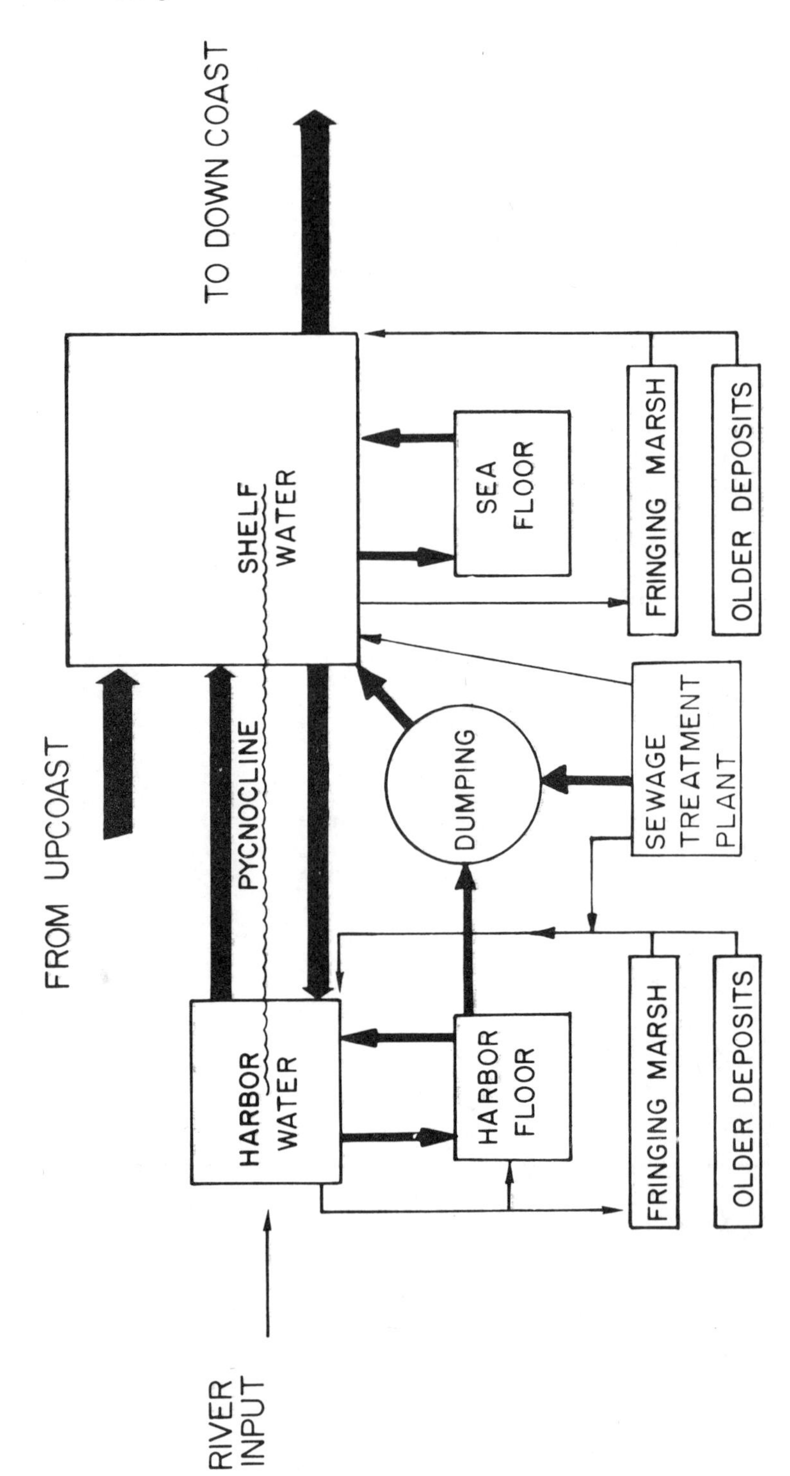

Figure 11
Schematic diagram of pathways and reservoirs controlling suspended-sediment budget of the New York Bight apex. From Freeland and Swift, in press b.

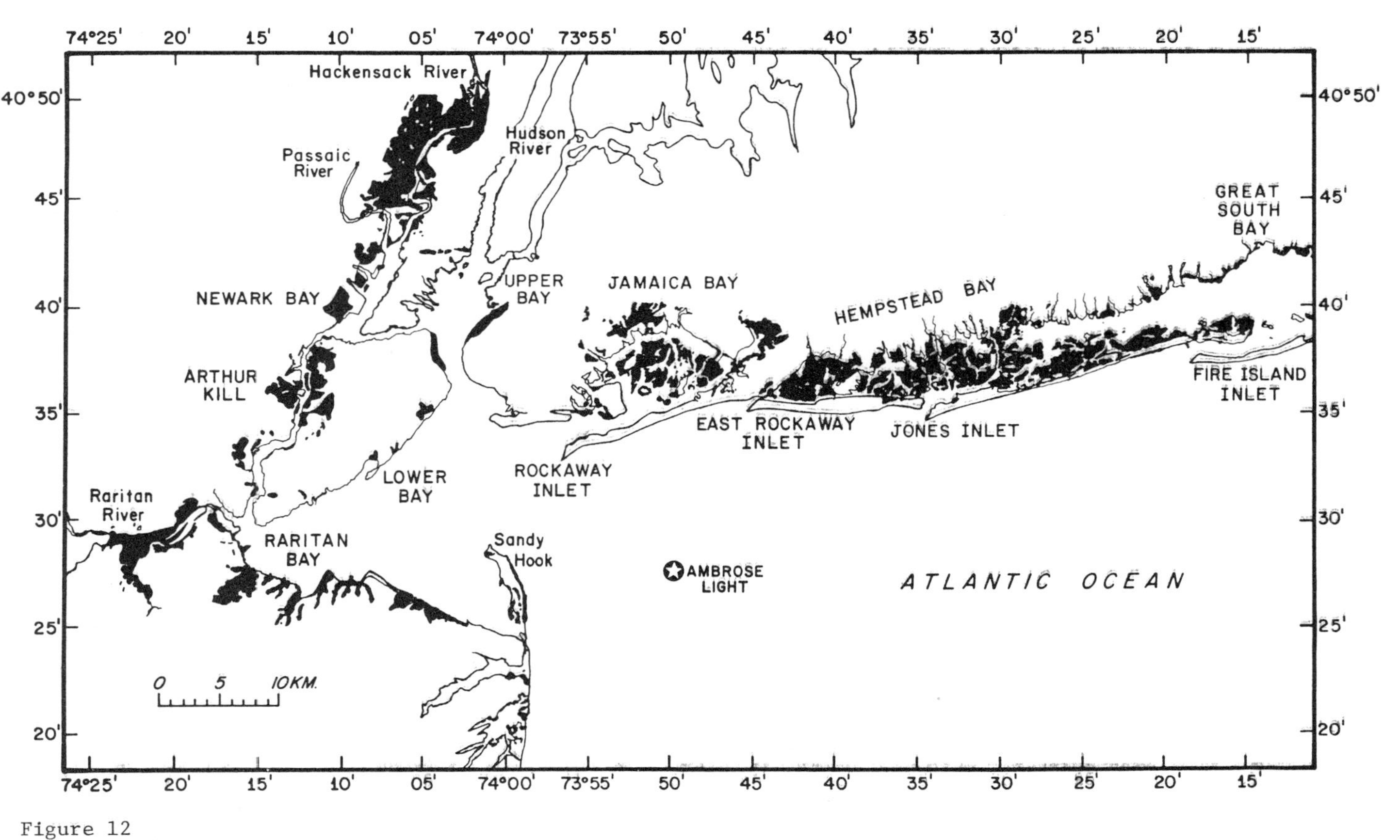

Figure 12
Wetlands of the New York Bight apex.

pattern known as the "Bight apex gyre" is in effect (Charnell and Mayer, 1975; Drake, 1977; Gadd and others, 1978). The fine-sediment transport budget approaches equilibrium in that nearly as much material escapes as is added to this system. Except in the immediate vicinity of the dredge-spoil dumpsite, the rate of accretion or erosion of the floor of the Christiaensen Basin is so low that during 37 years between bathymetric surveys, consistent trends cannot be distinguished from noise in the data (Fig. 13; Freeland and Merrill, 1976).

The entry of dumped waste into the fine-sediment transport system in the vicinity of the Christiaensen Basin mud deposit deserves special comment. Calculations indicate that of the sand, mud, and

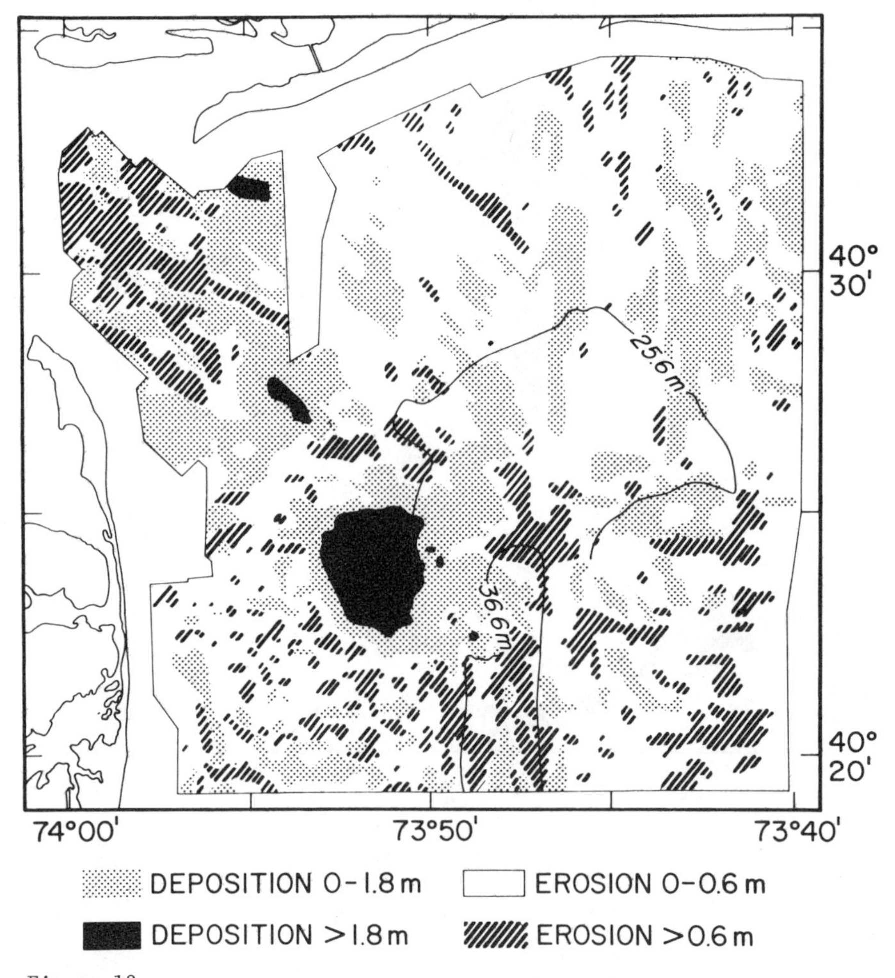

Figure 13
Net bathymetric change in the New York Bight apex between 1936 and 1973. From Freeland and Merrill, 1976.

gravel dumped at the dredge-spoil dumpsite since 1936, 87% has remained as a lag deposit, while the remainder has been dispersed (Freeland and Merrill, 1976). The nearly neutrally buoyant sewage sludge is considerably more mobile. The floor of the Christiaensen Basin beneath the sewage-sludge dumpsite proper is composed of sand containing negligible amounts of fine material (Fig. 2). Recent acoustical studies by Proni and others (in press) indicate that a large fraction of a sewage-sludge dump remains as a vertical column in the water, with high particulate concentrations, for several hours. Sludge particles are fed into the bottom nepheloid layer as a base surge of heavier particles during dumping. Less dense particles spread laterally at intermediate depths at the main thermocline and at secondary density discontinuities, and at the surface. Thus dispersion of sewage sludge through the transport pathways of the Bight apex begins even during the dumping process.

Our model for coastal fine-sediment transport suggests that the mud deposits of the western Long Island shoreface (Fig. 5) are controlled by the relationship between near-bottom hydraulic activity and near-bottom sediment concentration in turbid inner-shelf water. Major suspended-sediment inputs to this water mass are: 1) tidal exchange with inlets into Jamaica, Hempstead, and Great South Bays; 2) the Hudson River surface plume, driven eastward along the Long Island coast when blown by south and west winds; and 3) landward bottom drift from the Christiaensen Basin. The dominant flow component of this nearshore-water mass is constrained to an east-west direction by the proximity of land (Charnell and Mayer, 1975; Lavelle and others, 1976; Lavelle and others, in press; Gadd and others, 1978), but reverses its direction with a frequency of hours to days depending on tide and wind conditions. The sense of the weak two-layer flow (upwelling versus downwelling) also reverses on a similar time scale (Scott and Csanady, in press, especially Fig. 7). Thus, the exchange of sediment among the nearshore sediment sources, and between them and the shoreface mud deposits may be approximated as a stochastic process.

The shoreface mud patches seaward of Atlantic Beach (Fig. 5) that led to fears of a "creeping sludge front" have a complex distribution in time and space. In addition to being controlled by the model just described, they lie in the troughs of sandwave-like bedforms which are characteristic of the shoreface of the Middle Atlantic Bight and the Brazilian inner shelf (Swift and Freeland, in press; Figueiredo and Kowsmann, 1976). These bedforms appear to be flow-transverse features (therefore, "sandwave-like"), but their height (less than 1 m) is negligible with respect to spacing (greater then 200 m). Their orientation, with steeper flanks facing westward and slightly offshore, suggests that they are responses to the strong westward flows associated with winter "northeaster" storms (Beardsley and Butman, 1974; Lavelle and others, 1976, in press), which have a slight offshore component of bottom flow due to downwelling (Swift and others, 1975, Fig. 13; Magnell, 1976, p. 6-114 to 6-119, Figs. 6.5-1e, 6.5-5c). The shallow, gravelly troughs between sandwaves apparently become hydrodynamically sheltered zones during the quiescent summer months when net water movement is a slow eastward drift modulated by the surge of low amplitude swells. Mud deposited in troughs during the summer is buried or eroded (resuspended) during the winter season of repeated storms when the bedforms are reactivated.

FINE SEDIMENT BUDGET

The distribution of mud deposits in the New York Bight apex therefore appears to be controlled primarily by natural forces rather than by the activities of man except in the immediate area of the dredge-spoil dumpsite. There appears to be no compelling reason to believe that the distribution of muds is significantly different than when Henry Hudson's ships first visited the area, although the total area covered by mud may be somewhat greater. The 1845 chart of the apex (Survey of the Coast of the United States) shows "black mud", "mud", and "blue mud" mixed with "gray sand" in the Christiaensen Basin and the upper Hudson Shelf Valley at depths below 85 ft (26 m), a considerable area of "gray sand black specks" and "yellow sand black specks" on Cholera Bank at and around the present sewage-sludge dumpsite, and "mud", "muddy", "mud shells", and "black and white sand" to within 500 m of the shore off Rockaway, Long, and Jones Beaches.

Anthropogenic activities are more likely to change the composition of modern muds than their areal distribution. The degree of contamination of natural muds by dumped fine-grained wastes depends not only on the rate of dumped-waste input relative to the rate of natural mud input, but also on the rates of exchange of fine sediment along the pathways diagrammed in Figure 11 relative to the total fine sediment input. The exchange rates determine the extent to which dumped materials will be diluted with natural materials and dispersed. The concentration of suspended particulates and the level of hydraulic activity of the water masses determine the extent and duration of mud deposition and resuspension.

A few recent studies have provided preliminary numerical data on exchange rates and contamination. McLaughlin and others (1975), in their conceptual model of the New York Bight ecosystem, estimate sediment flux (both for fine and coarse sediment) into and out of the Bight (Table 1, this paper). An outstanding aspect of their budget is the importance of the biological subcycle. Biological assimilation (uptake) removes 14.3×10^6 tons/yr. of organic matter from the system, becoming unused dissolved matter. Schaeffer (1977) measured organic $^{13}C/^{12}C$ ratios in 14 Bight apex sediment samples, one midshelf Hudson Shelf Valley core, and two sewage treatment plant sludge samples. The results indicate that treatment plant organic matter (coming from both dumping of sludge and through outfalls) varies from 0% of the organic fraction in three samples off Rockaway Beach and Sea Bright, New Jersey, to 70-100% in the Christiaensen Basin near the sewage-sludge dumpsite. Total organic matter (based on the Wards Island treatment plant sample being 100% organic matter) reached a maximum of only 6.98% in a sample taken east of the dredge-spoil dumpsite and was less than 1% in all of the nearshore samples. Therefore, while contamination is occurring, the vast majority of all organic matter (including sludge) is being both biologically decomposed and physically dispersed. The figures presented in Table 1 also suggest that physical processes would tend to result in low ambient values of carbon due to dilution and dispersion.

An examination of the assumptions made by McLaughlin and others

Table 1
Estimates of fine sediment flux through New York Bight in 10^6 tons/yr. dry weight. From McLaughlin and others, 1975.

Pathway	Inflow	Outflow
Atmospheric input	0.1	
Dumped waste	2.1[a]	
Estuaries to shelf (includes river input)	0.5	
Shelf to estuaries		0.3
Biological production[b]	14.0	
Biological uptake[b]		14.3
Shelf advection		
From northeast	1.8	
To southwest		1.8
To shelf edge		1.0
Total	18.5	17.4
Net	1.1	
Total non-biological	4.5	3.1
Net non-biological	1.4	

[a]Does not include 4.0 x 10^6 tons/yr. of coarse dumped material (Mueller and others, 1976).

[b]Estimates for coarse sediment and fine sediment combined, marine organic materials only. See text.

(1975) suggests that their estimates for rates of exchange of fine sediment between the shelf and the estuary, and between the New York Bight and adjacent shelf compartments, may be too low. Even at these low estimates, the rate of transfer of fine sediment between the shelf and adjacent reservoirs results in a net inflow of only 1.1 x 10^6 tons/yr., or 6% of the total inflow. It may, therefore, be concluded that the system is a well-mixed one. Of all dumped fine materials which have not been destroyed by chemical or biological activity, about 50% will escape the Bight altogether, and the remaining will be uniformly dispersed through the natural mud deposits of the Bight apex--Hudson estuary system and its fringing bays, marshes, and secondary estuaries.

CONCLUSIONS

Although the input of fine-grained particles into the New York Bight apex is almost equally divided between anthropogenic and natural materials, relatively small amounts (by weight) of sewage sludge are causing the greatest environmental concern. Because of their low density, many of these organic particles do not settle rapidly to the bottom during a dumping operation, but are held in suspension in the water column to be dispersed and settle elsewhere. Most fine particles, including those from sewage-sludge, dredge-spoil, fluvial, estuarine, and shelf sources, settle temporarily in the low topography of the Christiaensen Basin and the Hudson Shelf Valley, or are carried by bottom currents into the estuaries which are semi-permanent sinks. Small patches of muddy sand also occur off the south shore of Long Island west of Fire Island Inlet.

Current knowledge of the behavior of these fine particles indicates that they settle out of the water column only during conditions of low bottom current speed and high concentration. This generally occurs in the Bight apex during the spring, summer, and early fall, from April to September. Because of shoal water depths throughout the apex, storms, particularly when winds are from the northeast, create bottom currents fast enough to erode muddy sediments even in the deeper low areas of the Christiaensen Basin and Hudson Shelf Valley along with more-easily-erodible sand. While sand particles redeposit rapidly during slowing bottom currents, fine particles stay in suspension considerably longer and are widely dispersed.

Water-mass studies in the nearshore shelf and estuary indicate that fine-grained matter is cycled between shelf and estuary, within the inner-shelf zone, and between deposits on shelf and estuary floors and the overlying water in time scales from diurnal to decades. Contaminated sediments that may be found during any one sampling are subject to transport at any time and cannot be considered as evidence of permanent accumulation.

ACKNOWLEDGMENTS

This work was done as part of NOAA's Marine Ecosystems Analysis (MESA) Project in the New York Bight. Thanks are due to the officers and crew of the NOAA Ship KELEZ, to students of Brooklyn College for aid in collection of samples, and to Messrs. John W. Kofoed and H. B. Stewart, Jr., for manuscript review.

REFERENCES

1. Baylor, E. R., 1973. Final Report of the Oceanographic and Biological Study for Southwest Sewer District No. 3, Suffolk County, New York. Report prepared for Bowe Walsh Associates by Marine Sciences Research Center, State University of New York at Stony Brook, V. 1, text, 695p.
2. Beardsley, R. C., and Butman, B., 1974. Circulation on the New England Continental Shelf: Response to Strong Winter Storms. Geophys. Res. Letters, 1:181-184.
3. Beardsley, R. C., Boicourt, W. C., and Hansen, D. V., 1976. Physical Oceanography of the Middle Atlantic Bight. p. 20-34, In: Gross, M. G., ed., Middle Atlantic Continental Shelf and the New York Bight. Special Symposia V. 2, American Society of Limnology and Oceanography, 440p.
4. Charnell, R. L., and Mayer, D. A., 1975. Water Movement Within the Apex of the New York Bight during Summer and Fall of 1973. NOAA Tech. Memo. ERL MESA-3, 29p.
5. Drake, D. E., 1976. Suspended Sediment Transport and Mud Deposition on Continental Shelves. p. 127-158, In: Stanley, D. J., and Swift, D. J. P., eds, *Marine Sediment Transport and Environmental Management*. New York, John Wiley and Sons, 602p.
6. Drake, D. E., 1977. Suspended particulate matter in the New York Bight Apex, Fall 1973. Jour. Sed. Petrol., 47:201-228.

7. Duedall, I. W., O'Connors, H. B., and Irwin, B., 1975. Sewage Sludge: Its fate in the New York Bight Apex. Jour. Water Pol. Control Fed.
8. Figueiredo, A. G., Jr., and Kowsmann, R. O., 1976. Interpretacao dos Registros de Sonar de Varredura Lateral Obtidos na Plataforma su riograndense Durante a Operasao GEOMAR VII. Rio de Janeiro, Brazil, Companhia de Pesquiesa de Recursos Minerais, 9p.
9. Freeland, G. L., and Merrill, G. F., 1976. Deposition and erosion in the Dredge Spoil and other New York Bight Dumping Areas. p. 936-946, In: Krenkel, P. A., Harrison, J., and Burdick, J.C., eds., *Dredging and its Environmental Effects*. New York, Amer. Soc. Civil. Eng., 1037p.
10. Freeland, G. L., Cok, A. L., and Swift, D. J. P., in press. Surficial Sediments of the New York Bight Apex. Boulder, Colorado, NOAA Tech. Report.
11. Freeland, G. L., and Swift, D. J. P., in press a. Surficial Sediments of the Long Island Inner Shelf. Boulder, Colorado, NOAA Tech. Memo.
12. Freeland, G. L., and Swift, D. J. P., in press b. Surficial Sediments. New York Bight Atlas Monograph 10, NY Sea Grant Institute, Albany, NY.
13. Gadd, P. E., Lavelle, J. W., and Swift, D. J. P., 1978. Calculations of Sand Transport on the New York Shelf Using Near Bottom Current Meter Observations. Jour.Sed.Petrol. 48:239-252.
14. Gross, J. G., 1972. Geologic Aspects of Waste Solids and Marine Waste Deposits, New York Metropolitan Region. Geol. Soc. Amer. Bull., 83:3163-3176.
15. Harris, W. H., 1976. Spatial and Temporal Variation in Sedimentary Grain Size Facies and Sediment Heavy Metal Ratios in the New York Bight Apex. p. 102-123, In: Gross, M. G., ed., Middle Atlantic Continental Shelf and the New York Bight. Soc. Limnol. Oceanogr. Spec. Symp. V. 2, 441p.
16. Hatcher, P. G., and Keister, L. E., 1976. Carbohydrates and Organic Carbon in New York Bight Sediments as Possible Indicators of Sewage Sludge Contamination. p. 240-248, In: Gross, M. G., ed., Middle Atlantic Continental Shelf and the New York Bight. Soc. Limnol. Oceanogr. Spec. Symp. V. 2, 441p.
17. Hatcher, P. G., Keister, L. E., and McGillivary, P. A., 1977. Steroids as Sewage Specific Indicators in New York Bight Sediments. Bull. Environ. Contam. Toxicol., 17:491-498.
18. Lavelle, J. W., Brashear, H. R., Case, F. N., Charnell, R. L., Gadd, P. E., Haff, K. W., Han, G. A., Kunselman, C. A., Mayer, D. A., Stubblefield, W. L., and Swift, D. J. P., 1976. Preliminary Results of Coincident Current Meter and Sediment Transport Observations for Wintertime Conditions, Long Island Inner Shelf. Geophys. Res. Letters, 3:97-100.
19. Lavelle, J. W., Swift, D. J. P., Gadd, P. E., Stubblefield, W. L., Case, F. N., Brashear, H. R., and Haff, K. W., in press. Fair Weather and Storm Transport on the Long Island Inner Shelf. Submitted to Jour. Geophys. Res.
20. McCave, I. N., 1972. Transport and Escape of Fine-Grained Sediment from Shelf Areas. Chapter 10, p. 225-248. In: Swift, D. J. P., Duane, D. B., and Pilkey, O. H., eds., *Shelf Sediment Transport: Process and Pattern*. Stroudsburg, PA, Dowden, Hutchinson, and Ross, 656p.

21. McLaughlin, D., Elder, J. A., Orlog, G. T., Kibler, D. F., and Buenson, D. E., 1975. A Conceptual Representation of the New York Bight Ecosystem. NOAA Tech. Memo. ERL MESA-4. National Oceanic and Atmospheric Administration, Boulder, Colorado, 359p.
22. Magnell, B., 1976. Summary of Oceanographic Observations in New Jersey Coastal Waters near 39°28'N Latitude and 74°15'W Longitude During the Period May 1974 through May 1975. EG&G Inc., Waltham, Massachusetts.
23. Manheim, F. T., Meade, R. H., and Bond, G. C., 1970. Suspended Matter in Surface Waters of the Atlantic Continental Margin from Cape Cod to the Florida Keys. Science, 167:371-376.
24. Mueller, J. A., Jeris, J. S., Anderson, A. R., Hughes, C. F., 1976. Contaminant Inputs to the New York Bight. NOAA Tech. Memo. ERL MESA-6, 347p.
25. Pearson, H., 1974a. Riding a Sludge Boat, Hearing of Alternatives. Newsday, August 1, 1974, Garden City, NY.
26. Pearson, H., 1974b. The Sludge Story: Muddy Waters. Newsday, August 1, 1974, Garden City, NY.
27. Pearson, H., 1974c. Buckley, Move Sludge Dump. Newsday, August 3, 1974, Garden City, NY.
28. Postma, H., 1967. Sediment Transport and Sedimentation in the Marine Environment. p. 158-180, In: Lauff, G. H., ed., *Estuaries*. Washington, Amer. Assoc. Advancem. Sci., 757p.
29. Proni, J. R., Newman, F. C., Young, R. A., Walter, D., Sellers, R., McGillivary, P., Duedall, I., Stanford, H., and Parker, C., in press. Observations of the Intrusion into a Stratified Ocean of an Artificial Tracer and the Concomitant Generation of Internal Oscillations. Jour. Geophys. Res.
30. Schaeffer, O. A., 1977. Anthropogenic Fluxes of Carbon into the Sediments of the New York Bight. Final Report, ERL-MESA Grant 0462244013, S.U.N.Y., Stony Brook, NY, 16p.
31. Schubel, J. R., 1972. Distribution and Transportation of Suspended Sediment in Upper Chesapeake Bay. p. 151-168, In: Nelson, B. W., ed., *Environmental Framework of Coastal Plain Estuaries*. Geol. Soc. Amer. Memoir 133, 617p.
32. Schubel, J. R., and Okubo, A., 1972. Comments on the Dispersal of Suspended Sediment Across Continental Shelves. p. 333-346, In: Swift, D. J. P., Duane, D. B., and Pilkey, O. H., eds., *Shelf Sediment Transport: Process and Pattern*. Dowden, Hutchinson, and Ross, Stroudsburg, PA, 656p.
33. Scott, J. T., and Csanady, G. T., in press. The COBOLT Experiment, First Current Meter Observations. Jour. Geophys. Res.
34. Segar, D. A., and Cantillo, A. Y., 1976. Trace Metals in the New York Bight. p. 171-198, In: Gross, M. G., ed., Middle Atlantic Continental Shelf and the New York Bight. Amer. Soc. Limnol. Oceanogr. Spec. Symp. V. 2, 441p.
35. Smith, R. N., and Ali, S. A., 1973. Sediment - Physical Aspects. p. 223-246, In: Baylor, E. R., ed., Final Report of the Oceanographic and Biological Study for the Southwest Sewer District #3, Suffolk County, New York. State University of New York, Stony Brook, NY, Marine Sciences Research Center, V. 1, 695p.
36. Soucie, G., 1974. Here Come de Sludge. Audubon, 76:108-113.

37. Stubblefield, W. L., Permenter, R. W., and Swift, D. J. P., 1977. Tiem and Space Variation in the Surficial Sediments of the New York Bight Apex. Estuarine and Coastal Mar. Sci., 5:597-607.
38. Swift, D., Cok, A., Drake, D., Freeland, G., Lavelle, J. W., McKinney, T., Nelson, T., Permenter, R., and Stubblefield, W., 1975. Geological Oceanography. p. 16-61, In: Charnell, R., ed., Assessment of Offshore Dumping: Technical Background. NOAA Tech. Memo. ERL MESA-1, 83p.
39. Swift, D. J. P., and Freeland, G. L., in press. Mesoscale Current Lineations on the Inner Shelf, Middle Atlantic Bight of North America. Estuarine and Coastal Marine Science.
40. Veatch, A. C., and Smith, P. A., 1939. Atlantic Submarine Valleys of the United States and the Congo Submarine Valley. Geol. Soc. Amer. Spec. Paper 7, 101p.

Depositional Characteristics of Sediments at a Low Energy Ocean Disposal Site, Savannah, Georgia

George F. Oertel

ABSTRACT

Since the early 1900's, the bar-channel system of the Savannah navigation channel has been dredged for maintenance and improvement. Between 1967 and 1977, approximately 10.8 x 10^6 yds^3 (8.3 x 10^6 m^3) of this dredge material was deposited at a predetermined ocean disposal site. The prominent sources of material dredged from the navigation channel are fine-grained river deposits, littoral sands, Pleistocene river gravels, and submarine Miocene outcrops. Material that is released from hopper dredges is hydraulically sorted as it settles through the water column. Coarse material is deposited in relatively thick mounds corresponding to the disposal path of the dredge, thin layers of fine material accumulate in deeper areas adjacent to mounds.

After deposition, the fine and coarser material is relatively stable during non-storm conditions. During storm conditions, coarse material in the upper parts of the mounds are reworked by waves and storm currents.

INTRODUCTION

The southeast coast of the United States is bordered by a broad shallow continental shelf that experiences intermediate tidal ranges between 2 and 3 m. The average water depth at 10 and 40 kilometers offshore is 9 m and 18 m, respectively. Large waves are refracted by the topography of the shallow shoreface and wave energy is "dampened" before it reaches the shore. Measurements of sea swell and period reveal monthly averages of 0.6-1.2 m (2-4 feet) and 5-9 seconds, respectively (based on measurement made 22 km offshore at the Savannah Light Tower by the U. S. Army Corps of Engineers, (1973). Data from the U. S. Army Corps of Engineers Littoral Environment Observation Program (LEO) illustrated similar findings. Breaker heights were between 0.6-1.2 m (2 and 4 feet) and mean periods were 7-8 seconds (U. S. Army Corps of Engineers, 1976).

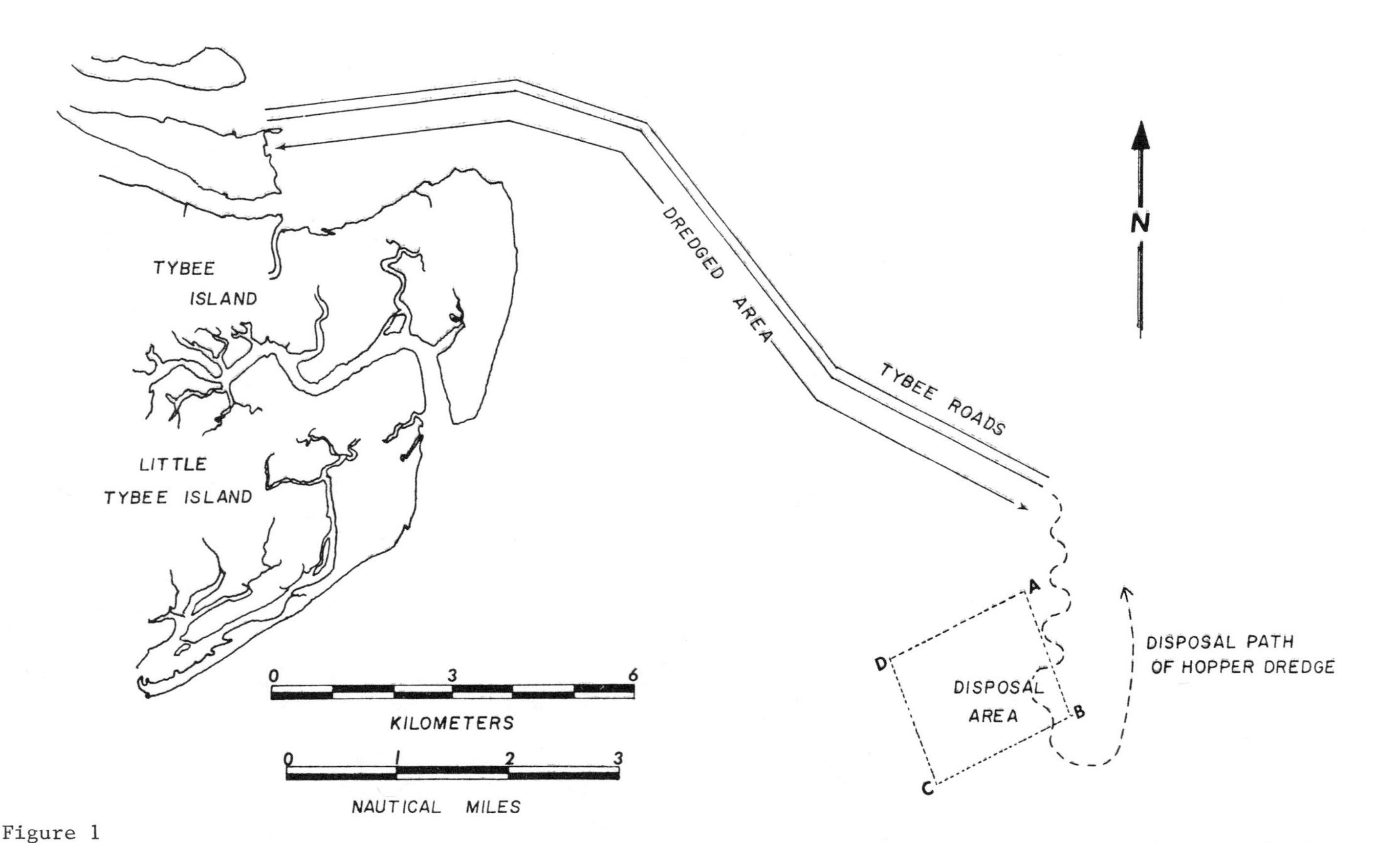

Figure 1
Location map of dredged area and the disposal area for dredged material. The hopper disposed material along a traverse between buoys A and B and buoys B and C.

The Georgia shoreline is interrupted by numerous inlets, some of which have rivers that transect the Coastal Plain and head in the upland provinces of the Piedmont and Blue Ridge. Inlets that head in marsh-lagoons of the Coastal Plain exchange approximately 10^9 cubic feet of water with the ocean during ebb and flood tides. These large prisms of water are accelerated as they pass through the inlet and inlet water is trajected 10 to 20 kilometers offshore during the 6 hour duration of the ebb. The excursion of the lagoonal water mass is clearly marked by plumes of turbid water that are injected into blue-green shelf waters. Wave energy that is dissipated on the inner 10 kilometers of the sea floor also contributes to the turbidity of this zone by resuspending fine sediments into the water column.

An obvious seasonal pattern exists in the character of wave energy; during the spring and summer low energy waves are from the south, during the fall northeast storms are very prevalent, and during the winter wave approach is variable. While all waves have some influence on shore processes, only the storm waves effect the seabed at depths greater than 10 meters.

SUMMARY OF DREDGING HISTORY

The material dumped at the Savannah ocean disposal area was dredged from the outer limits of the Savannah River navigation channel between the entrance and sea buoy (Figure 1). In 1722, the Savannah River was in a natural state, however, as early as 1773 some navigation maintenance was performed in the upper harbor with the removal of several sand bars. From 1775 to 1822, a 13-foot channel (MHW) existed between Savannah and the Atlantic Ocean. As shipping increased, the demand for better navigational channels became obvious. Between 1867-1872, the U. S. Army Engineering Department was charged with maintaining a 10.5 foot channel (MLW) by 1888. Material dredged from the outer part of the channel was dumped in deeper water on the seaward side of the outer bar (distal shoal). As navigation demanded, the channel across the bar was widened and deepened. In 1902, 1930, and 1945, the controlling depth of the channel was increased to 26, 30, and 34 feet, respectively. In each case, the dredged material was dispersed in the deeper water on the seaward edge of the outer bar. In 1964, a greater demand for control was put on the location of dredge material disposal and a site was marked for future disposal operations (Figure 1).

A 5.0 sq km disposal site was located 15 km seaward and southeast of the Savannah River entrance. The site is in a hummucky area characterized by large sand waves that were present prior to the utilization of the site for a disposal area. Since 1967, disposal in the area was performed by hopper dredges that maintained and improved the channel between the entrance jetties and the sea buoy (Young, 1976).

Since 1967, the mid channel depths for the navigation channel have been increased from 36 to 42 feet. In 1976, improvements called for the creation of a 600 foot wide channel with 40 foot controlling depths. Improvements to the navigation channel that involve widening and deepening have required the dredging of Miocene and occasionally Pleistocene deposits. Studies by Woolsey (1977) indicate that Holocene sediments in the vicinity of the navigation channel and seaward of the jetties are between 1 and 6 m thick. Pleistocene deposits are generally missing, however, some Pleistocene cut and fill is present in lenses up to 5 meters thick. Miocene is present just below the Holocene deposits and varies in depth from 5-10 m below the surface of the sediment. "Relict" Miocene and Pleistocene sediments are considerably coarser (less than 2ϕ) than the Holocene sands (greater than 2ϕ, Gorsline, 1963) and often contain fossil shells and sharks teeth.

Maintenance dredging is required primarily to remove littoral sands spilling into and being trapped in the deeper navigation channel. This material is transported and deposited by inlet tidal currents and wind driven wave currents.

GENERAL SETTING

Although the Savannah disposal site is located approximately 15 km offshore, the average depth of the sea floor is only 14 m and is subject to the reversing inlet tidal currents of the Savannah River. The turbid plume of the Savannah River crosses the disposal area approximately 1 hour before and after low water. Water velocities over the disposal area are variable and depend on river runoff, synodic and semi-diurnal tides, and wind and wave currents. Generally, velocities at the seabed are lower than 20 cm/s and grain transport is sporadic and often related to storms or oscillatory wave action.

The disposal area is located at the margin of the Recent-Relict boundary described by Gorsline (1963) and Pilkey and Frankenberg (1964) and contains a mixture of relatively coarse relict sediments and fine modern sediments.

SEDIMENTATION

Sediment deposition and retention in the disposal area are controlled by mechanics of deposition, physical forces, and sediment characteristics. Mechanics of deposition at the dredge material disposal area are both natural and man induced. Between 1967 and 1977, approximately 10.8 x 10^6 cu yd of sediment were transported to the area by hopper dredges. If all of the material stayed within the limits of the disposal area, the sea floor should have increased approximately 1.8 m in elevation. Surveys revealed that 95% of the material deposited during the 1973 improvements could be accounted for following that operation. However, a sediment blanket of uniform thickness was not spread over the entire area. Coarse-grained sediments (probably from Miocene and Pleistocene deposits) formed a mound extending 6 m above the mean depth of the sea floor and correspond to the depositional path of the hopper dredge (Figures 1, 2, 3).

Figure 2 (at right)
Sketch of hopper dredge releasing dredged material into the water column. Coarse material is deposited rapidly and fine material is spread laterally in a plume as it settles to the bottom. Finer material is sorted away from the disposal path of the dredge.

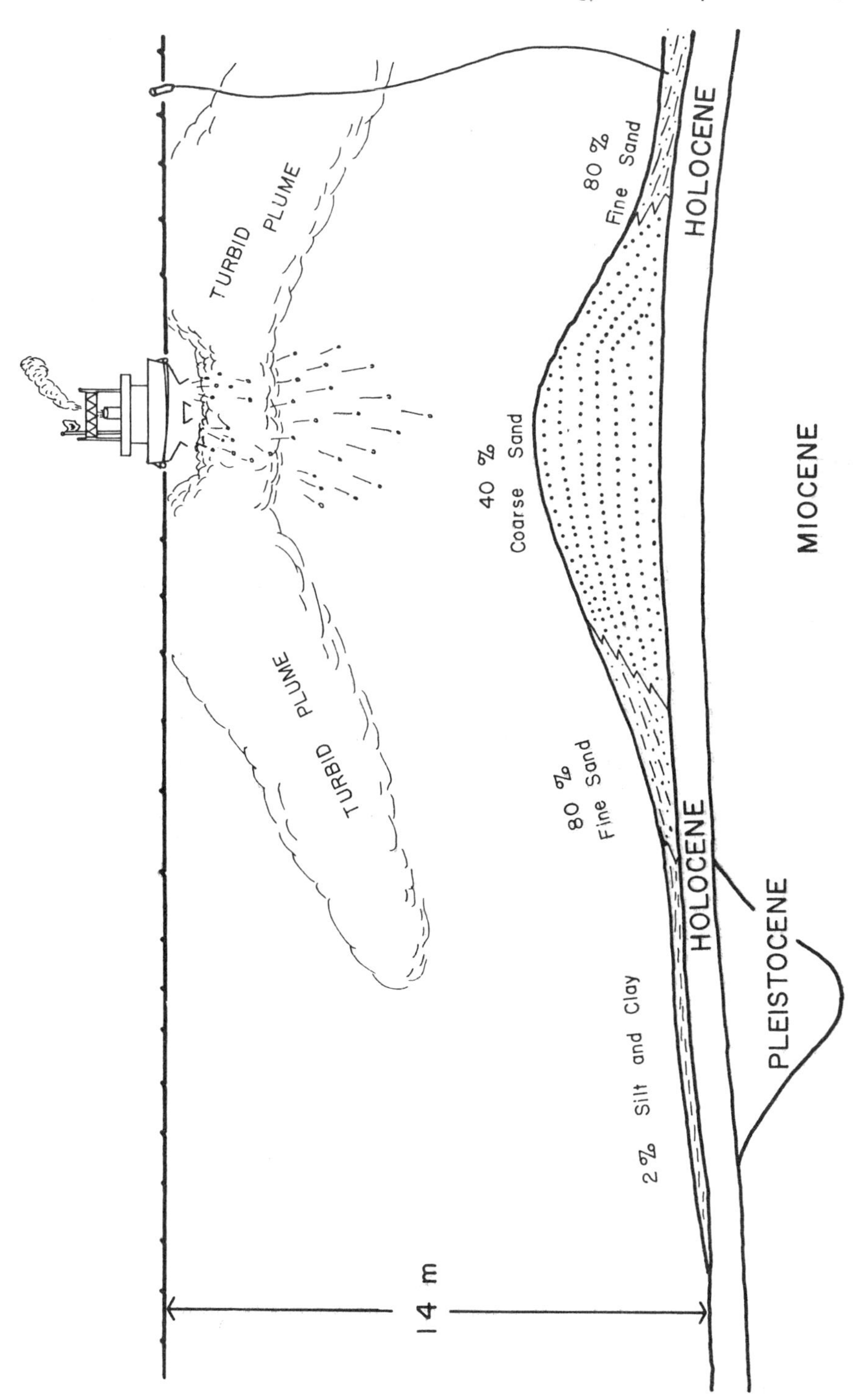

TURBID PLUME
TURBID PLUME
80 % Fine Sand
40 % Coarse Sand
80 % Fine Sand
2 % Silt and Clay
HOLOCENE
HOLOCENE
MIOCENE
PLEISTOCENE
14 m

The shallowest part of this mound was approximately 9 m deep. Fine sediments were dispersed greater distances by tidal currents as they settled through the water column and accumulated in areas approximately 14 meters deep (Figure 2).

In general, dredged material was sorted laterally away from the disposal track in accordance with grain size. Coarse Miocene and Pleistocene sediments were deposited in mounds whereas fine Holocene and Recent sediment were deposited in troughs between mounds. Sediments were transported in a southeast and northwest direction during the ebbing and flooding tide, respectively.

In 1974, the average current speed measured 1 m above the sea floor of the disposal area was 13.5 cm/s. All sediment coarser than very fine sand reached the bottom in less than 20 minutes and was displaced less than 175 m (approximately 575 ft) away from the disposal track. Finer grained deposits (silts and clays) which represent less than 5 percent of the hopper material were transported up to 3 km before reaching the bottom. However, as weak tidal currents are generally reversing and sometimes rotary in nature, the flow trajectory is often helicoidal and most of the very fine hopper material remains within the boundaries of the disposal site. Inlet currents from the Savannah River often increase surface velocities over the disposal area and may enhance the initial displacement distance of sediments. These currents are present only occasionally and attenuate rapidly 3 m below the water surface.

Periodically, increases in turbulence caused by waves or high velocity currents may keep silts and clays in suspension for very long periods. Turbulent conditions during two-day storms may cause silts and clays to be transported tens of miles away from the disposal area. When storms produce a mass transport of water in a landward direction, these fines contribute to the turbidity of the nearshore zone. However, the contribution to turbidity by dredge disposal is minor when compared to that caused by resuspension in the nearshore zone and by turbid plumes from inlets (Brokaw and Oertel, 1976; Oertel, 1976).

Once the disposed material accumulates on the sea floor, it remains relatively stable. Diver observations in the disposal area revealed a variety of small bedforms and surface features. During mid summer, when the mean sea swell and period was 0.4 m and 4.3 seconds, respectively, no sediment movement was observed at the sea floor. During this low energy period a film of brown algae cemented surface grains at the sea floor into a thin crust.

During intervals of 0.3-1.0 m sea swell, with 8 second wave periods, sediment suspension was noted above ripple crests. The shallow part of the sea floor appeared to have a high concentration of medium to coarse sand with numerous shell fragments with black stains. Small scale topography was characterized by longitudinal megaripples

---

Figure 3 (at right)
Contour maps of the disposal area illustrating the percentage of coarse sand, silt and clay and bioturbation. Data for maps were collected during the 1973 disposal operations. The path of the hopper is highlighted by high percentages of coarse sand and low percentages of silt and clay. Heavily bioturbated areas were associated with relatively high concentrations of silt and clay.

PERCENTAGE
COARSE SAND
(0.0 Ø)

PERCENTAGE
SILT & CLAY
(>4.0 Ø)

PERCENTAGE
BIOTURBATION

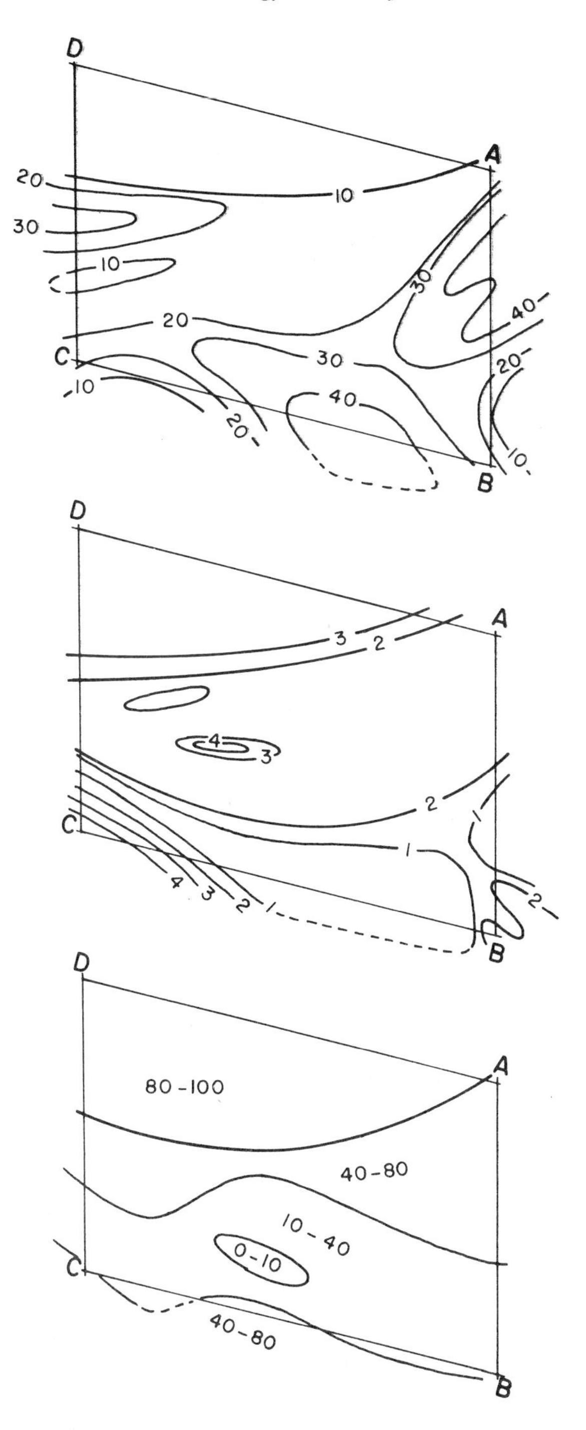

which were apparently a response to the oscillatory motion of waves. Aggregate pebbles of sandstone cemented by a brown algal matrix were common in troughs between megaripples and apparently were remnants of the algal crust observed during dives in low energy periods.

Deeper portions of the disposal area were characterized by finer grained sediments in linguoidal megaripples. These bedforms apparently were not influenced by oscillatory wave motion, but rather were formed by low velocity currents flowing near the bottom. Sediment analysis from undisturbed box cores revealed that renewed biological activity in this disposal material was rapid and related to mean grain size and water depth. Biological activity, indicated by bioturbation, was generally related to the distribution of silt and clay in the deeper parts of the disposal area (Figure 3). Sampling of the benthic infauna revealed a distinct assemblage of approximately 743 organisms per square meter, 62% of which were polychaetes and 31% were amphipods. The record of this relatively intense biologic activity is preserved in bioturbated sediments.

The shallower portions of the disposal area also had a distinct assemblage of benthic infauna, with a density of approximately 177 organisms per square meter. Several families of polychaetes described as active burrowers associated with coarse grained shoals (Day, 1955; Pettibone, 1963) comprised approximately 40% of the sample population.

Of the 67 species encountered only 4 percent were common to the shallow mounds and the deeper areas. The different faunal assemblages were most probably related to differences in sediment type and sediment stability (as influenced by wave activity).

In contrast to the deeper part of the disposal area, the substrate of the shallow mound did not have a record of intense biological activity of the benthic infauna. At the seabed, dwelling tubes of the worm Diapatra sp. were common, and biological mixing by heart urchins was apparent from x-ray radiographs of several cores. The relatively low degree of bioturbation reflects increased physical reworking produced by wave activity. Although the disposal area is located 15 km offshore, the mean depth of the sea floor is only 14 m and the coarse-grained mound extends to a depth of 9 m. The shallower depth is often effected by the orbital motion of storm waves. Sand grains are resuspended during the passage of long period waves that "touch" bottom. As grains settle to the sea floor they are displaced by weak currents that are lower than the normal threshold velocity for the grains. This pattern of rhythmic transport occurs with the passage of each wave crest. Ripples on this relatively active sediment carpet produce foresets in the coarse substrate on the disposal mound.

Figure 4 (at right)
Generalized sketch of the disposal area between D and C. The dredge material occurs in mounds over thin units of Holocene sand that were deposited on Miocene sediments. Pleistocene cut and fill structures are occasionally present in the Miocene. During the 3-month period between the 1973 disposal operation and the 1973 northeast storm period, the coarse grained sand ridge was displaced 0.62 km in a southerly direction.

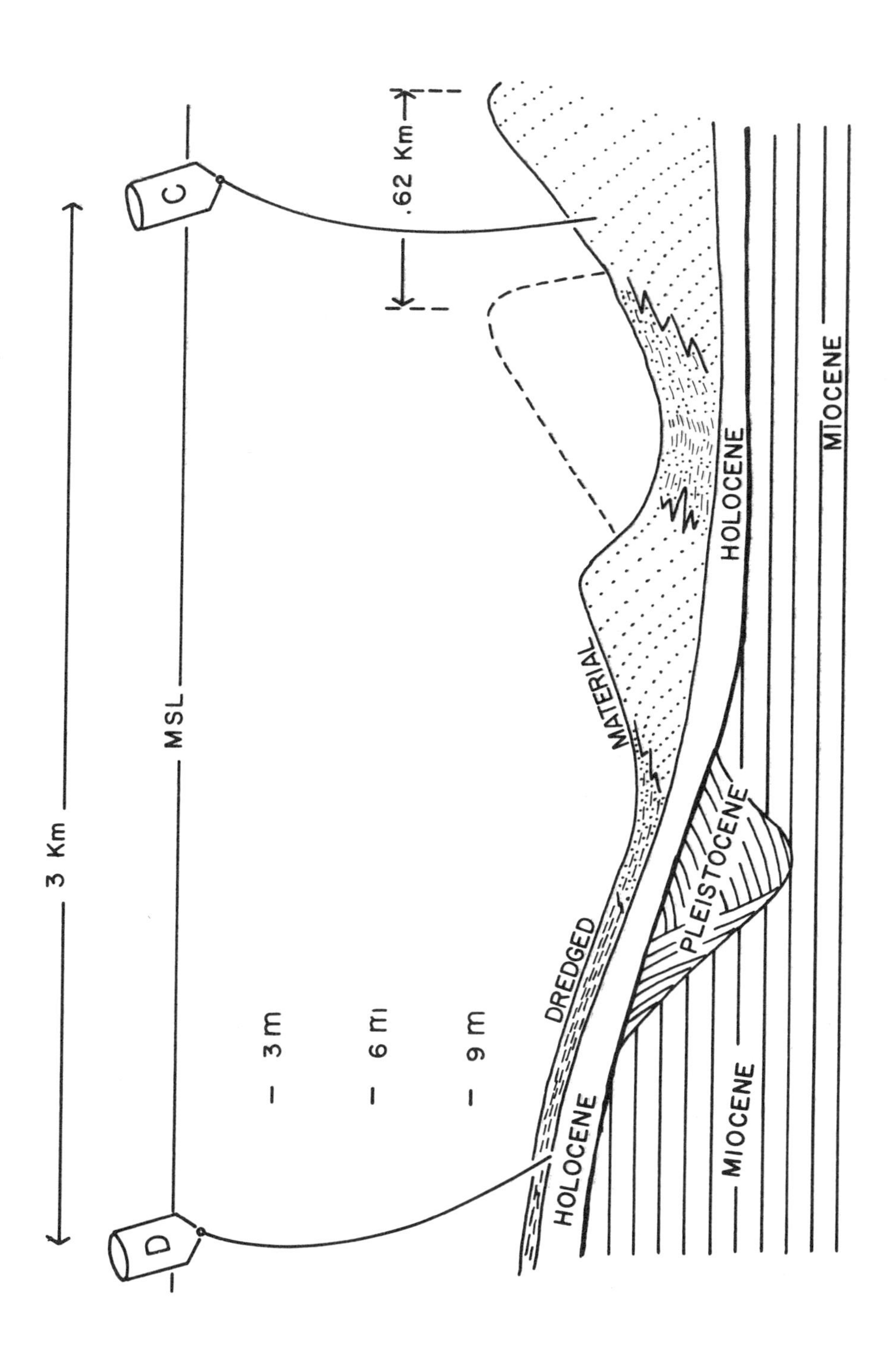
3 Km
MSL
C
D
.62 Km
- 3 m
- 6 m
- 9 m
DREDGED
MATERIAL
HOLOCENE
HOLOCENE
PLEISTOCENE
MIOCENE
MIOCENE

The shape of a coarse-grained mound is to a large extent controlled by the rhymthic pattern of sediment movement by waves and weak currents. Water flow observations at the mound illustrate a decrease in current speed with depth. Measurements of mean speeds near the top of the mound were 15 to 30 percent greater than measurements at the base of the mound. Thus, in water greater than 9 m deep the rhythmic transport of sediment decreases with the attenuation of orbital motion and currents. During storms when sediment transport is pronounced, the rate of sediment movement at the top of a mound is often greater than at its base. As a result, mounds become asymmetrical during storms. An asymmetrical mound (ridge) was observed in the area before 1973; however, a series of northeast storms in the fall of 1973 caused the ridge to move out of the disposal area. Within a 3 month period, the ridge was displaced approximately 0.33 nautical miles (0.6 km) (Figure 4). During the prior 6 months, there were no apparent changes in the ridge position.

CONCLUSIONS

The Savannah ocean disposal area is located 15 km offshore, where the average depth of the shoreface is relatively shallow (14 m). Tides, wind-generated waves, and thermohaline currents which influence the entire water column are major agents causing sediment dispersion. However, the ultimate depositional pattern of disposal material is determined by the hydraulic characteristics of sediment. Since dredge material from maintenance work is usually fine-grained littoral sands and material from improvement operations is coarse-grained relict sands, contrasting patterns of sedimentation may be related to source. Coarse grained material generally settles through the water with minor lateral dispersion, and as a result coarse-grained mounds of Pleistocene and Miocene sand form along the disposal path of the hopper dredge. Fine-grained sands are dispersed greater distances and therefore accumulate in broader carpets around the mounds. During non-storm conditions, the magnitude and duration of currents permit all disposal material to be deposited within the designated disposal area. However, during storms, fine-grained material may be transported tens of miles out of the disposal area.

Once deposited on the sea floor, fine-grained material remains relatively stable and although a veneer of coarse-grained material is reworked by orbital wave motion, lateral migration of sediment is only minor. Relative degrees of sediment stability are indicated by the intensity of bioturbation. Fine-grained material in deep areas generally illustrate intense bioturbation, whereas coarse-grained mounds are characterized by structures produced by physical processes.

During the fall, northeast storms occur frequently along the southeast coast of the United States. Because of rapid attenuation of current and wave energy with depth, the mounds of coarse material often receive a greater portion of storm energy than the deeper fine-grained area. Major storms are capable of causing the upper portions of mounds and ridges to migrate leaving behind a platform that is approximately equal in depth to storm-wave base. Although mounds may become asymmetrical following storms, they do not migrate as a result of sediments avalanching over one major slip face but rather move by differential rates of megaripple migration with depth.

ACKNOWLEDGMENTS

Portions of this project were supported by contract DACW21-73-C-004 with the Savannah District, U. S. Army Corps of Engineers and Grant R/EE-4 with the Georgia Sea Grant Program. I also acknowledge with appreciation the assistance of J. L. Harding for reviewing the manuscript and D. W. Menzel for editing the manuscript.

REFERENCES

1. Brokaw, R. and G. F. Oertel. 1976. Suspended sediment data from nearshore waters of Georgia. Georgia Marine Science Center Technical Report Series, No. 76-3.
2. Day, J. H. 1955. Polychaeta of South Africa. Part II, Sedentary species from Cape shores and estuaries. Journ. Linn. Soc. London Zool., 42:494-817.
3. Gorsline, D. S. 1963. Bottom sediments of the Atlantic shelf and slope off the southern United States. Journ. Geology, 71: 422-440.
4. Oertel, G. F. 1976. Characteristics of suspended sediments in estuaries and nearshore waters of Georgia. Southeastern Geology, 18:107-118.
5. Pettibone, M. 1963. Marine polychaete worms of the New England region. Part I, Families phroditidae through Trochochaetidae. Bull. U. S. National Museum, 227:1-356.
6. Pilkey, O. H. and D. Frankenberg. 1964. The relict-recent sediment boundary on the Georgia continental shelf. Georgia Acad. Sci. Bull., 22:38-42.
7. U. S. Army Corps of Engineers. 1973. Sea swell and period from the Savannah Light Tower. (Unpublished data)
8. U. S. Army Corps of Engineers. 1976. Littoral Environmental Observations. Data from Savannah District LEO Program.
9. Woolsey, J. R. 1977. The neogene stratigraphy of the Georgia coast and inner-continental shelf. Ph.D. Dissertation, University of Georgia, Athens, Georgia. (Unpublished data)
10. Young, W. 1976. Personal communication with Mr. William Young, Navigation Division, Savannah District Corps of Engineers, Savannah, Georgia.

Containment of Particulate Wastes at Open-Water Disposal Sites

H. J. Bokuniewicz
R. B. Gordon

ABSTRACT

The degree of retention of particulate silt-clay size waste at open-water disposal sites depends primarily on the sedimentary regime at the site and the height of the waste deposit. In estuarine, lacustrine and coastal localities particulate waste released at the water surface is found to deposit on the bottom within a radius of about 200 m from its impact point. This radius is insensitive to water depth and the volume of material released. After deposition the waste deposit self-consolidates and expels interstitial water. Bio-aggregation decreases the erosion resistance of the deposited material; bioturbation keeps a surficial layer of the waste exposed to the ambient water. As the deposit is made higher it becomes more susceptible to disturbance by waves and currents and may itself alter the local hydraulic flow field. A necessary condition for the retention of particulate waste at a disposal site is that it be below the wave-affected zone. In marine localities the tidal stream will disperse waste placed in a deposit rising above natural topographic features of the bottom unless its surface is armored by erosion-resistant material or the natural sedimentation rate exceeds a critical value that depends on pile height.

These factors are evaluated for dredged material disposal sites in Long Island Sound. Silt-clay is found to be dispersed at the Cornfield Point site, where the natural sedimentation rate is zero, but retained at the New Haven site, where natural sedimentation is 0.8 kgm/(m^2yr). At the latter site 1.2×10^6 m^3 of dredged material was placed in a conical mound 8 m high. Self-consolidation occurred over several weeks and caused an outward advection velocity of interstitial water as great as 5×10^{-5} cm/sec. The top of the waste pile penetrates the wave-affected zone but, because it is armored with sand and clods of cohesive silt, it is not dispersed. The lower part of the waste pile is unarmored. Tidal resuspension of the neighboring muddy bottom and the waste deposit occurs every tidal cycle and is increased by as much as a factor of ten during severe storms. A layer of wastes less than 1 cm thick is intermixed with sediments from the neighboring sea floor within a few weeks after deposition is completed; subsequent entrance of wastes into the water column is due to biological mixing in the top 10 cm of the waste deposit. The results show that permanent waste deposits may be constructed in coastal localities with favorable natural sedimentary regimes provided close control over the deposition process is attained.

INTRODUCTION

When particulate waste material is to be placed on an open water disposal site, it is often desired to minimize contact between the waste and the ambient water. We examine here the question of how effectively particulate waste materials can be contained--kept out of the water column--for long periods of time at a coastal disposal site. By "particulate" we mean material composed principally of mineral grains and containing relatively small amounts of organic matter. This excludes sewage sludge and paper mill effluents, for example, but includes most dredged wastes and many wastes from offshore drilling operations. Attention is focussed on wastes where the mineral content is largely in the silt-clay range ("mud") because pollutants such as heavy metals are associated with the muddy rather than the sandy waste (Benninger, Lewis and Turekian, 1976).

Particulate waste can be placed at a desired location on the bottom in coastal waters with good accuracy using existing marine technology (Bokuniewicz et al., 1977a). The retention of the material so placed depends on the internal stability of the deposit, the susceptibility of the deposited material to erosion, the natural sedimentary regime at the disposal site, and the activities of benthic animals that colonize the deposit. We present observations on each of these factors obtained during studies of three disposal sites in Long Island Sound, a large estuary on the Atlantic coast (Figure 1). Two of the sites (Eatons Neck and New Haven) are capable of retaining particulate waste; dispersion occurs at the third (Cornfield Point) site. This study area is typical of many coastal waters surrounding the North Atlantic, and the results can be generalized to other localities.

RESULTS

Placement of Waste on the Bottom

Particulate waste released at the water surface is deposited on the bottom in a three step process (Gordon, 1974). After falling through the water column as a descending density current, the material impacts with the bottom and then spreads radially outward from the impact point as a bottom surge. The spread of the waste over the bottom after impact is determined by the rate at which the kinetic energy of the waste flow is dissipated during descent, impact and lateral spread. These processes have been studied during the disposal of dredged sediment in six coastal and lacustrine localities (Bokuniewicz et al., 1977a). Water depths ranged up to 65 m and currents to 4 knots.

After release at the surface, the descending material quickly acquires the lateral velocity of the current flowing over the disposal site; its point of impact on the bottom is then displaced

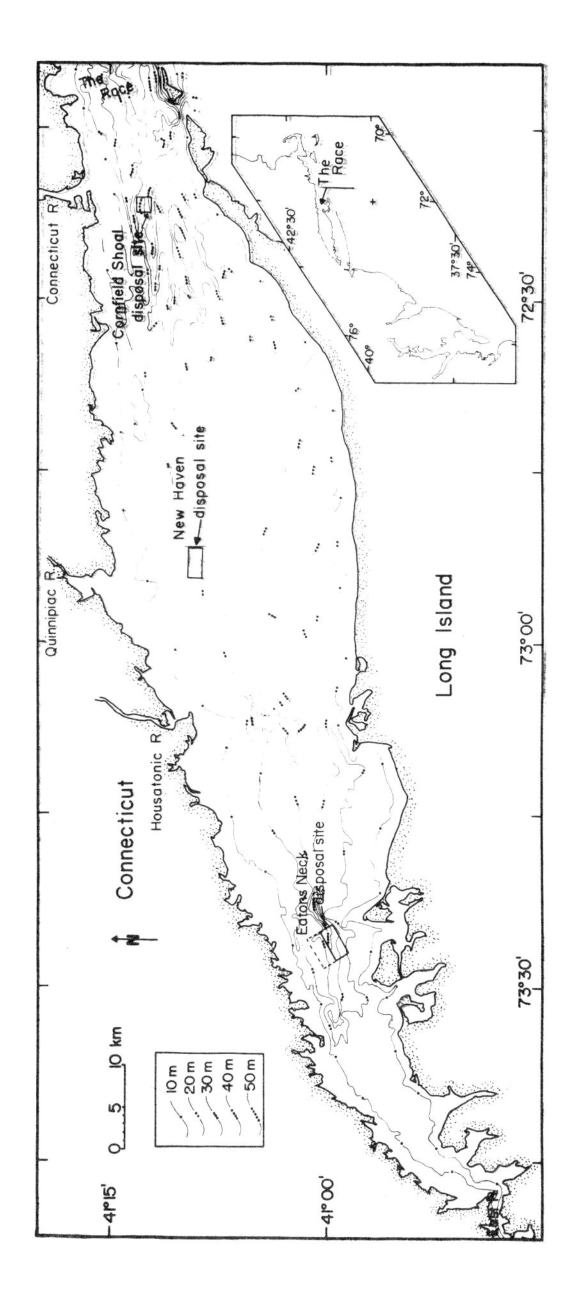

Figure 1
Long Island Sound showing bathymetric contours and the location of the Cornfield Point, New Haven, and Eatons Neck disposal sites.

relative to the surface release point. The bottom surge formed at impact does not cover an area of more than about 200 m radius under any conditions likely to be encountered in the disposal of waste from barges, scows and hopper dredges. (The maximum water depth in which the density current of waste will reach the bottom is yet to be determined but it is at least 65 m for barge-load quantities of particulate wastes.) Consequently, if the location of the release point at the surface is carefully controlled, the waste deposited in still water can be confined to a circular area of about 200 m radius and displacement due to ambient water velocity can be easily calculated. The configuration of the waste deposit can be controlled by selecting appropriate release points. At the New Haven disposal site in Long Island Sound (Figure 1), for example, all material was discharged within a circle of 250 m radius at the surface and the greatest displacement due to flow of the ambient water was 270 m. A conical deposit on the bottom having a radius of ∿750 m was formed, as shown in Figure 2 (Bokuniewicz et al., 1976).

Mechanical Stability

Changes in the configuration and properties of the waste deposit may result from mechanical instability within the deposited pile. When cohesionless waste is deposited from a bottom surge, a deposit with small side slopes is formed and slumping on the pile flanks is unlikely. (If the material is cohesive and descends as clods, most of the material would not be deposited from a bottom surge; steep, unstable slopes may be formed on the deposit.) The initial density of the deposited waste is expected to be lower than that of the material released at the surface because of water entrained during the placement process. Consolidation of both the deposit and the marine sediment under the deposit occurs with the explusion of interstitial water. A theory of this process may be constructed using the stress-strain relations for a linearly compressible, porous medium (Biot, 1941).

Physical constants required for the calculation are the compressibility and the permeability of the sediment. A one-dimensional model of the deposit consists of a layer of spoil of thickness h_1, deposited on a layer of marine sediment of thickness h_2. The final settlement, w, of the surface depends on the final compressibility, $\bar{a}$, as defined by Biot and averaged from the surface to the substrate. The settlement is given by:

$$w = \bar{a}\rho g h_1 \, (h_2 + h_1/2)$$

where ρg is the submerged unit weight of the spoil.

Necessary parameters have been determined for a Long Island Sound silt and used to calculate the expected settlement of the spoil deposit at the New Haven site. Changes in the thickness of this spoil deposit have been measured by repeated bathymetric surveys. These are illustrated by the sequence of cross-sections shown in Figure 2. The observed settlement compares well with the theoretical estimates; the ratio of the final settlement to the deposit thickness is measured as 0.33 ± 0.13 while the calculated ratio is about 0.24. In this instance, the water content of the deposited spoil was about

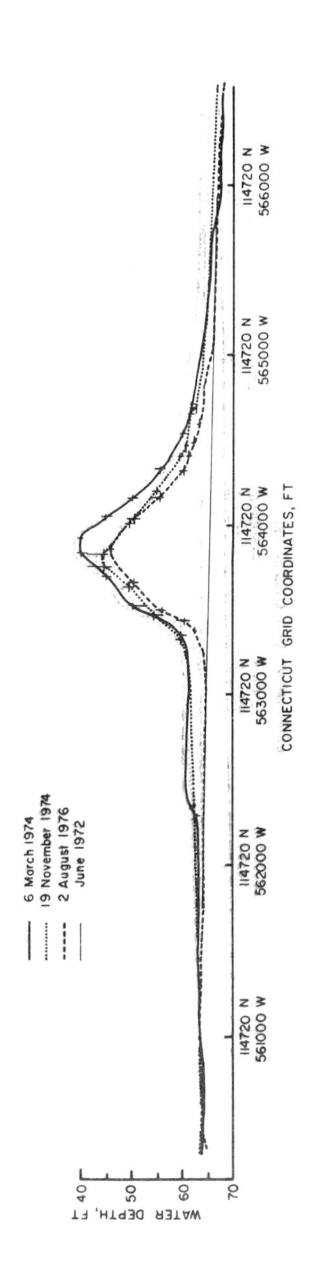

Figure 2
Successive cross-section profiles of the spoil pile on the New Haven disposal site at times T after deposition (T = 0).

double that of the sediment before dredging. The average strain rate of a column of the deposited material was about 0.004/day for the first 28 days and was too small to be measured after 200 days. The calculated advective flux of interstitial water across the sediment-water interface decreases from 5 x 10^{-5} cm/sec during self-consolidation. As a result of the consolidation, the height of the waste deposit is reduced and it presents less of an obstruction to the water flow. In addition, as the water-content of the surficial sediment is decreased, the material becomes more resistant to erosion (Terwindt and Breusers, 1972).

Biological Processes

In many marine and lacustrine environments the mechanical properties of natural silt-clay sediments are modified by biological processes (Rhoads, 1974). These processes include the ingestion of mineral particles by benthic animals and the production of organic binding materials by bacteria. In areas with active benthic communities, like Long Island Sound or Buzzards Bay, nearly all of the sediment resuspended in the water column is in the form of organically bound fecal pellets (Benninger, 1976). Other areas, such as the Severn estuary of Great Britain, have very limited benthic fauna and biological processing of muddy sediment is relatively unimportant.

Where benthic communities are active the properties of the surficial layer of sediment are largely determined by the processing by the animals. The silt bottom of Long Island Sound is covered by a layer of fecal pellets about 1 cm thick. The binding of silt into mineral-organic aggregates coupled with increased water-content at the sediment-water interface due to bioturbation, results in a critical erosion velocity (at 1 m above the bed) for the processed sediment which may be only one-sixth of that for the unprocessed mineral grains (Rhoads, 1973). The fall velocity for the aggregated material may be as much as 100 times faster than the fall velocity of the constituent silt particles (Haven and Morales-Alamo, 1968).

Rhoads, Aller, and Goldhaber (1976) have examined the biological activity in Long Island Sound and specifically at the New Haven disposal site. They have found that dredge spoil placed on the silt bottom of Long Island Sound is recolonized by benthic animals over a period of about six months. The benthic animals resident on the surface of the spoil deposit cause bioturbation; material is overturned and reworked by deposit feeding animals and is thereby kept in continuous communication with the water at the sediment-water interface. They find that this biological mixing extends to a depth of about 10 cm at the New Haven disposal site.

Resuspension and Dispersion

Waste material on the bottom may be subject to resuspension and transportation by currents and by waves. Resuspension may be followed by mixing of the waste material with sediment resuspended from the surrounding sea floor and subsequent redeposition over an area larger than that originally occupied by the waste deposit. It is helpful to review the processes by which muddy sediment subjected to periodic disturbance by a tidal stream is resuspended. The processes occurring at the sediment-water interface are shown schematically in Figure 3. Both erosion and deposition occur simul-

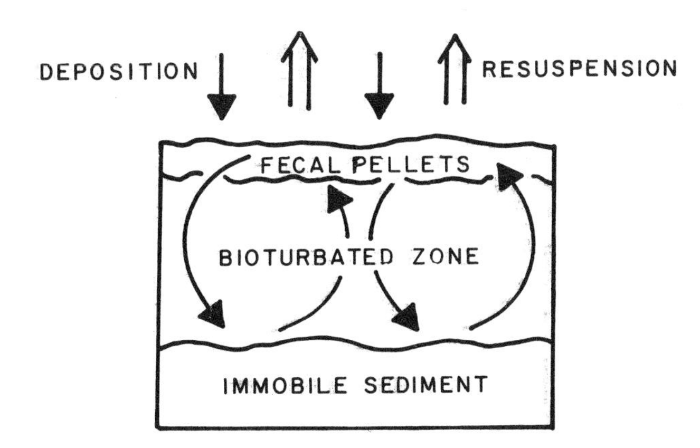

Figure 3
Processes active at the sediment water interface.

taneously and their relative rates determine whether there is net erosion or deposition. At times of high tidal current speeds material may be eroded, intermixed in the water column with material from surrounding areas of the sea floor, and advected by the current. When the current is slack deposition predominates and the intermixed material in the water column settles into the bioturbated zone. Sediment particles at the surface of this zone are subject to continuous intermixing with material deeper in the deposit, down to a depth reached by the burrowing, benthic animals.

Because of the inhomogeneous distribution of bottom stress in a turbulent flow and the patchy distribution of benthic populations, erosion, deposition and bioturbation do not occur uniformly over the bottom. Each fluctuates with characteristic time and length scales. When, for example, turbidity measurements in the water column show that an average of a few millimeters of bottom sediment has been eroded, this erosion could be concentrated in a few small areas and so penetrate much more deeply, perhaps through the bioturbation layer. These length and time scales, if known, would define appropriate averaging intervals for process measurements. Many monitoring programs now in use do not deal with this problem adequately.

The quantity of suspended sediment in the water column depends both on characteristics of the water flow and of the sediment particles. Most theories relate the sediment transport rate to a quantity that varies as the power dissipated in friction by the transporting fluid. We shall use the power of tidal and storm-generated flows as a measure of their relative importance in the disturbance of deposited waste. The deposition rate depends on the settling velocity of suspended particles. If resuspension is due to the tidal stream and the time required for particles to fall through the water column is small compared to the half-tidal period, the concentration of suspended sediment would be expected to show a regular variation with the ebb and flood of the tide. Such a variation has been observed in Chesapeake Bay (Schubel, 1972) and in Long Island Sound (Bokuniewicz *et al.*, 1976). The high fall velocity of suspended material in these areas is attributed to bioaggregation, as discussed earlier.

To the resuspension and deposition due to the tidal stream considered above must be added effects due to the net flow of the estuarine circulation, perturbations in the flow field due to storms, and alterations of the flow due to the presence of the waste deposit itself. Thus, when there is a net flow superimposed on the tidal oscillations, the drift of resuspended material must be added to the mixing caused by the tides in calculating the transport of waste from the deposition site. At estuarine and coastal disposal sites the mean flow is likely to be variable and is often small compared to the tidal stream. Consequently, care must be taken that current measurements are made over a long enough time and a large enough area to sufficiently define this flow (e.g. Weisberg, 1976; Bokuniewicz *et al.*, 1977b). In central Long Island Sound, the net current velocity is about 2% of the maximum tidal velocity, producing an advective flux comparable to the diffusional flux of resuspended material. From an examination of current meter records obtained at thirty locations over several years, Gordon and Pilbeam (1975) determined that at least 10 to 20 days of observations were necessary to define a reliable mean flow. This mean flow is identified as the estuarine circulation. The current meter data show the bottom water at depths greater than 18 m to be flowing into the Sound parallel to the shoreline. At depths less than 18 m there is a shoreward flow of bottom water into a zone where saline bottom water and fresher surface water are intermixing. Over the disposal sites where the water is greater than 18 m deep, the mean flow advects suspended sediment parallel to the coastline.

In many environments, fluctuations in the flow velocity occur over a wide range of frequencies. These fluctuations may be large compared to the mean flow and comparable to the tidal flow. In the description of geophysical flows resort must be made to a statistical representation of the components having short length and time scales. Sometimes a natural break in the fluctuation spectrum suggests the scales at which the statistical description should apply (Webster, 1969), but no such natural dividing point is found in Long Island Sound. Hence, we separate from the total flow both those variations which have tidal frequencies and the net flow (such as might be due to an estuarine circulation). All remaining components are described as fluctuations.

The fluctuating velocity component at one location in Long Island Sound is shown in Figure 4. The largest fluctuations are comparable in amplitude to the strongest tidal flows. It has been shown that in the Sound the fluctuating velocity distribution is broadened during windy periods (Bokuniewicz, Gordon, and Pilbeam, 1976). The occurrence of greater velocities over the bottom is expected to cause increased erosion of bottom sediment, and, hence, an increase in suspended sediment concentrations. Measurements were made of the amount of sediment resuspended during a gale in March, 1975, at a location one mile south of the New Haven Disposal Site, where the water is 23 m deep. The 3 hr average of the fluctuating velocity measured 2 m above the bottom and the total amount of resuspended sediment in the water column is shown in Figure 5. When measurements were terminated because of severe weather conditions, the amount of resuspension material was equivalent to the erosion of a layer about 2 mm thick from the bottom. This is about twice the amount resuspended by the tides during calm weather.

Since direct measurements of sediment transport are costly, and

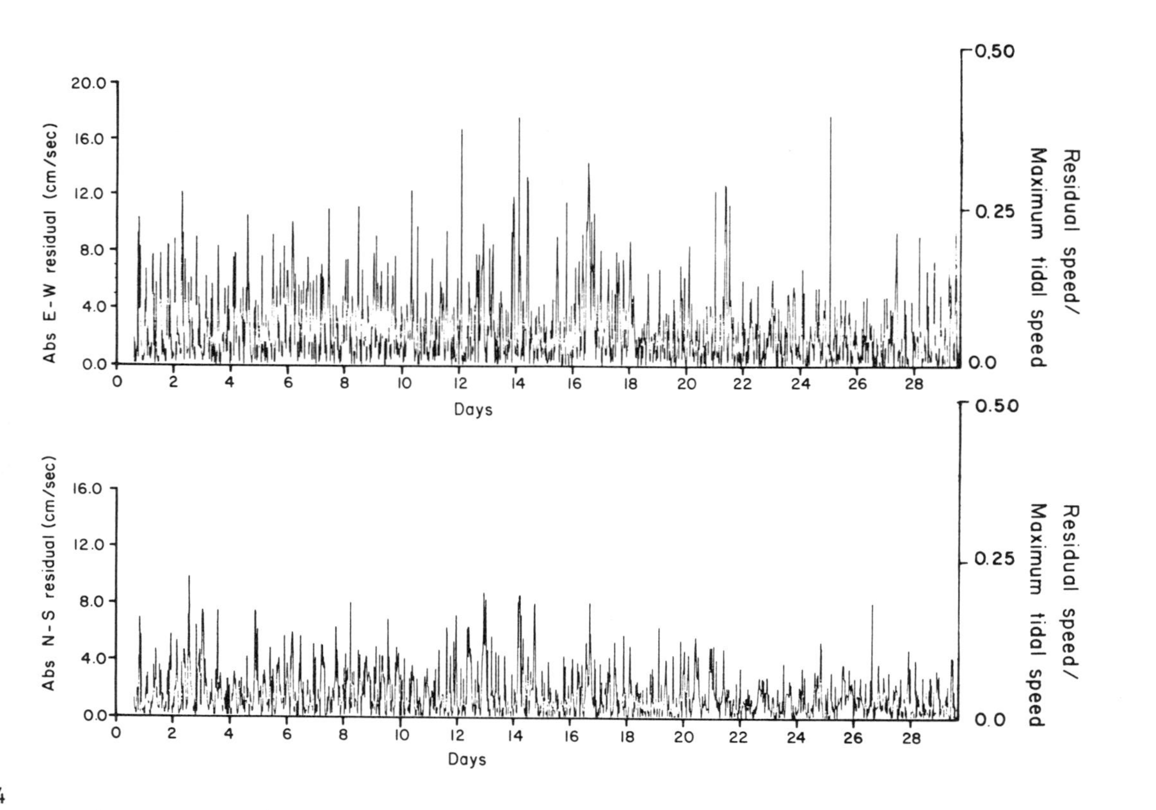

Figure 4
Components of the current velocity fluctuations measured 2 m above the bottom at the Eatons Neck disposal site.

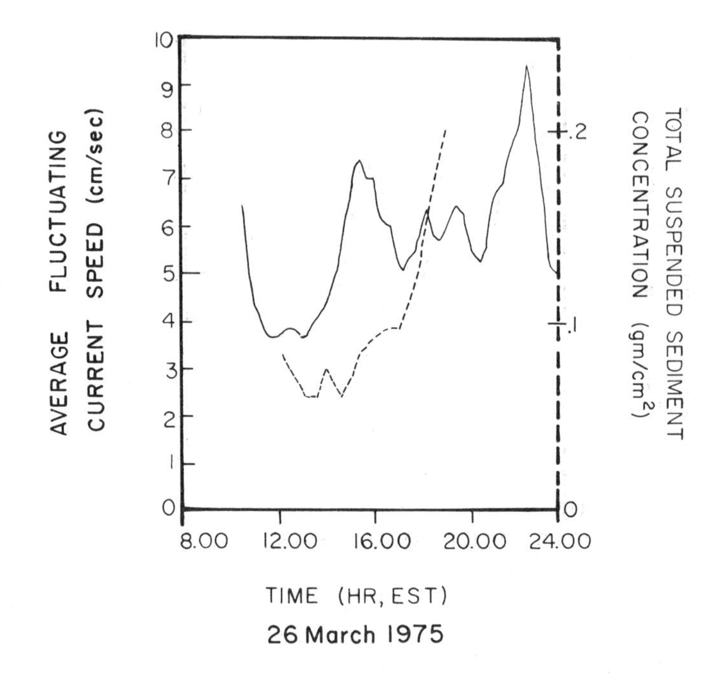

Figure 5
One-hour average fluctuating current speed (calculated every 15 minutes) at 2 m above the bottom and the total suspended sediment concentrations measured near the New Haven disposal site in central Long Island Sound.

intense storms are infrequent, it is useful to be able to estimate the effectiveness of storms in disturbing the bottom material from long-term indirect measurements of, for example, wind velocity or water level. During stormy periods a large amount of work is done by the wind against friction at the sea surface. Part of this energy goes into raising the sea state, part into a change in water level and part into an increase in the flow of water over the bottom.

Water level data were used to determine the probability of storm disturbances at the disposal sites in Long Island Sound. Winds strong enough to appreciably alter the distribution of fluctuating velocity cause changes in the water level. The non-tidal water level deviation, δh, is defined as the difference between the observed water level and the water level of the predicted tide. Non-tidal water level deviations may be as great as the tidal amplitude, and usually persist over several tidal cycles. If the mean water depth is much greater than δh, the additional energy stored is proportional to δh (Taylor, 1919); this energy is available to resuspend sediment. The recurrence interval for any given storm intensity may be found from a time series of δh. The graph of recurrence intervals shown in Figure 6 was calculated from hourly water level data collected over the last 38 years at New London, Connecticut; these are representative of the water level of the whole of Long Island Sound. During this period,

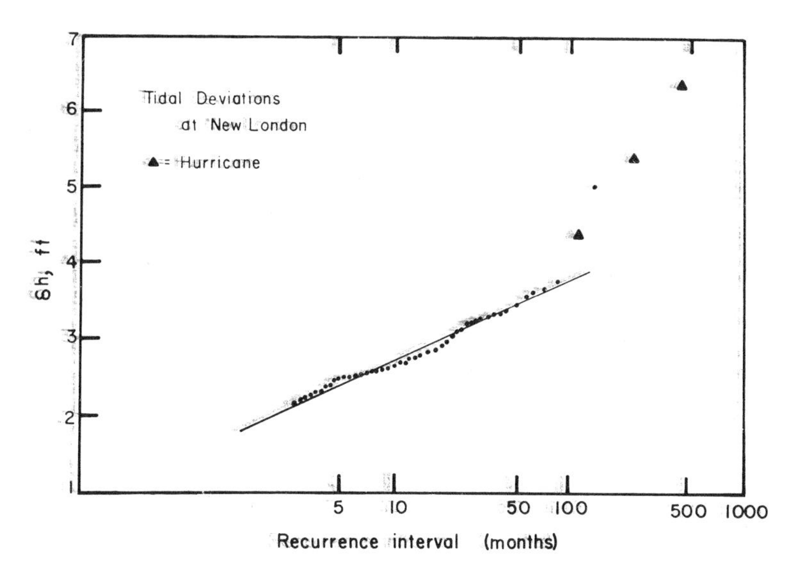

Figure 6
Recurrence interval of extreme deviations of water level at New London from the predicted tide over 38 years.

storms equal in energy to the gale of March, 1975 discussed above have a recurrence interval of about one month. The most severe storms had an energy of about 5 times the energy of this gale. The energy dissipated by the March gale is estimated to be about 30% of the typical tidal power. Since this storm resuspended a layer of sediment 2 mm thick, we estimate that under the most severe winter storm conditions resuspension will extend about 1 cm deep. This is comparable to the thickness of the mantle of fecal pellets observed on the bottom of the Sound, but is much less than the thickness of the bioturbated zone.

The average δh for the winter months may be used as a measure of the storminess of the season. Seasons of greater and lesser storminess are easily recognized in the water level records from New London. Winter storm energy trends vary by as much as a factor of 1.5 between different years. These data provide a basis for deciding how representative any given winter season is.

In areas where muddy sediment is accumulating the bottom is likely to be nearly flat and any large deposit of waste may alter the bathymetry significantly compared to the natural roughness. If the top of a waste pile reaches a height which is greater than the boundary layer thickness of the flow, the shear stress at the top of the mound may be several times greater than on the surrounding sea floor. In addition, the flow must converge both laterally and vertically over the waste deposit. This convergence increases the flow velocity, but the increase is expected to be important only if the height of the deposit is comparable to the depth of the water. These effects were measured at the New Haven study site, where a conical pile of waste material rose to about half the water depth.

A current meter was placed 772 meters east of the center of the New Haven site spoil pile and a second was placed on its top. These meters recorded simultaneously; both were set 2 m above the sediment-water interface. In comparing the results obtained at the two stations, allowance must be made for the large fluctuating component of velocity, which is observed throughout Long Island Sound, by comparing records of at least 10-days duration. The number of times current velocities fall within successive velocity intervals, is plotted against the magnitude of the velocity in the distributions shown in Figure 7. The greatest expected tidal velocity at the New

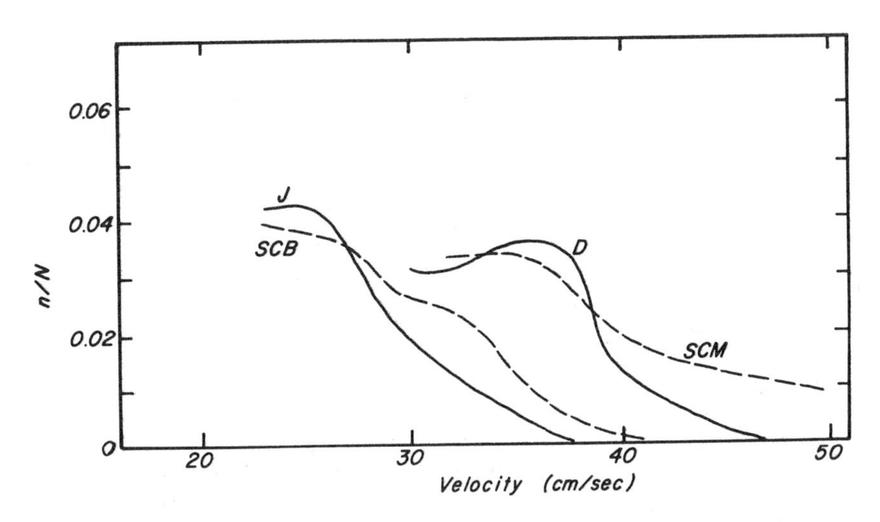

Figure 7
Frequency of occurrence (n/N) of current speeds measured 2 m above the bottom at the edge (J), the top (D), and one mile south (SCB), of the spoil pile on the New Haven disposal site. Also shown is n/N recorded at mid-depth in the water column one mile south of the disposal site (SCM).

Haven site is ∿ 30 cm/sec; the occasional occurrence of much higher current speeds is evident in the distributions. Figure 7 shows that the current speed over the top of the spoil pile (curve D) is about 30% greater than that on the surrounding sea floor: Tidal speeds reach about 40 cm/sec and the range of velocity fluctuations is correspondingly increased.

The increased current observed over the New Haven spoil pile may be due to alteration in the flow regime by the pile, or it may be due to the meter on the top of the pile sampling higher in the water column, where velocities would be greater even in the absence of the pile. This may be tested by comparison with data from a vertical array of three current meters operated one mile south of the disposal site. The bottom and middle meters in this array were placed at the same relative depths in the water column as the meters at the top and side of the spoil pile. The distributions calculated from these records are also shown in Figure 7; curve SCB is for the bottom meter and curve SCM is for the middle meter. The increase in current speed at mid-depth is comparable to that observed at the top of the spoil pile. Hence, it appears that the spoil pile, despite its size and considerable height, has no measurable effect on the speed of water movement over the disposal site.

Suppose that a deposit of muddy waste is placed on an existing muddy bottom. As the elevation of the waste deposit is raised, the resuspension rate illustrated in Figure 3 is increased (because of the higher current speed over the surface) and the deposition rate decreased (because of the decrease in suspended sediment concentration with height above the bottom). Hence, the deposit is expected to be unstable; the ongoing processes of resuspension and deposition will disperse it over the surrounding sea floor. (The fact that mud bottoms of estuaries tend to be flat shows that they are stable against local perturbations in elevation.) How, then, can a stable waste deposit be formed? One way is to armor its surface against erosion by covering it with material having a higher critical erosion velocity than the neighboring bottom, either by allowing a lag deposit of coarse material to accumulate on its surface or by placement of erosion-resistant material on top of less-resistant waste during disposal operations. The latter situation obtains at the New Haven disposal site. Although the pile contains silt-clay sediment from maintenance dredging operations, a photograph of the waste pile, Figure 8, shows that the surface is composed of sand and clods of cohesive mud. This material is sufficiently erosion resistant that little or no change in the configuration of the top of the pile has occurred even though it is high enough to be exposed to currents about 30% greater than those that run over the surrounding muddy bottom.

Even in the absence of an armored surface a deposit of waste at an elevation above the surrounding bottom may resist dispersion if placed in an area in which the natural sedimentation is sufficiently high. Suppose the waste deposit is built up above the surrounding muddy bottom and is then recolonized by benthic animals so that a surficial layer of granules is formed in a time short compared to the time that would be required for the net erosion to remove the deposit. Because of the higher elevation, erosion is expected to remove the layer of granules and the waste material that is being incorporated into them. However, if the deposit is in an area of net deposition so that new mineral material is being added to the bottom, this may replenish the loss from the layer of granules rapidly enough that the waste below remains unexposed. This will be the case if the rate of supply of new material equals or exceeds the difference between the erosion and redeposition rates that would obtain in the absence of the supply of new mineral matter. The accumulation of new sediment on the waste deposit will be, then, less than on the surrounding bottom but sufficient to prevent net erosion.

Erosion and resuspension rates as a function of elevation are not easily measured and we need to find some other indicator of the stability of elevation differences on the bottom to use in evaluating the capacity of disposal sites to retain waste. The same stability argument applies to naturally occurring differences in bottom elevation, so the magnitude of natural bottom slopes may be taken as a measure of the greatest acceptable slope for a deposit of waste which is to be retained at its disposal site. For example, the greatest observed slope on the natural mud bottom over length-

Figure 8
Photograph of a 20 cm span of the surface of the spoil pile at the New Haven disposal site.

scales comparable to the size of the New Haven disposal site and under the same hydraulic regime is about 1:1000. This sloping surface occurs in a region of net deposition and appears to be stable. The composition of the bottom sediment and the benthic animal population here are also the same as at the disposal site. Hence, it is expected that material that can be contained under a surface slope 1:1000 will be retained at the disposal site even though the inherent critical erosion velocity of the waste is as low as that of the natural bottom.

Containment Capacity of a Disposal Site

Natural sedimentary processes set a limit to the amount of waste that can be retained at any given disposal site. As a waste deposit is built upwards from the natural bottom its top will be exposed to higher water velocities and, eventually, it may enter the wave-affected zone. Increased erosion and reduced deposition rates will result. Hence, any open water disposal site has a capacity which, if exceeded, results in unacceptably large losses of material as judged against some established dispersion criterion. This capacity depends on the hydraulic and sedimentary regimes at the disposal site, the properties of the waste and the configuration of the waste deposit. There is an optimum configuration, the one which gives the greatest contained volume of waste for the smallest acceptable loss rate. Careful placement of the waste is required if this optimum is to be attained.

CONCLUSION: THE MANAGEMENT OF A DISPOSAL SITE

The description of the physical processes of erosion, transport and deposition of fine-grained sediments discussed above can be used as a guide in the selection and operation of a site for the containment of particulate waste material.

Disposal Site Location

The first condition to be met is that the water depth be sufficiently great that the bottom not be exposed to disturbance by waves. This appears to be the criterion used many years ago in the selection of some dredge spoil disposal sites along the east coast of the U.S. In Long Island Sound, for example, spoil grounds are usually at the closest spot to the source of spoil where the water is 20 m deep, the greatest depth of the wave-affected zone (Bokuniewicz *et al.*, 1977b). Sites in a similar depth in less protected waters (such as the one just south of Newport, R.I.) may not be out of the zone where waves will disturb the bottom. Once out of the wave-affected zone there is no advantage *per se* to going to deeper water; there are potential disadvantages. Accurate placement of the waste at the desired location on the bottom in the presence of strong currents becomes less probable as the water depth increases. Haulage costs and navigational difficulties are also increased.

A second important factor in the selection of a disposal site is the natural rate of sediment accumulation. Not only does the accumulation of natural sediment suggest by analogy that deposited waste of comparable properties may be retained, but, as shown above, a deposit of unarmored waste which forms at a higher elevation than the natural bottom can only be contained at a site of natural deposition. The long-term, average accumulation rate of the muddy bottom of Long Island Sound has been measured (Bokuniewicz, Gebert and Gordon, 1976) and is shown in Figure 9 along with the locations of the spoil grounds in the Sound that have been used in the past. All but two of the designated disposal sites are in areas of natural deposition but several of the sites need be moved only short distances to be in more favorable locations. The natural deposition rate at the Cornfield Point disposal site is zero and it is not expected that this site will retain particulate waste. The water depth is 57 m but the tidal stream reaches speeds of 1 m/sec. Over a four week interval in the Fall of 1976 120,000 m^3 of silt-clay dredged spoil was released at this site. Despite the great depth and strong current this material reached the bottom. At the end of the four week interval a bathymetric survey showed that not more than 60% of the spoil remained on the site.

If it can be established that the sedimentary environment at a selected disposal site is favorable, the current regime at the site is not of direct concern. Current measurements can be helpful in evaluating the stability of the site, however. The tidal stream characteristics at the site can be determined from a relatively short run of data but much longer records may be required to adequately define the net water movement and the perturbation of the flow under storm conditions. We have presented methods for evaluating such data in another report (Bokuniewicz et al., 1977b).

A final factor to be considered in the selection of a disposal site is the strength properties of the bottom sediments. Part of the kinetic energy of descent of the waste causes plastic deformation of the bottom sediments at the impact point. The greater the energy dissipated this way, the less will be the energy available to disperse the waste. Thus, cohesive blocks of waste may come to rest without breakup on a soft bottom but be broken up and dispersed at an otherwise equivalent site with a hard bottom. The amount of energy dissipated this way can be calculated if the bottom hardness is known (Gordon, 1972).

Method of Disposal

Successful retention of waste at a containment site requires that appropriate care be taken with the method of disposal. Among the factors that must be considered are the speed of the vessel at the time of discharge, navigational accuracy in the location of the discharge points, the rate at which waste is deposited, and the season of the year during which disposal operations are carried out. The method of disposal must be designed to produce the optimum configuration of the waste pile.

When waste is released from a vessel held in a fixed position the material descends quickly through the water column and is deposited within a small area on the bottom. Sufficiently slow release from a moving vessel will inhibit formation of a descending density current; the waste will settle as individual particles and widespread dispersion will result.

Accurate placement of waste on a containment site requires discharge from a stopped vessel or release at a rate sufficiently fast

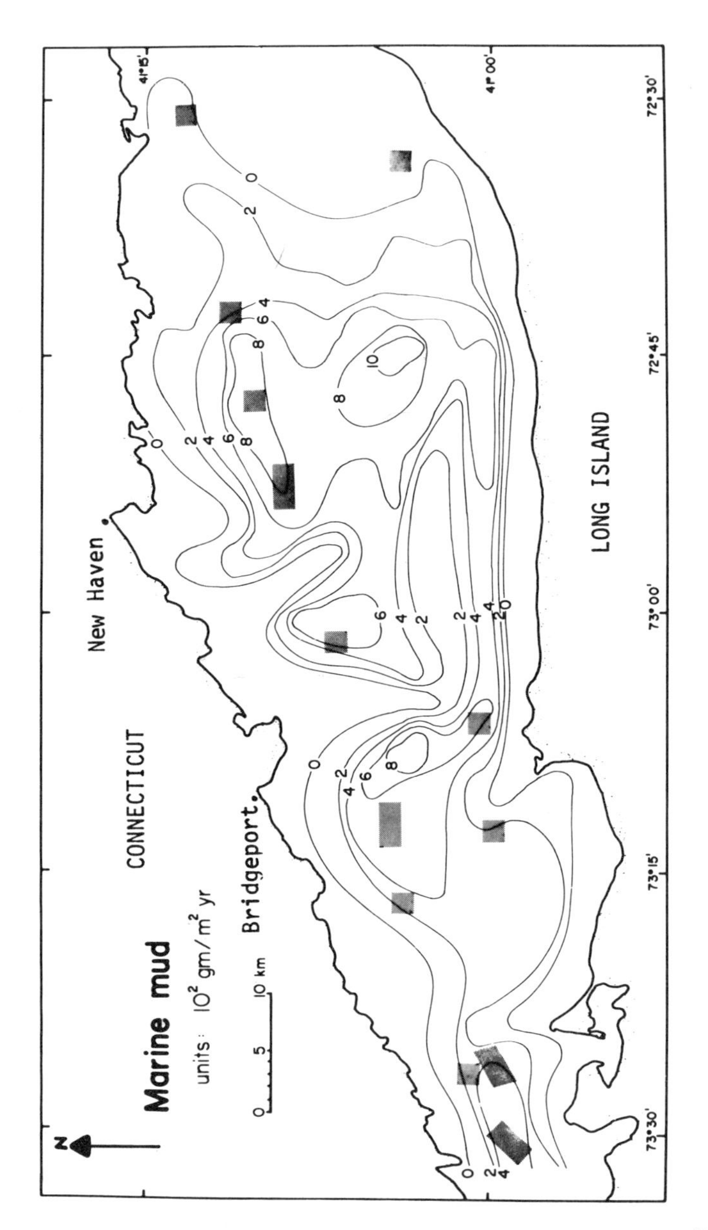

Figure 9
Locations of disposal sites in Long Island Sound superimposed on contours of the mean natural rate of sedimentation.

relative to the speed of a moving vessel to insure formation of a descending density current. The area of the bottom covered by waste released at the surface is determined by the range of the bottom surge, which is a few hundred meters. Construction of a waste deposit that covers the bottom uniformly with surface slope held below the maximum acceptable for the site requires release of the waste on a pattern of predetermined points within the disposal area. The positioning accuracy required is that the error radius must be small compared to the range of the bottom surge. The radius of the positioning error circle should be less than ∿ 50 meters. Such placement, though well within the capability of modern marine technology, has not been customary in disposal operations. When there is a strong current running over the disposal site, allowance must be made for the lateral deflection of the descending jet of waste when selecting discharge locations.

Bioturbation of the waste deposit will increase its susceptibility to erosion. Consequently, it is desirable to minimize the time that freshly-deposited material is exposed at the sediment-water interface. Waste material should thus be applied to the disposal area at the maximum possible rate to prevent recolonization during construction. A few large disposal operations or, perhaps, a combination of several smaller projects are preferable to a multiplicity of operations spread over a long period of time.

The time at which waste is placed on the disposal site should be chosen with regard to biological conditions at the site. During the winter benthic animal activity, and the resultant bioturbation rate, is low. At this time bacterial exudates bind the sediment particles and increase the resistance of the particles to erosion. Disposal operations carried on between November and February not only bury the fewest benthic animals, but also allow bacteria populations to increase (D. C. Rhoads, Yale University, personal communication). In addition to making the sediment less erodible, bacterial exudates also provide a food source for macro- and microfauna that may colonize the waste pile in the spring.

Of the waste material placed on the bottom, that in the bioturbation layer will be put in direct contact with the sea water above where it is subject to dispersion. In New England waters the bioturbation layer is typically about 10 cm thick. Material below this depth will remain undisturbed at a disposal site where waste is retained. The minimum possible radius of a waste deposit is set by the range of the bottom surge generated at the impact point. Hence, there is a minimum size disposal operation below which all material placed on the bottom will be dispersed by resuspension and deposition. As the amount of waste material placed on the bottom is increased, the fraction of the waste that will be dispersed is decreased. The spread of the bottom surge superimposed on a variation of variable position of the discharge point of 200 m leads to the construction of a conical pile with a radius of about 400 m. Assuming that this is the minimum attainable radius for a multi-barge disposal operation, that a conical mound will be constructed and that there is no net erosion, the fraction of material deposited that will suffer dispersion is illustrated in Figure 10. A large disposal operation results in relatively less dispersion than a small one, so long as the total volume of waste remains below the site capacity.

It was shown above that this capacity can be estimated for the

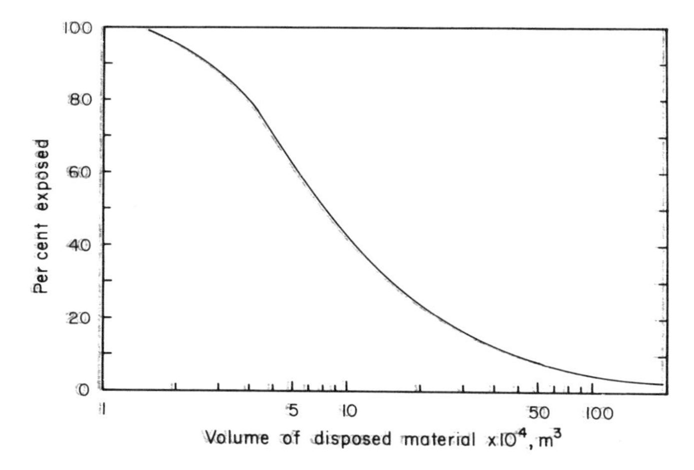

Figure 10
Percentage of disposed material that will undergo dispersion as a function of the size of the disposal project. Calculations are for a conical deposit 400 m in radius assuming a 10 cm bioturbated zone, a 20% deflection of the bottom under the mound, and no long-term, net erosion.

New Haven site on the condition that the slope of the surface of the waste deposit relative to the natural mud bottom be less than about 1:1000. The capacity of the one square mile site is then about 1.3×10^6 m^3, allowing for a 20% deflection of the bottom under the weight of the pile.

The capacity of a site can be substantially increased if the surface of the waste is covered by more erosion-resistant material which does not provide a suitable habitat for deep-burrowing, deposit-feeding animals. Such a cover was attained at the New Haven disposal site more by accident than design. The cohesive character of the silt-clay spoil resulted in formation of the stable substrate for the sand subsequently deposited. If sand is placed on a thinner deposit of less cohesive waste, intermixing or even dispersion of the waste may result from the impact. Generally applicable methods of capping deposits of silt-clay wastes at open water disposal sites have not yet been developed.

If waste material is to be contained on a coastal disposal site, the disposal operation should be managed as a construction project intended to build a permanent bottom feature from the spoil. Adequate placement can be attained with the equipment now in common use in the dredging industry, but much closer control over the placement operation than has been customary in dredging projects will be required.

Acknowledgments

Data used in this paper were obtained in studies made for the United Illuminating Company of New Haven, the New England Division of the U.S. Army Corps of Engineers and the U.S. Army Engineer Waterways Experiment Station.

REFERENCES

1. Benninger, L. K., 1976. The uranium-series radionuclides as tracers of geochemical processes in Long Island Sound, Ph.D. Dissertation, Yale University, New Haven, Connecticut, 151 pp.
2. Benninger, L. K., D. M. Lewis and K. K. Turekian, 1976. The use of natural Pb-210 as a heavy metal tracer in the river-estuarine system, ACS Symposium Series No. 18, Marine Chemistry in the Coastal Environment.
3. Biot, M. A., 1941. General theory of three-dimensional consolidation, J. Appl. Phys. 12, 155-163.
4. Bokuniewicz, H. J., J. A. Gebert and R. B. Gordon, 1976. Sediment mass balance in a large estuary, Est. Coast. and Mar. Sci. 4, 523-536.
5. Bokuniewicz, H. J., J. A. Gebert, R. B. Gordon, P. Kaminsky, C. C. Pilbeam, M. W. Reed and C. L. Tuttle, 1976. Environmental consequences of the disposal of dredged materials in Long Island Sound, Phase III; Geophysical Studies, April 1975 - April 1976, report to the U. S. Army Corps of Engineers New England Division, Waltham, Massachusetts, 95 pp.
6. Bokuniewicz, H. J., J. A. Gebert, R. B. Gordon, J. L. Higgins, P. Kaminsky, C. C. Pilbeam, and M. W. Reed, 1977a. Field study of the mechanics of the placement of dredged material at open water disposal sites, report to the U. S. Army Crops of Engineers, Waterways Experiment Station, Vicksburg, Mississippi.
7. Bokuniewicz, H. J., J. A. Gebert, R. B. Gordon, P. Kaminsky, C. C. Pilbeam, M. W. Reed, and C. B. Tuttle, 1977b. Field study of the effects of storms on the stability and fate of dredged material in subaqueous disposal areas, report to the Army Corps of Engineers, Waterways Experiment Station, Vicksburg, Mississippi.
8. Bokuniewicz, H. J., R. B. Gordon and C. C. Pilbeam, 1976. Stress on the bottom of an estuary, Nature 257, 575-577.
9. Gordon, R. B., 1972. Hardness of the sea floor in nearshore waters, Jour. Geophys. Res. 77, 3287-3293.
10. Gordon, R. B., 1974. Dispersion of dredged spoil dumped in near-shore waters, Est. Coast. Mar. Sci. 2, 349-358.
11. Gordon, R. B. and C. C. Pilbeam, 1975. Circulation in central Long Island Sound, Jour. Geophys. Res. 80, 414-422.
12. Haven, D. and R. Morales-Alamo, 1968. Occurrence and transport of fecal pellets in suspension in a tidal estuary, Sed. Geol. 2, 141-151.
13. Rhoads, D. C., 1973. The influence of deposit-feeding benthos on water turbidity and nutrient recycling, Am. J. Sci. 273, 1-22.
14. Rhoads, D. C., 1974. Organism-sediment relations on the muddy sea floor, Oceanogr. Mar. Bio. Ann. Rev. 12, 263-300.
15. Rhoads, D. C., R. C. Aller and M. B. Goldhaber, 1976. The influence of colonizing benthos on physical properties and chemical diagenesis of the estuarine sea floor, Belle Baruch Symposium on Ecology of Marine Benthos, in press.
16. Schubel, J. R., 1972. Distribution and transportation of suspended sediment in upper Chesapeake Bay, Geol. Soc. Am. Mem. 133, 151-167.
17. Taylor, G. I., 1919. Tidal friction in the Irish Sea, Phil. Trans. Roy. Soc. London A 220, 1-33.

18. Terwindt, J. H. J., and H. N. C. Breusers, 1972. Experiments on the origin of flasher, lenticular and sand-clay alternating bedding, Sed. 19, 85-98.
19. Webster, F., 1969. Turbulent spectra in the ocean, Deep-Sea Res. 16, 357-368.
20. Weisberg, R. H., 1976. A note on estuarine mean flow estimation, Jour. Mar. Res. 34, 387-394.

Dredging and Disposal in Chesapeake Bay, 1975–2025

M. Grant Gross
W. B. Cronin

In Chesapeake Bay frequent dredging is required in three locations: Bay entrance, harbors, and the turbidity maxima at the fresh-/salt-water interfaces. Between 1975 and 2025, 560×10^6 m^3 (740×10^6 yd^3) are scheduled to be dredged to maintain existing channels and to deepen them to accommodate larger vessels. One third of the dredging is associated with new construction projects; two thirds comes from maintenance dredging. Approximately half the maintenance dredging will be done in port areas where much of the dredged material may be mixed with industrial wastes and sewage discharges.

Sources of deposits in channels and port facilities in Chesapeake Bay are poorly understood. River sediment discharge to Chesapeake Bay is probably 5 to 10×10^6 tons (15 to 30×10^6 m^3/yr) per year, in years without major floods. Smaller but unknown amounts of sediment are contributed by sewage treatment plants, urban runoff, industrial processes, and shoreline erosion.

Available data suggest mostly localized and transitory biological effects for those disposal operations involving unpolluted materials that do not alter the physical or chemical character of the bottom. Long-term effects of dredged material disposal in the Bay such as altered current regimes or sediment transport can not be predicted with confidence.

Because of shortages of wetland disposal sites, opposition to open-water disposal, and high costs of diked disposal areas, new disposal sites and techniques will likely be needed in the next 50 years to handle the disposal of materials from needed dredging in Chesapeake Bay. Dredging and disposal of dredged materials will continue to be major geological processes in Chesapeake Bay.

INTRODUCTION

Availability of low-cost waterborne transport has controlled economic development and growth in Chesapeake Bay since colonial times[1]. Combined with the building of the railroads in the nineteenth century and highways in the twentieth century, waterborne transport has determined the location and growth of the Bay's cities and suburbs. Today Baltimore and Hampton Roads (Norfolk) remain

*Chesapeake Bay Institute Contribution 250

among the nation's top five ports, but their future growth and vitality depend on maintaining navigation channels through the Bay for ocean-going vessels.

Because of shoaling of navigation channels, dredging has been a continuing activity. Building port facilities and navigation channels (Figure 1) as well as maintaining them has required extensive

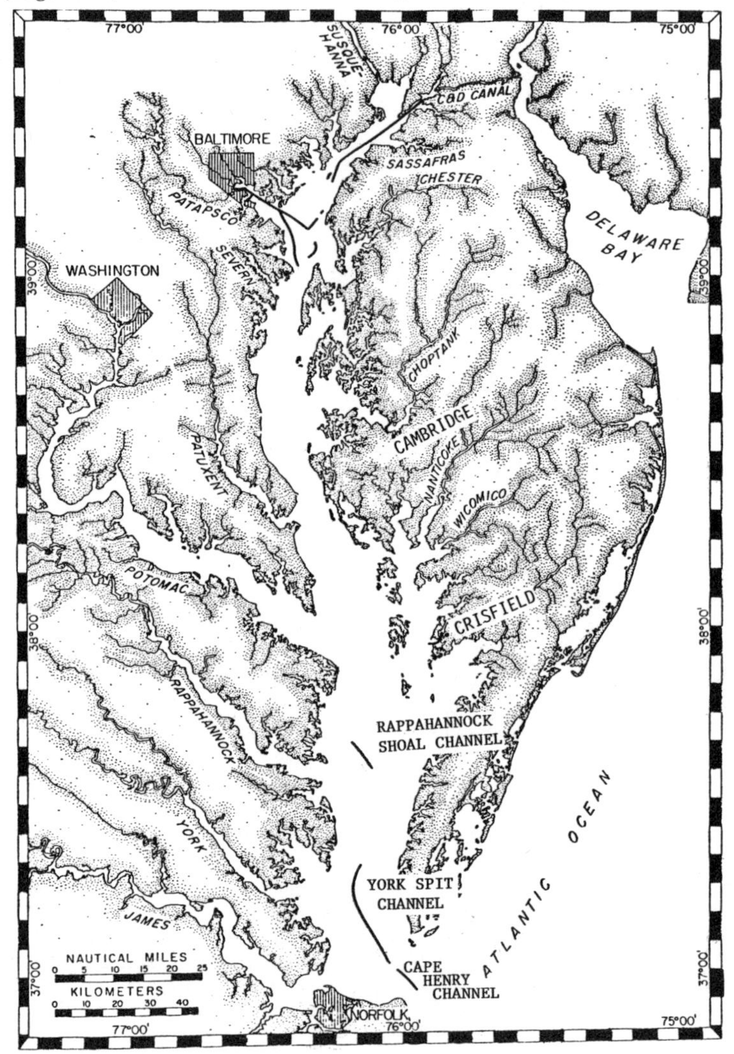

Figure 1
Major navigation channels in Chesapeake Bay

dredging since the 1790s. Dredging and disposal of dredged materials will remain a high priority item for the Chesapeake Bay region during the next 50 years, especially as navigation channels are deepened to accommodate larger vessels used to transport bulk cargoes--petroleum, iron ore, and coal.

Dredging of Baltimore Harbor began as early as 1798[2] and by 1804 at Georgetown near Washington, DC[3]. Extensive erosion caused by primitive agricultural practices and the soil depletion resulting from tobacco farming[4] and coal mining and washing in the Susquehanna River draining caused widespread shoaling in many tributaries. Harbors for several important tobacco ports on smaller tributaries of the Bay completely silted in and today are several miles from the Bay shoreline[3,5]. And it is highly probable that the soils eroded during colonial times are still moving into the Bay[6].

In short, shoaling of ports and navigation channels is a continuing problem in Chesapeake Bay and dredging remains the only practicable solution to maintain access for ocean-going vessels to the Region's ports. But concern about the immediate problems of dredging badly shoaled channels and disposal of the dredged materials has obscured the need for planning on a regional scale for long-term needs for dredging and acceptable disposal of the dredged materials.

Several questions must be answered when we consider the Region's dredging and disposal problems over the next 50 years.

How much dredging will be required and where?

What are the sources of the materials deposited in the channels and what can be done to reduce the amount of sediment reaching the Bay?

What are the physical and chemical characteristics of the dredged materials?

What can be done to dispose of the dredged materials at acceptable costs?

DREDGING REQUIREMENTS

Dredging is required to maintain needed channel depths in Baltimore and Norfolk harbors, in Northern Chesapeake Bay, and at the ocean entrance to the Bay (Table 1). Between 1925 and 1975, approximately 115×10^6 m^3 of materials were removed from Baltimore Harbor (Figure 2). A comparable amount (110×10^6 m^3) was removed from Hampton Roads projects (Figure 3) between 1900 and 1974.

Data on disposal of the dredged materials are scarce. Most of the dredged material went into disposal sites near the channels. Apparently little of the material went into landfill sites.

Looking ahead to 2025, the amount of material scheduled for dredging to deepen the channels in the Bay (Table 2) amounts to about 200×10^6 m^3. About three fourths is to be done in the Norfolk-Hampton Roads area, at nearly double the projected annual rate for FY 1977-1981. Maintenance dredging is also scheduled to increase markedly (Figure 2), amounting to 360×10^6 m^3 (Table 3).

Most of these materials will be deposited in disposal sites in Chesapeake Bay because of the scarcity of acceptable disposal sites on land, the long haul distances to ocean disposal sites, and the national and international efforts (such as the London Convention of 1973) to control waste disposal in the ocean.

Table 1
Annual maintenance requirements of principal Chesapeake Bay navigational channels in Chesapeake Bay[7]

	$10^6 m^3$	$(10^6 yd^3)$
NORTHERN CHESAPEAKE BAY		
Chesapeake and Delaware Canal (to MD boundary)	0.15	(0.2)
Western approach channel to the C&D Canal	0.9	(1.2)
Baltimore Harbor Approaches and connecting channels to the Canal western approach channel	0.6	(0.8)
Baltimore Harbor	0.2	(0.3)
Subtotal	1.85	(2.5)
SOUTHERN CHESAPEAKE BAY		
Port of Hampton Roads	2.9	(3.8)
James River	0.9	(1.2)
Rappahannock River	0.04	(0.05)
Baltimore Harbor Channels (VA portions)	0.8	(1.0)
All other Federal projects in Norfolk D.T.	0.8	(1.0)
Subtotal	5.44	(7.05)
TOTALS	7.2	(9.55)

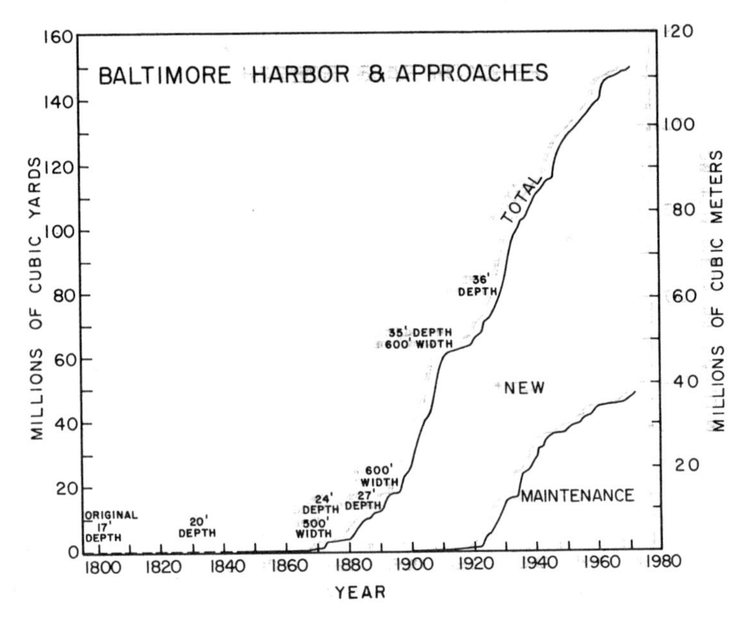

Figure 2

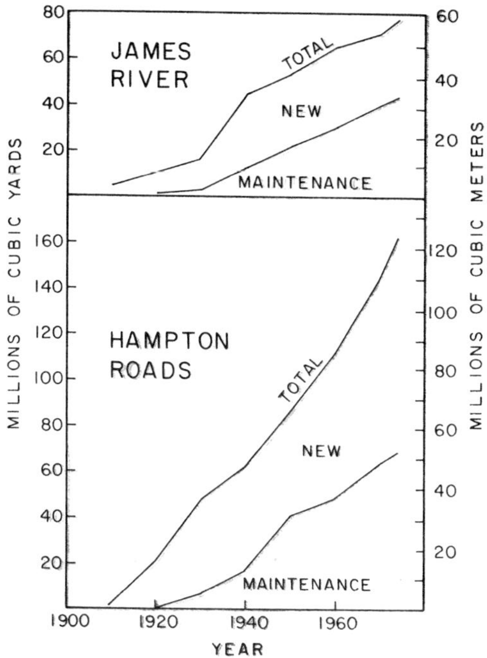

Figure 3
Volumes of materials dredged from Hampton Roads and James River by Federal projects, 1900–1974. (Data compiled from Annual Reports of the Chief of Engineers.)

Table 2
New channel dredging, principal navigational channels in Chesapeake Bay[7]

	$10^6 m^3$	($10^6 yd^3$)
NORTHERN CHESAPEAKE BAY		
Western approach channel to the Canal(to 35')	7.6	(10.0)
Baltimore Harbor Approaches (to 50') connecting channels to the Canal western approach channel (to 35')	26.3	(34.3)
Baltimore Harbor (to 50')	11.2	(14.7)
Crisfield Harbor	0.15	(0.2)
Subtotal	45.2	(59.2)
SOUTHERN CHESAPEAKE BAY		
Port of Hampton Roads	95.0	(125.0)
York River	16.0	(21.0)
Southern Branch, Elizabeth River	2.4	(3.2)
Baltimore Harbor Channels (VA portions)	40.0	(52.0)
Subtotal	153.4	(201.2)
TOTALS	198.6	(260.4)

Table 3
Projections of dredging requirements (in millions of cubic meters) for Chesapeake Bay, 1975 - 2025 [7]

	Maintenance		New Construction
NORTHERN CHESAPEAKE BAY			
Turbidity maximum (Approach Channels)	82		34
Harbors			
Baltimore	10		11
Crisfield	--		0.2
Subtotal	92		45.2
SOUTHERN CHESAPEAKE BAY			
James River	45		--
York River	--		16
Baltimore Harbor Channels	40		40
Harbor			
Hampton Roads, Elizabeth River	185		97
Subtotal	270		153
TOTALS	362	+	198 = 560

SOURCES OF SEDIMENT

Sediments deposited in navigation channels and port facilities of Chesapeake Bay come from:

Riverborne sediments eroded from farms, mines and construction sites,

Urban runoff,

Sewage discharges,

Industrial plants,

Shoreline erosion,

Marine organisms that grow in the Bay waters,

Continental shelf sediments outside the Bay mouth.

The relative importance of each source varies depending on the location of the dredging site.

Specification of the importance of each source requires detailed study of the materials dredged from each area as well as the local circulation patterns and biological processes. No detailed studies have been made of Chesapeake Bay areas that require routine dredging aside from the James River estuary[8].

Data on sources of sediments deposited in Chesapeake Bay are scarce and most of the available data deal with riverborne solids. Limited measurements of the sediment discharge of the Susquehanna and Potomac Rivers and estimates for other rivers indicate an annual sediment discharge of more than 6.5 million tons/year in the absence of major floods (Table 3); Wolman[5] "guestimated" the discharge of 8 million tons/year. Sediment discharge during two floods on the Susquehanna River in 1966-1975 was 40 million metric tons out of 50 million tons discharged during the decade (Gross, et al., in press). No data are available on suspended solid discharges through the Region's municipal and storm sewers or from industrial operations. Continued suburbanization of the Bay's shores will doubtlessly increase local sediment discharges[9].

Shoreline erosion is also an important sediment source. Schubel[10] estimated that erosion along approximately 235 miles of Northern Chesapeake Bay shoreline contributed about 100,000 tons/year of silt and clay to the Bay. If we assume that the entire Bay shoreline of 4600 miles contributes sediment even at rates only half those of Northern Chesapeake Bay, this would amount to approximately one million metric tons per year or comparable to the annual sediment discharge of the Susquehanna River (Gross et al., in press).

The Region's cities and industries also discharge suspended sediment to the Bay, primarily in the Baltimore, Washington, and Hampton Roads areas. These discharges combined with runoff from construction sites are locally important sources of sediment to the Bay[11].

Investigations of sources of materials dredged from shoaling areas in northern Delaware Bay[12] showed that three sources contributed 88.5 percent of the 6.8 million tons of sediment dredged each year (Table 4):

Erosion of estuary bed and banks	39.0%
Riverborne solids from watershed	23.9
Diatom production	25.6

Table 4
Contributions of suspended and dissolved solids in northern Delaware Bay, in millions of tons per year (after Neiheisel[12])

Source	Solids (10^6 tons/year)		
	Suspended	Dissolved	%
Erosion of drainage	1.4	---	19.7
Erosion of estuary bed and banks	2.3	---	32.4
Diatom production	1.5	---	21.1
Dredging	0.4	---	5.6
Sewers	0.1	0.4	7.0
Industrial	0.06	0.8	12.1
Atmospheric fallout	0.1	---	1.4
TOTALS	5.9	1.2	

But only 86% of the shoaling could be accounted for unless it was assumed that the dissolved solids in sewage and industrial discharges were incorporated in the sediment (see Table 4).

Similar processes may be active in Chesapeake Bay but definitive data are lacking. We do not, for example, have any information on the contribution of diatoms to sediments deposited in Northern Chesapeake Bay. Nor do we know the contributions to the dredging problems arising from discharges of sanitary and storm sewers or waterfront industries.

PHYSICAL AND CHEMICAL COMPOSITION OF DREDGED MATERIALS

Sediments dredged from Chesapeake Bay harbors such as Baltimore or Norfolk, and from the turbidity maximum in the Northern Bay are usually fine grained, have high water contents, compact slowly, and often have high concentrations of organic carbon and industrial, metal-rich wastes[13]. These dredged materials are undesirable for landfill and waterfront disposal sites are not often readily available. For such material diked enclosed areas have been built to contain the materials dredged from Baltimore Harbor[13] and Norfolk Harbor[14]. Large areas of the upper Potomac estuary have been filled, and used for such purposes as Washington's National Airport, and several parks[15].

Clay minerals dominate Chesapeake Bay sediments. Illite, chlorite, and kaoline were identified in all sediments from Northern Chesapeake Bay[16,17] and in the James River[8].

Organic carbon contents of the Bay sediments are generally high. In the turbidity maximum of the James River, organic carbon contents range from 0.3 to 5.9 percent, averaging 1.2 percent[8]. In the silty sediments from the deeper portions of central Chesapeake Bay (near Patuxent River), organic carbon contents were typically between 2.0 and 2.6 percent[16]. We have no data on the composition of sediments from the Chesapeake Bay turbidity maximum or from the Bay mouth channel. Sediments from the turbidity maximum of Delaware Bay (Marcus Hook) contained more carbon than the average for the Bay (Table 5).

Table 5
Principal constituents in Delaware Bay sediments and dredged materials from Marcus Hook (after Neiheisel[12])

Constituent	Delaware Bay*	Marcus Hook	Bay Mouth
Quartz	62.6%	24.5%	93.5%
Clay minerals	12.0	26.4	1.4
Diatoms	8.1	18.0	0.1
Feldspar	8.1	14.1	3.4
Organic matter	2.6	8.0	0.3
Coal	2.9	5.4	0.0
	96.3%	96.4%	98.7%

*excl Marcus Hook

Because of concern about hazardous metals in sediment deposits in harbors, most of the data on chemical composition of Chesapeake Bay sediment come from possibly polluted deposits near Baltimore, Washington, and Norfolk (Table 6). In the Potomac River deposits near the Blue Plains Wastewater Treatment Plant, the deposits were unusually rich in copper, chromium, silver, and cadmium[18]. Substantial seasonal variability in concentrations of potential pollutants such as zinc and nickel were observed but not explained.

Comparison of metal concentrations in sediment deposits in Chesapeake Bay with its industrialized and urbanized harbors indicates that the deposits in Baltimore Harbor and the Elizabeth River are distinctly enriched in common industrial metals[18,19]:

Enrichment with respect to unpolluted Bay sediments

chromium	20 times
copper	50 - 30 times
lead	3 - 30 times
zinc	3 - 7 times
cadmium	> 7 times
mercury	15 times

No data are available on the physical or chemical characteristics of materials from navigation channels at the entrance to Chesapeake Bay. These deposits are likely to be sands, distinctly different from the fine-grained materials dredged from Northern Chesapeake Bay.

Table 6
Metal concentrations (mg/kg) in sediment deposits from Baltimore Harbor, Delaware River, Potomac River, James River, Chesapeake Bay and Elizabeth River (after Villa and Johnson[19]).

Metal	Baltimore Harbor	Delaware River	Potomac River	James River	Chesapeake Bay	Elizabeth River
Chromium						
Low	10	8	5	NO	18	NO
Average	492	58	25		25	
High	5745	172	90	DATA	42	DATA
Copper						
Low	<1	4	10	NO	<1	20
Average	342	73	30*		6.4-7.0	900
High	2926	201	730	DATA	22	1500
Lead						
Low	<1	26	20	4	9	10
Average	341	145	35*	27	27	100
High	13890	805	100	55	86	500
Zinc						
Low	31	137	125	10	33	80
Average	888	523	275*	131	128	350
High	6040	1364	1000	708	312	1300
Cadmium						
Low	<1	<1	<0.10	NO	<1	NO
Average	6.3-6.6	2.9-3.1	<0.10*		<1	
High	654	17	0.60	DATA	<1	DATA
Nickel						
Low	12	NO	3	NO	5	NO
Average	36		25		12	
High	94	DATA	50	DATA	27	DATA
Manganese						
Low	121	NO	500	NO	218	NO
Average	739		--		690	
High	2721	DATA	4800	DATA	1608	DATA
Mercury						
Low	<0.01	<0.01	0.01	0.02	<0.01	0.30
Average	1.17	1.99	--	0.32	0.061-0.067	0.90
High	12.20	6.97	0.03	1.00	0.31	3.00

*Median values

EFFECTS OF DISPOSAL OPERATIONS

Disposal of the large volumes of dredged materials in Chesapeake Bay projected for 1975 - 2025 will cause long-term problems because of unavoidable alteration of Bay shorelines and bottom configurations, especially if diked enclosures are used to contain polluted materials. For example, the amount of materials to be dredged from Baltimore Harbor is about two-thirds the amount removed from the Harbor since 1794 or about one-sixth the entire volume of water in the Harbor at low tide[20]. And most of the more acceptable disposal sites have already been used and apparently have little capacity left.

The situation is slightly more favorable in the Approaches to the Chesapeake and Delaware Canal than in Baltimore Harbor. Here there is a much larger area of the Bay potentially usable for disposal sites, including some land sites. Also the dredged materials are less polluted than deposits removed from Baltimore Harbor, providing more disposal options.

In Southern Chesapeake Bay, the disposal situation appears to be less troublesome than in the Baltimore area. In addition to existing diked disposal areas, there appear to be more possibilities for open water disposal. More attention must be given to regional planning for disposal of dredged materials if we are to avoid the long-term --and probably irreversible--changes reported from the James River[21]. There disposal of dredged materials in the James Estuary has built up shoals which are shallow enough to be eroded by wave action and redistributed by tidal currents.

Short-term effects of dredged material disposal have been studied extensively although many questions remain unanswered. Some of the results are summarized here.

Dredging and disposal operations temporarily alter distributions and abundances of benthic organisms. Available data, usually of a qualitative, descriptive nature, indicate temporary destruction through removal of the benthic communities during (dredging) or through burial during disposal operations. If the physical character and composition of the bottom is not drastically altered, recovery begins within a few months[22,23,24]. Resettlement occurs by migration of mobile animals and--more importantly--by settling of larval or juvenile forms[22].

Effects of dredging and disposal of silty sediment from Northern Chesapeake Bay were studied by Cronin and others[23]. The materials hydraulically dredged were discharged from a pipeline to shallow areas in Northern Chesapeake Bay near the dredged channels with the following observed results:

- Fine grained materials dredged from the channel were released as a semi-liquid mixture in shoal water over sediments having similar physical and chemical properties.
- Turbidity increased over an area of 1.5 - 1.9 square miles (4 - 5 square kilometers) around the disposal site. Over most of the area, the suspended sediment load was within the range of natural variation observed, but at a different season from observed natural maxima.
- Suspended sediments (in the top of 10 feet or 3 meters of water) were carried in a tide-related plume to a maximum distance of about 3.1 miles (5000 meters), and virtually disappeared within two hours after disposal operations ceased.
- Concentrations of total phosphate and nitrogen in the water increased near the dredge discharge by factors of about 50 and 1000, respectively, over ambient levels. Limited field experiments did not show any detectable effects on photosynthesis by phytoplankton.
- The dredged material deposited on the bottom covered to at least 1 foot (0.3 meters) an area at least 5 times as large as that of the defined disposal site.
- Approximately 12% of the deposited sediment disappeared from the spoil "pile" in 150 days after deposition.
- No obvious effect of dredging or spoil disposal was observed on phytoplankton primary productivity, zooplankton, fish eggs and larvae, or fish.
- There was a reduction of about 70% in the average number of benthic organisms per square yard and of about 65% in the

benthic biomass in the spoil disposal area, accompanied by a marked reduction in the number of species present. After one and a half years, numerical abundance, biomass, and species diversity had recovered to approximately the predisposal levels. Individual species varied greatly in susceptibility to damage and in recovery patterns.
- At the site of dredging in the channel, an erratic series of species fluctuations occurred. After one year, the channel had about the same number of individuals as during the predredging period, but fewer species were present.

MANAGEMENT ISSUES FOR DISPOSAL OF DREDGED MATERIALS IN CHESAPEAKE BAY

Data available for the period 1975 - 2025 indicate that dredging is a long-term problem in Chesapeake Bay and will likely continue at levels well above those projected for FY 1977 - 1981. The increase in the volume of materials to be disposed of and the increasing shortage of acceptable disposal sites and techniques indicate that environmental management and regulatory agencies in the region should adopt strategies to cope with dredging and disposal problems.

The present technique for screening materials for acceptability for open water disposal is the elutriate test[25] which involves analyzing waters that have been mixed with established volumes of sediment for a prescribed amount of time. The waters are then analyzed and their concentrations of selected constituents are compared with those observed in waters taken from the proposed disposal site. Its utility in estuarine situations like Chesapeake Bay remains to be demonstrated.

The test seems ill-suited for application in Chesapeake Bay when the waters and organisms are routinely subjected to large volumes of sediment from river discharge, especially floods (see Table 5) from erosion of the shoreline and bottom especially after storms and from carbon-rich sediments resulting from the abundant growth of phytoplankton in the Bay waters. Such materials although unaffected by human activities or waste discharges are likely to be judged unacceptable to the elutriate test. Furthermore, there is substantial uncertainty about the water sample with which the elutriate test results are to be compared. The suspended sediment concentrations and presumably the sediment associated constituents from the eroded deposits are subject to wide variation with depth in the waters as a consequence of erosion of the bottom by tidal currents.

While there is yet no generally accepted test for identifying troublesome deposits that would be inappropriate for unrestricted disposal operations, it is clear that such a test should be developed that would successfully classify fine-grained deposits such as those found in Chesapeake Bay. A test or series of tests should be devised to indicate which dredged materials are appropriate for open water disposal including covering over older waste deposits or materials too polluted to permit removal by dredging operations. Some of the materials dredged from the ocean entrance channels may be useful for landfill or for beach replenishment if they were found to be free of contamination.

Strategies to control the amount of material to be dredged have particular appeal. The argument is that if we reduce the amount of sedimentary material coming into the estuary, we could reduce the amount of dredging required to remove those materials from navigation channels. But on the basis of this survey, this strategy seems to be an attractive option for only one of the three major areas requiring frequent maintenance dredging. Clearly it would have little to offer for materials dredged during construction activities, and new construction dredging accounts for nearly two-thirds of the materials to be moved during the next fifty years in Chesapeake Bay.

In harbors, more vigorous enforcement of sediment pollution controls might reduce the amount of sediment washed into the harbor through urban runoff from construction sites. Better treatment of industrial wastes and possibly more advanced treatment of municipal sewage might also reduce the amount of sediment discharged to the harbor from these sources. While the volume of dredged materials affected is relatively small, it is nonetheless important. Such materials are among the most heavily polluted of the materials dredged and are usually the most difficult to find adequate disposal sites for.

In the turbidity maximum, there seems to be little that can be done directly to reduce the amount of dredging. Most of the material apparently comes from river discharge and is first deposited in the estuary and then reworked by tidal currents and wave action to be eventually deposited in the navigation channels. The amount of material available for reworking in the estuary is large. Furthermore, the lower course of the rivers still contain large volumes of materials disturbed by agriculture and other human activities and only about 5% of that material has yet reached the estuary. So better agricultural practices are not likely to have a dramatic effect on the amount of sediment from such sources for decades after control measures are instituted.

Much of the sediment deposited in the Bay entrance channels apparently comes from littoral drift--the movement of beach and nearshore sands along the coast under the influence of the waves. These materials do not contribute a large volume and have generally been found to be unpolluted and potentially acceptable for either ocean disposal or for such uses as beach replenishment.

CHANGES IN SEA LEVEL

Because of the continued long-term adjustment of sea level since the last ice age, the level of the average sea surface is rising at 1.5 mm (0.005 ft) per year[26]. In addition to the world-wide sea level change, Chesapeake Bay is also subsiding between 1.5 and 2.3 mm per year, giving a net change in sea level of about 3.0 mm per year in Northern Chesapeake Bay and 2.3 mm per year in the Southern Bay (see Table 7). This means that the Bay is able to accommodate some additional deposits without overall changes in depth.

Table 7
Changes in yearly mean sea level in Chesapeake Bay, in millimeters per year[26]

Port	Entire Series of Data	1940-1972
Baltimore	3.39 (1903-1972)	2.94
Annapolis	4.23 (1929-1972)	3.49
Washington, DC	3.28 (1932-1972)	3.26
Solomons, MD	3.87 (1938-1972)	3.83
Hampton Roads, VA	4.63 (1928-1972)	3.84
Portsmouth, VA	3.81 (1936-1972)	3.87

Over the 50 years of this projection the Bay will be able to accommodate about 2×10^9 m^3 or about 4×10^7 m^3/year. This is about twice the volume of sediment ($\sim 20 \times 10^6$ m^3/yr) transported to the Bay each year by the rivers.

Thus the continued rise in sea level offers some hope that the Bay will be able to accommodate some large volume discharges of dredged materials if they were widely dispersed.

CONCLUSIONS

1. Between 1975 and 2025, approximately 450×10^6 cubic meters (580×10^6 cubic yards) of sediment deposits are scheduled for dredging from Chesapeake Bay, 31% for construction of deeper channels and 69% for maintenance of existing and new channels and other port facilities.

2. Dredging in Chesapeake Bay occurs primarily in three locations: harbors (Baltimore and Norfolk-Hampton Roads), the turbidity maximum where most riverborne sediment is deposited near the head of the Bay and its major tributaries, and in the ocean entrance channels.

3. Dredging of Chesapeake Bay normally involves fine grained sediment (silts and clays) with high water contents, high organic matter, high nitrogen and high chemical oxygen demands. Dredging of new channels will probably involve coarser grained materials with less organic matter and lower oxygen demands.

4. Sediments dredged from harbors at Baltimore, Norfolk, and Washington will contain varying amounts of potentially hazardous metals and hydrocarbons from industrial and municipal discharges, as well as urban runoff.

5. Materials dredged from the entrance channels will primarily involve sands with relatively low concentrations of metals and organic matter.

6. Potential long-term effects of dredged material disposal involve changes in current regimes and creating sediment deposits that are easily eroded by waves and tidal currents and cause locally increased problems of sediment accumulation.

7. Biological and chemical consequences of dredged material disposal are relatively short-term and usually reversible, especially where the physical and chemical characteristics of the bottom are not permanently altered.

8. Owing to the lack of knowledge about sources of sediments that accumulate in navigational channels, it is not possible to develop management strategies or to evaluate the cost effectiveness of those already proposed.

9. One attractive sediment management proposal is to control the amount of sediment discharged through municipal sewers and coming from industrial discharges.

10. There are no acceptable methods of identifying potentially troublesome materials and to classify them for disposal by the most cost effective means.

REFERENCES

1. Boorstin, D.J. 1958. The Americans: The Colonial Experience. Random House, New York. 434 p.
2. Kanerek, H.K. 1976. A monument to an engineer's skill: William P. Craighill and the Baltimore Harbor. Baltimore District, U.S. Army Corps of Engineers, Baltimore, MD.
3. Gottschalk, L.D. 1945. Effects of soil erosion on navigation in Upper Chesapeake Bay. Geogr. Rev. 35:319-338.
4. Eaton, C. 1961. The growth of southern civilization. Harper & Row, New York. 357 p.
5. Wolman, M.G. 1968. The Chesapeake Bay: geology and geography. p II-7 to II-48. In Proceedings of the Governors' Conference on Chesapeake Bay, September 12-13, 1968.
6. Trimble, S.W. 1975. Denudation Studies: Can we assume steady state. Science 188:1207-1208.
7. CRC. 1977. Proceedings of the Bi-State Conference on Chesapeake Bay. Chesapeake Research Consortium Publ. 61. Baltimore, MD.
8. Nichols, M.M. 1972. Sediments of the James River Estuary, Virginia. p. 169-212. In B.W. Nelson (ed) Environmental Framework of Coastal Plain Estuaries. Geol. Soc. Amer. Mem. 133, 619 p.
9. Wolman, M.G. 1967. A cycle of sedimentation and erosion in urban river channels. Geograf. Ann. 49:385-395, Series A.
10. Schubel, J.R. 1968b. Shore erosion of the northern Chesapeake Bay. Shore and Beach. April 1968, p. 22-26.
11. Wolman, M.G. and A.P. Schick. 1967. Effects of construction on fluvial sediment, urban and suburban areas of Maryland. Water Resources Research 3:451-464.
12. Neilheisel, J.D. 1973. Long range spoil disposal study, Part III, Sub-study 2. Nature, source and cause of the shoal. U.S. Army Corps of Engineers, Philadelphia District. 140 p.
13. Kolessar, M.A. 1965. Some engineering aspects of disposal of sediments dredged from Baltimore Harbor, p. 613-618. In Proceedings of the Federal Interagency Sedimentation Conference, 1963. Agricultural Research Service Misc. Publ. 970. 933 p.
14. Norfolk District, U.S. Army Corps of Engineers. 1974. Norfolk Harbor, Virginia - Report of Survey Investigations: The Craney Island Disposal Area. 2 volumes.
15. Meade, R.H. 1969a. Errors in using modern stream-load data to estimate natural rates of denudation. Geol. Soc. Amer. Bull. 80:1265-1274.
16. Biggs, R.B. 1967. The sediments of Chesapeake Bay. p. 239-260. In G.H. Lauff (ed) Estuaries. Am. Assoc.Adv. Science. Publ. 83. 757 p.
17. Schubel, J.R. 1968a. Suspended sediment of the Northern Chesapeake Bay. Chesapeake Bay Institute Tech. Report 35. The Johns Hopkins University, Baltimore, MD. 264 p.
18. Pheiffer, T.H. 1972. Heavy metals analyses of bottom sediment in the Potomac River Estuary. U.S. Environmental Protection Agency, Annapolis Field Office Tech. Report 49.
19. Villa, O. and P.G. Johnson. 1974. Distribution of metals in Baltimore Harbor sediments. Env. Protection Agency, Tech. Rept. 59. Annapolis Field Office.
20. Garland, C.F. 1952. A study of water quality in Baltimore Harbor. Chesapeake Biol. Lab. Dept. Res. and Ed. Publ. 96. 132 p.

21. Nichols, M.M. 1978. The problem of misplaced sediment. In this volume.
22. Harrison, W., M.P. Lynch, and A.G. Altschaefel. 1964. Sediments of Lower Chesapeake Bay with emphasis on mass properties. J. Sed. Petrol. 34:727-755.
23. Cronin, L.E. (ed) 1970. Gross physical and biological effects of overboard spoil disposal in Upper Chesapeake Bay. Natural Resources Institute Special Rept. No. 3, University of Maryland. 66 p.
24. Stickney, R.R. and D. Perlmutter. 1975. Impact of intracoastal waterway maintenance dredging on a mud bottom benthos community. Biol. Conserv. 7:211-226.
25. EPA/COE. 1977. Ecological evaluation of proposed discharge of dredged material into ocean waters. Environmental Protection Agency/Corps of Engineers Technical Committee on Criteria for Dredged and Fill Material. U.S. Army Engineers Waterways Experiment Station, Vicksburg, MS.
26. Hicks, S.D. and J.E. Crosby. 1975. An average long-period sea-level series for the United States. NOAA Tech. Mem. NOS 15. Rockville, MD. 6 p.

The Problem of Misplaced Sediment

Maynard M. Nichols

ABSTRACT

Disposal of estuarine dredged material is a major depositional process with important sedimentologic and hydraulic consequences. Large-scale dredging in U.S. East Coast estuaries has cut channels far below natural equilibrium depths and induced rapid sediment accumulation. In turn, the increased frequency of dredging has created a massive disposal problem and a need for ocean dumping.

An estuary channel attains maximum stability by adjusting its bed geometry and hydraulic regime. When a channel is deepened, tidal currents are reduced and sediment deposition accelerates. Deepening causes more salty water to penetrate landward and shifts the zone of maximum shoaling upstream. Landward density currents accelerate the potential for return of dredged material from seaward reaches, while higher stratification enhances the potential for entrapment. Therefore, dredging is partly self-perpetuating.

Long-continued open water disposal along channel shoulders of the Upper James Estuary, Virginia, has built up mounds close to the surface where they are subject to wave erosion and current scour. Mounds of misplaced sediment with slopes greater than 6° allow return of sediment to the channel and they restrict tidal flow. By reducing the cross sectional area, they supress the tide range and tidal discharge in landward reaches. Consequently, dispersion of pollutants is reduced and accumulation of fine-grained sediment is promoted. As a consequence of dredging and disposal, many estuaries are losing their capacity to absorb more misplaced sediment.

INTRODUCTION

Every year about 19 million tons of sediment is dredged from coastal estuaries, lagoons and harbors of the U.S. East Coast. Dredging is required to maintain sufficient water depth in shipping channels to Philadelphia, Savannah, and other major port cities. As tankers and other ocean vessels become larger, greater channel depths are required to accomodate deeper draft ships. Deeper channels induce faster sediment accumulation, which in turn, necessitates a greater frequency of dredging. A major problem is the disposal of huge amounts of dredged material. Combined dredging and disposal have altered the geometry of some estuaries to such an extent that the hydraulic regime is altered and natural processes are impaired.

This paper aims to show how dredging and disposal affect the hydraulic and sediment regime of an estuary. Important questions bearing on waste disposal in the ocean are: How are estuaries affected by long-continued channel deepening and ocean disposal? Can estuaries absorb more dredged material and thus provide an alternative to ocean dumping?

Misplaced sediment consists of waste dredged material deposited in a foreign environment of deposition. It is unevenly or heterogeneously distributed and often creates anomalous bed topography. The maldistribution of sediment is in disequilibrium with the prevailing energy regime. Eventually, misplaced sediment is returned to the channel from which it was dredged.

THE PROBLEM

Dredged material from estuaries is disposed of in the most expedient and economical way. About 42 percent of the maintenance material dredged by the Corps of Engineers from the U.S. East Coast is dumped in confined areas behind dikes (Boyd, et al., 1972). Small amounts are dumped on uplands or used as landfill. However, most dredged material consists of cohesive mud which does not make good fill or foundations. Waterfront disposal sites are scarce, especially around industrialized areas. Elsewhere, salt marshes formerly have been used for disposal, but now their use is often restricted because of their value as productive habitats. Material disposed on land or behind dikes disrupts natural drainage and alters vegetation patterns. Because the material often retains large quantities of water, it remains soft and unstable for years rendering the area unsuitable for development. Many potential upland disposal sites are not available within economic distances from dredged channels.

Much dredged material from U.S. East Coast estuaries, about 47 percent of total maintenance material dredged by the Corps of Engineers, is dumped in open water and unconfined areas either in deep holes or along estuary channels and bordering tidal flats and marshes. When waves and tidal currents disperse the dredged material, the muddy suspensions often create excess turbidity or release nutrients and toxic pollutants into the water. When the suspended material settles out or disperses as a mud flow, it threatens benthic habitats including clam and oyster grounds. Some dredged material is carried back to the dredged channel. Material retained in disposal areas as mounds or protruding banks and islands distorts the estuarine geometry and in turn, may change the hydraulic regime. Long-continued disposal can drastically reshape an estuary.

Since adequate disposal areas are often lacking near dredge sites, large amounts of dredged material are taken by barge or hopper dredge and dumped at sea. An estimated 18.4 M tons of dredged material, including both maintenance material and material from new cuts, was dumped off the Atlantic Coast in 1973 (EPA, 1974). Dredged wastes reportedly make up 80 percent by weight of all wastes dumped at sea (CEQ, 1970). Off New York over 7 M tons are dumped annually and the rate of supply exceeds the natural sediment influx for the entire mid-Atlantic region (Gross, 1972). Most ocean disposal sites are located on the floor of the continental shelf, 1.8 to 3.0 km from shore, in water depths less than 20 meters (Fig. 1). Greatest

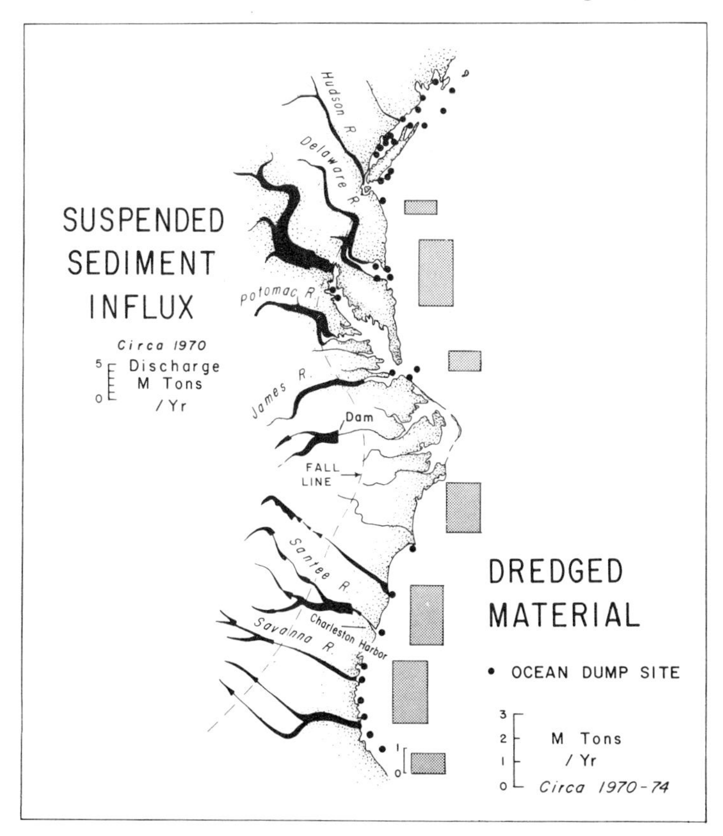

Figure 1
Average annual suspended sediment influx by major rivers to U.S. East Coast estuaries; figure and data modified from Meade and Trimble (1974). Also, average annual quantities of dredged material taken from major estuaries; data from U.S. Army Corps of Engineers annual reports, 1970-74. Location of ocean dump sites from CEQ (1970).

amounts of dredged material come from the Delaware River estuary where 2.84 M tons are taken annually on the average to maintain the shipping channel (Fig. 1). Additionally, 2.48 M tons are dredged annually from Charleston Harbor and 2.74 M tons from the Savannah River estuary. As channels in these estuaries have been deepened below natural depths, about 6 to 7 meters, sedimentation rates have increased. Consequently, dredging and disposal are a growing problem.

Time trends in the production of maintenance dredged material in selected estuaries since the early 1900s are shown in Figure 2. The greatest increase occurred in Charleston Harbor between 1934-44

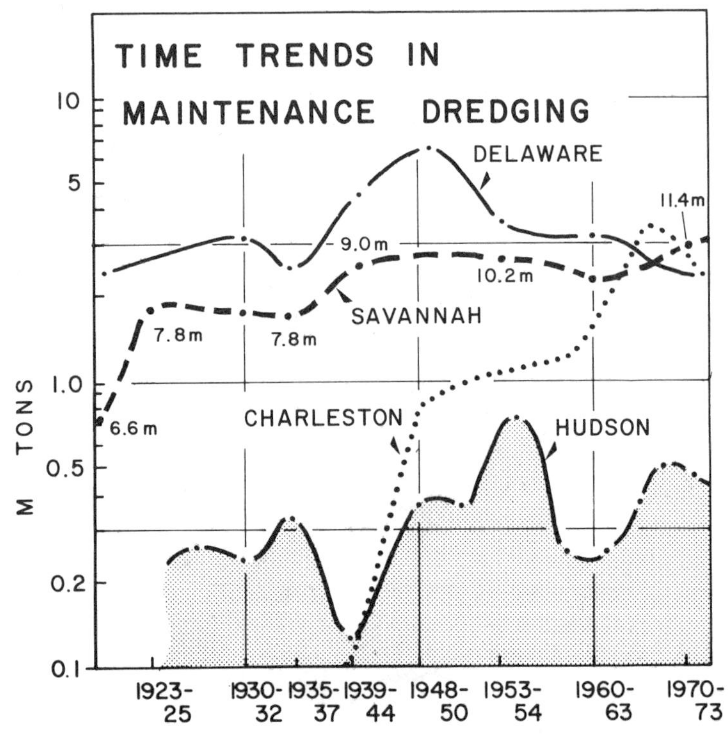

Figure 2
Time trends in the production of maintenance dredged material in selected U.S. East Coast estuaries. Data from annual reports of the Corps of Engineers. Value associated with the Savannah curve represents dredged channel depths at various times.

and 1966-68, a time when construction of a dam diverted drainage from the Santee River into Charleston Harbor and created a four fold increase in sediment influx. Because the increased freshwater inflow shifted harbor circulation from a well-mixed to a partly-mixed or two-layered regime, most of the river-borne sediment load was effectively trapped in the harbor. Deepening of the shipping channel by nearly two meters further enhanced entrapment.

In the Hudson River estuary, maintenance dredging markedly increased following World War II when much material was dumped in open water of seaward reaches, the Narrows (Fig. 2) (Panuzio, 1963). In the Delaware River estuary, dredging and disposal quantities increased 2.8 times between 1935-37 and 1948-50, a time when the channel was deepened nearly 2 meters and maintenance material was dumped in open water. After 1948-50 when dredged material was dumped in diked areas, dredging rates diminished to 2.5 M tons per year, the level in 1920-30. In the Savannah, between 1915-17 and 1939-44, each successive increase in channel size or depth caused an increase in maintenance dredging (Rhodes, 1949). With greater channel deepening after 1939-44, shoaling and maintenance dredging leveled off. Although the rate of maintenance dredging has generally slowed down in recent years, the need for larger harbors and deeper channels to accomodate larger supertankers is increasing. At the same time, many existing disposal sites are filling up, new sites are more difficult to locate or more costly, and many potential sites are areas of environmental concern.

EQUILIBRIUM CONCEPTS

Like all natural systems, an estuary channel develops toward a state of maximum stability. To attain dynamic equilibrium, an estuary continually co-adjusts its tidal discharge and its bottom geometry. Its channel must be neither too deep nor too shallow for the amount of discharge and for the load of sediment that passes through it.

An ideal estuary channel is funnel-shaped. As the banks converge and the floor shoals headward, widths and depths diminish landward. When a progressive wave is propagated landward from the sea, its amplitude tends to be preserved. Friction with channel boundaries produces energy loss and an amplitude decrease, while landward convergence causes energy concentration and an amplitude increase. Therefore, for frictional dissipation to balance amplitude increase, there is an appropriate degree of convergency. Width and depth conform to an exponential rate of change with minimal deviation. For minimum work, the depth change must be constant between the mouth and landward limit of tidal influence (Wright, et al., 1973). This is attained by adjusting the channel geometry through erosion and deposition, or by changes in tidal characteristics including wave length, amplitude and longitudinal gradient of tidal discharge. When a natural estuary is dredged to depths greater than those dictated by the equilibrium regime, sediments accumulate to reestablish an equilibrium depth in accord with the tidal hydraulics.

In broad bays and shallow lagoons, scour by wind waves and currents is believed to establish an equilibrium with sedimentation and frictional resistance of the bed and shorelines (Price, 1947). Dredging of channels across a bay and the formation of islands or banks of dredged material may reduce the effective bay width and wind fetch. As a result, the bay is shoaled by sediment deposits which build up to the level of equilibrium appropriate to the ratio between average wind fetch and water depth. When an equilibrium depth is reached, river-borne sediments move through the bay and out to sea.

Superimposed on the oscillatory movement of the tide, there is a slow residual movement of water created by fresh-salt water density differences. The resulting estuarine circulation pattern is controlled by the relative volume of river inflow and tidal flow, as well as the width, depth and geometric configuration. In partly-mixed coastal plain estuaries common to the U.S. East Coast, net flow is seaward in the freshened upper layer, and landward in the relatively salty lower layer. Where landward flow and seaward river flow meet near the inner limit of salty water, shoaling rates are often high. Consequently, the rate of maintenance dredging in this zone is often higher than elsewhere.

When a channel is deepened by dredging, salty water in the lower layer progresses farther landward on the average whereas freshened river water in the upper layer extends farther seaward.

With greater stratification, mixing between the lower and upper layer is reduced, the volume transport is slowed down, and the level of no-net motion (i.e., the boundary between upper and lower layer) is lowered (Nichols, 1972).

Similar dynamic changes take place when an estuary is narrowed by diking and overboard disposal. When river inflow and tidal velocities are held constant, a narrowed cross section decreases the ratio of river flow to tidal flow. A decrease in cross section acts together with an increase in depth to reduce vertical mixing and shift the estuarine circulation pattern from type C towards type A (Pritchard, 1955). When sediments are dumped into seaward parts of a salt wedge system, they can be carried back to the shipping channel by the landward flow. By contrast, when sediments are dumped on banks or shoals bathed by the upper layer, they can be flushed seaward and escape the estuary. By redistributing misplaced sediment, these residual transport mechanisms are the means by which an estuary seeks to heal dredged cuts and to reestablish equilibrium. It is useful to examine several case histories that reveal how an estuary responds to dredging and disposal.

TAMPERING WITH THE BED AND FLOW

Although dikes have been commonly built in the expectation that they will train tidal currents and counter bank erosion, they are now used to enclose portions of shoals and harbors and to accomodate disposal of dredged material. In Hampton Roads, Virginia, a dike structure called Craney Island Disposal Area was constructed 2.4 m high between 1946 and 1957. It impounded 2,500 acres (11.1 sq. km) of harbor waters to accomodate maintenance dredged material. Although the disposal area eliminated a continued need for ocean disposal, it was not without consequences and side effects. More than 12 percent of protected harbor area was lost along with a loss of marine habitat and recreational opportunities. By narrowing the harbor cross section 23 percent (Fig. 3), tidal currents were speeded up and a large portion of the harbor floor scoured 0.2-1.2 m deep. River-borne sediment normally deposited in the section before diking was "throttled" through the reach and most likely deposited elsewhere, either farther upstream around oyster grounds or farther downstream in the Norfolk shipping channel. Because alternate disposal sites are lacking, life of the Craney Island Disposal Area most likely will be extended by building higher dikes (U.S. Army Engineer District, Norfolk, 1974).

In the Delaware Estuary, hydraulic engineers have tampered with channels and dikes, reach by reach since 1850, until the channel geometry of the entire upper estuary has been altered. Comparison of mean depths and widths along the estuary between 1878 and 1970 (Figs. 4A and 4B) indicate the magnitude of geometric alteration. Cross sectional areas along the estuary length in 1878 increase exponentially seaward (Fig. 4C) suggesting quasi-equilibrium between geometry and flow. However, in zones where the mean depth and width have been increased substantially, shoaling is greater than elsewhere. In many places, dikes have effectively trained tidal currents, increased channel depths and diverted sediment deposition from one reach to another. By realigning the direction of flow with the channel axis, the lateral supply of sediment has been reduced.

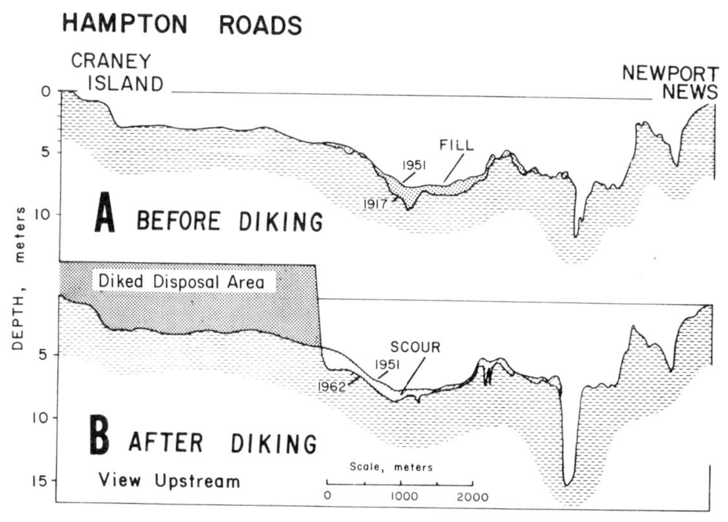

Figure 3
Bottom profiles across Hampton Roads Harbor showing changes related to fill and scour before and after construction of Craney Island Disposal Area.

After dredged material was placed in diked areas along the Delaware estuary instead of in unconfined open water sites, maintenance dredging decreased more than 2½ times. Dikes not only retained the dredged material, but they prevented its return to the channel. They are an obvious advantage in containing dredged material that is polluted.

The beneficial effects of dikes are often offset by formation of new shoals in other sections of the estuary. In turn, the longitudinal distribution of flow is changed and the hydraulic equilibrium is altered. A large contraction in the flow cross section by two dikes in the Delaware below Philadelphia in 1891-1904 and in 1912-15, caused an 18 percent reduction of tide range farther upstream (Wicker and Rosenzweig, 1950). Consequently, dikes are only a temporary expedient for coping with shoaling and disposal problems in estuaries. They do not flush sediment into the sea nor do they correct the basic cause of shoaling.

Although submerged mounds and banks of dredged material formed by open water disposal are less conspicuous than dikes, they have a potential for disrupting hydraulic equilibrium and changing the sedimentation regime. Like dikes they reduce the cross-sectional area for tidal flow. At the same time, the anomalous topography creates frictional losses, which in turn, affect the rate of tide propagation and the tide range.

In the Upper James River estuary, maintenance dredging is required every 1 or 2 years to align the channel course and to maintain a 7.5 m channel depth for shipping. Long-continued overboard disposal since 1878 has built up banks of misplaced sediment along the channel (Fig. 5) to a level where they are subject to wave erosion and scour by tidal currents. Upward building for a thickness of 2 to 3 meters has increased slopes from 1:170 to less than

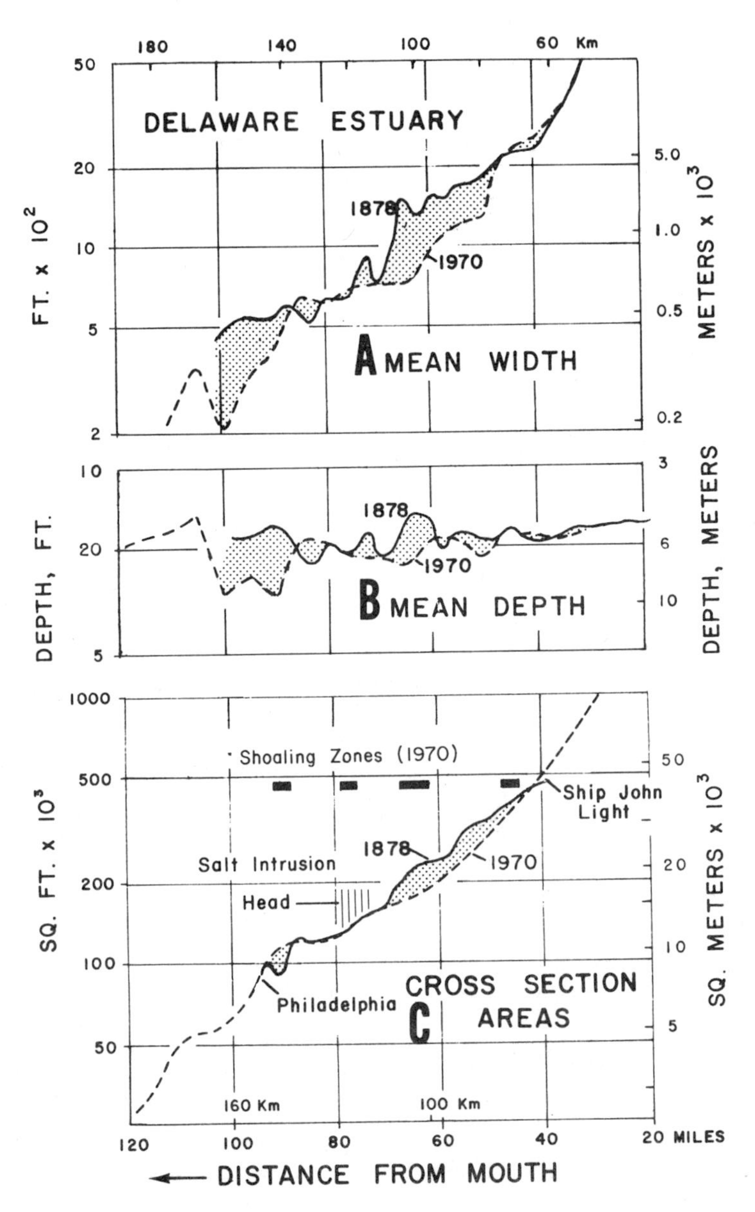

Figure 4
Comparison of (A) mean width, (B) mean depth and (C) cross section area along the Delaware estuary between Ship John Light and Philadelphia. From surveys of 1878 (U.S. Coast and Geodetic Survey Report, 1883) and 1970 (U.S. Army Engineer District, Philadelphia, 1973).

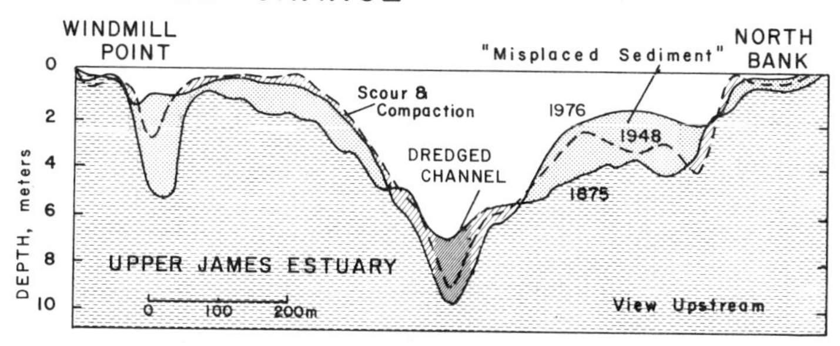

Figure 5
Bottom profiles across the Upper James estuary, Virginia, at Windmill Point illustrating changes in bottom topography by channel dredging and overboard disposal on adjacent shoals since 1875.

1:45, a slope which permits low density material to return directly to the channel (Fig. 5). Long-term disposal at Windmill Point, James River, reduced the cross-sectional area by 26 percent. Since 1878 the tide range at Hopewell, 13 km upstream of Windmill Point, decreased 8 percent. This trend suggests that narrowing the cross section by disposal caused a loss of tidal energy. This is partly substantiated by the fact that sediment texture is generally finer-grained and cohesive in the relatively young mounds than in the old mounds. The trend for more silt and clay accumulation in the channel with time amplifies the disposal problem since fine-grained material is difficult to utilize. Additionally, it tends to shoal the channel floor to a greater height, or regime depth, than sandy sediment. The equilibrium channel depth is less in silt than in sand because the shear force, which is proportional to the current speed gradient near the bed, must be greater to initiate particle movement. With a reduction in tidal range, the tidal prism and tidal discharge are also reduced. Such a trend also reduces the mixing capability of tidal currents to flush pollutants like Kepone through the reach. In short, it appears the Upper James is losing its capacity to absorb more dredged material.

In some partly mixed estuaries, dredging and disposal cause major changes in the density circulation with important sedimentologic consequences. Changes recorded in Savannah Harbor since 1917 illuminate a whole sequence as the channel was progressively deepened and extended upstream (Fig. 6). Simmons (1965) shows that more than two-thirds of the maintenance dredging is performed on a shoal that forms in the null zone. This is a zone of weak residual current near the bottom where seaward flowing river water meets landward flowing estuarine water. When the channel was progressively deepened from 7.8 to 10.2 m, between 1923-25 and 1953-54, shoaling rates increased 2.6 times. Moreover, it caused the null zone to shift landward 12 km and in turn, a landward shift in the locus of maximum shoaling (Fig. 6) (Simmons, 1965).

Shoaling in the Savannah increased, despite an overall reduction in sediment influx by river dams from 2.5 M tons annually in 1891 to 0.8 M ton annually in 1953-57. Meade (1969) suggests the apparent deficit in supply of shoaling material came from offshore

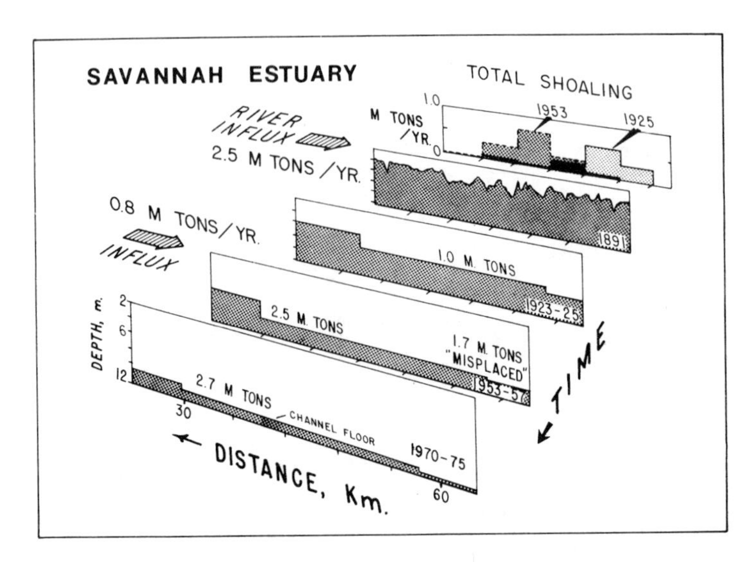

Figure 6
Changes in depth profiles along the Savannah estuary channel with time and with progressive deepening since 1891. Shoaling rates and river influx of suspended sediment are presented in average annually megatons.

and from previously dredged sediment. With additional channel deepening, between 1953-57 and 1970-75, the increase in harbor-wide shoaling was relatively small (Fig. 6). According to Simmons (1965), sediment trapping by the density circulation developed to such an extent that all potential shoaling material entering the harbor was trapped. The present site of major shoaling is now close to a port area where disposal areas are limited.

THAMES ESTUARY CASE

Tampering with the channel and banks of the Thames proceeded piecemeal for centuries until its consequences were understood and corrected (Inglis and Allen, 1957). In 1909, the Port of London initiated a major dredging program to improve shipping access to London. The channel was dredged through the Mud Reaches (Fig. 7) to 8.1 m below low water or about 1.8 m below the natural regime depth. In this zone, suspended sediment loads are higher than elsewhere and form a turbidity maximum. Sediment is entrapped by opposing river and estuarine bottom currents in the null zone. A channel was also dredged through the Gravesend Reach (Fig. 7) to 9 m below low water. In this reach, deposition is caused by a rapid seaward expansion in width which for the same discharge, results in a velocity reduction. Moreover, alignment of the shipping channel differs from the natural channel. Changes in mean depth and corresponding increase in area of flow, mainly caused by dredging between 1830-39 and 1939-49, are illustrated in Figure 8. The shoaling problem lies in reaches where there is a substantial supply of fine

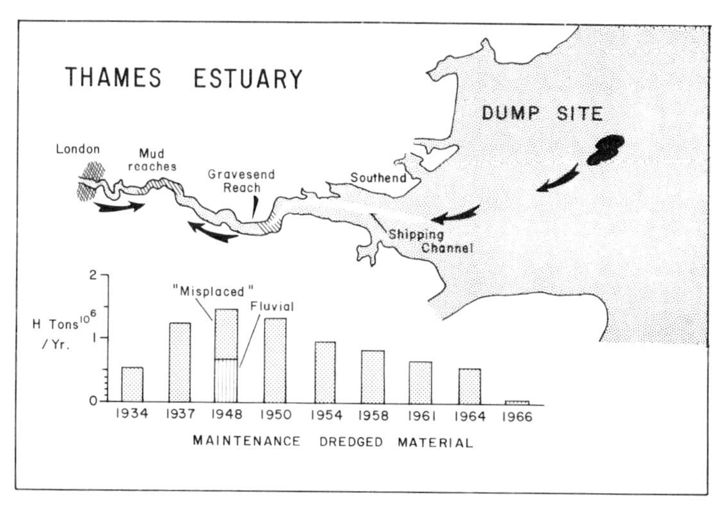

Figure 7
Chart of the Thames estuary showing location of principal shoals and the ocean dump site. Arrows represent direction of net drift near the bed. Lower graph gives the quantities of maintenance dredged material in average hopper tons per year from the Mud Reaches and Gravesend Reach. One hopper ton is a volume measure representing 0.902 cu yd in a hopper.

sediment and a large increase in mean depth and cross-sectional area of flow with distance seaward.

Most dredged material was dumped offshore in Black Deep approximately 48 km offshore from Southend (Fig. 7), an area of channels and sand banks swept by currents more than one meter per second. Maintenance dredging increased steadily in the Mud Reaches and Gravesend Reaches from about 0.6 million hopper tons in 1934 to 1.4 million hopper tons in 1948. Once initiated, dredging necessitated further dredging on a large scale to maintain a small increase in depth.

The problem was investigated by hydraulic engineers for 7 years between 1949-56 using historical records, hydraulic models, radioactive tracers and intensive field observations of current velocity, salinity and suspended sediment concentrations. The investigation aimed to determine the routes and rates of sediment supply, the cause of shoaling and the effect of dredging on the estuary regime.

The investigation (Inglis and Allen, 1957) revealed that not more than 45 percent of the sediment deposited in the shoals came from fluvial sources. The rest came from offshore. Sediment moved landward by a systematic landward drift along the bed to the Mud Reaches. Most dredged material dumped offshore in Black Deep re-entered the estuary and was carried landward by the net drift to areas of deposition. As sediment moved landward, it became progressively finer and hence more difficult to dredge and dispose of. When dredging was done at Gravesend in the middle estuary, less sediment was carried farther landward to the upper estuary. The investigation emphasized the importance of the regime, or equilibrium depth, to provide a self-maintaining channel.

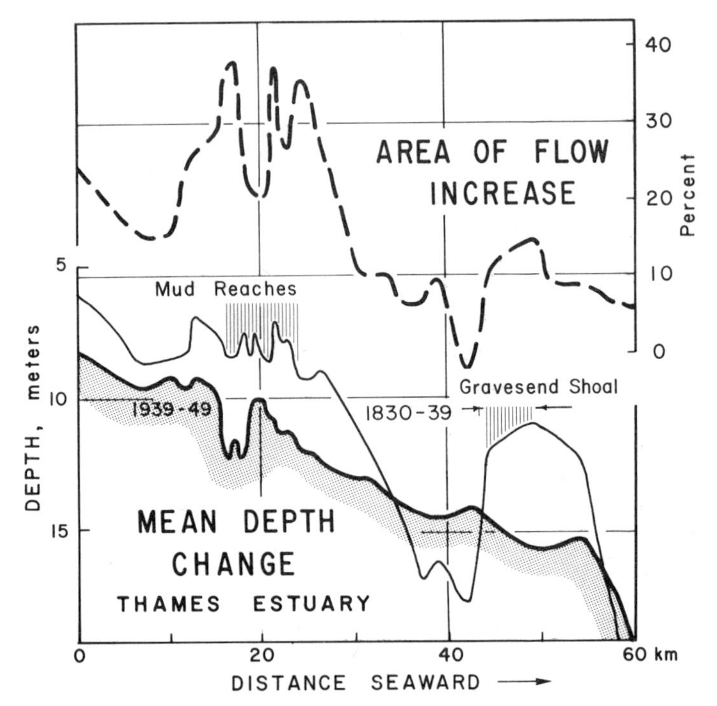

Figure 8
Distribution of mean depth change and percent area of flow increase between 1830-39 and 1939-49 along the Thames estuary. Based on data from Inglis and Allen (1957).

Following recommendations of the investigation (Inglis and Allen, 1957), the offshore dump site was abandoned in 1960. Instead, dredged material was dumped ashore and kept out of the estuary. Dredging of the Mud Reaches ceased in 1965 and the channel was allowed to revert by sedimentation to an equilibrium depth, which is only 0.3 m above the maximum controlling dredged depth. Shipping was partly scheduled to ride the high tide across the undredged shoal, and for another part, to use alternate docks in seaward reaches with deepwater access. The dual measure, i.e., pumping ashore and maintaining an equilibrium depth, reduced the amount of dredged material by 74 percent within 10 years (Fig. 7) (personal communication, R. Kirby and J.R. O'Donnell).

DISCUSSION

Although the physical forces that create and maintain landward density currents tend to break down seaward from an estuary mouth, substantial evidence exists for landward transport of natural sediment in many estuaries along the U.S. East Coast (Meade, 1969). It is likely that misplaced sediment is transported into U.S. East Coast estuaries via the net landward drift (Fig. 9). Escape from the dump site and return to the dredged channel is promoted by waves

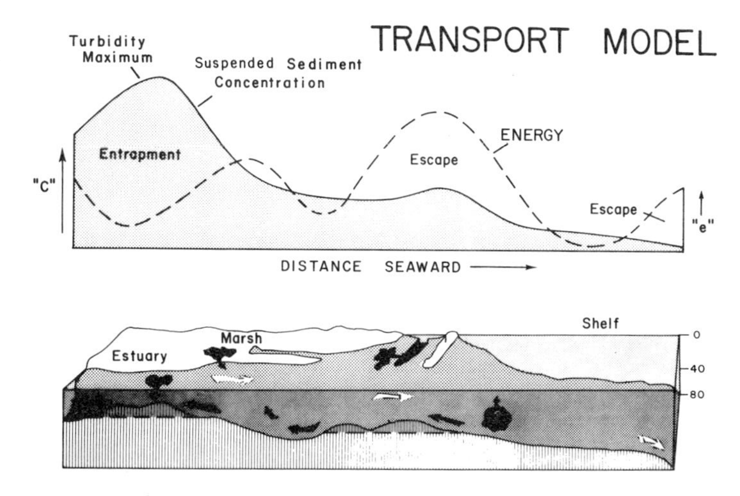

Figure 9
Upper figure displays hypothetical relationship between energy levels, "e", and the natural gradient of suspended sediment concentration, c, with distance seaward along an estuary. Recycling routes of dredged material in a hypothetical estuary. White arrows represent routes of escape, black arrows routes leading to entrapment.

and currents acting in estuary mouths and on the inner shelf floor. They can not only put dredged material into suspension but maintain it during transport. Other recycling routes include escape downslope from overboard sites on shoals or on marshland (Fig. 9). The Thames estuary case emphasizes the great care that must be taken to ensure that dredged material dumped offshore will not return to the estuary.

The link between dredging and ocean disposal is strengthened by dredging itself. Channel deepening not only accelerates landward flow but increases stratification and hence the potential for entrapment. Experience in the Thames shows that the circuit can be broken by disposal outside the estuary, on uplands or behind dikes. However, in many urbanized areas this means of disposal is limited as previously discussed. Such limitations emphasize the need for practical utilization of dredged material and improvement of engineering properties.

Another approach to the problem of disposal is to reduce the rate of shoaling and hence the quantity of material that requires dredging and disposal. The Thames case illustrates that a small increase in depth may trigger a large increase in sediment deposition. The further the bed is lowered from equilibrium conditions, the faster the sedimentation. As suggested by Price and Kendrick (1976), for major changes affecting the hydraulic character of an estuary, the chief aim is to obtain maximum improvement with minimum disturbance.

Shoaling can also be reduced by any measure or structure that restricts supply of sediment at its source in the watershed, on banks or in shoals. Since a substantial amount of dredged material may consist of older misplaced sediment, disposal in confined areas should reduce long-term shoaling and dredging commitments. Additional benefits may be gained by increasing the tidal discharge relative to river inflow such as increased fresh-salt water mixing which allows greater amounts of suspended sediment to escape an estuary. Tidal discharge could be increased by developing a progressive landward converging geometry that contracts tidal energy of a progressive wave. This effect is further enhanced by reducing frictional dissipation through smoothing of geometric irregularities and by developing a single axial channel which favors ebb flow.

The disposal capacity of some estuaries may be improved for short periods by dikes and hydraulic training works designed to shift shoals into areas where desirable disposal areas are available. As land use patterns change, new sites may become available. Shifting shoal sites may be desirable to reduce the pumping or hauling distance to disposal areas. Thus, the problem of misplaced sediment is complicated by many factors. Proper long-range dredging and disposal plans must be based on an understanding of the prevailing hydraulic regime and sedimentological effects.

ACKNOWLEDGEMENTS

The paper benefitted from ideas expressed by Dr. Richard Faas. Dr. John Munday and Dr. William Hargis of VIMS reviewed the manuscript. Recent information on the Thames was provided by Dr. Robert Kirby of the Institute of Oceanographic Sciences, Great Britain, and Mr. J.R. O'Donnell of the Port of London Authority.

REFERENCES

1. Boyd, M.B., Saucier, R.T., Keeley, J.W., Montgomery, R.L., Brown, R.D., Mathis, D.B., and Guice, C.J., 1972. Disposal of dredge spoil problem identification and assessment and research program development. U.S. Army Waterways Experiment Station, Tech. Rep. H-72-8, 121p.
2. Council on Environmental Quality, 1970. Ocean dumping a national policy; a report to the President, 45p.
3. Environmental Protection Agency, 1974. Administration of the ocean dumping permit program. Second Annual Report, EPA, Washington, D.C.
4. Gross, M.G., 1972. Geologic aspects of waste solids and marine waste deposits, New York metropolitan region. Geol. Soc. Am. Bull., 83:3163-3176.
5. Inglis, C.C. and Allen, M.A., 1957. The regimen of the Thames Estuary as affected by currents, salinities, and river flow. Proc. Inst. Civil Engineers, 7:827-878.
6. Meade, R.H., 1969. Landward transport of bottom sediment in estuaries of the Atlantic coastal plain. J. Sediment. Petrol., 39:222-234.

7. Meade, R.H. and Trimble, S.W., 1974. Changes in sediment loads in rivers of the Atlantic drainage of the United States since 1900. Proc. Paris Sympos., IAHS-AISH Pub. 113:99-104.
8. Nichols, M., 1972. Effect of increasing depth on salinity in the James River estuary. In: B.W. Nelson, Environmental Framework of Coastal Plain Estuaries. Geol. Soc. Am. Memoir, 133: 571-589.
9. Panuzio, F.L., 1963. Lower Hudson River siltation. Proc. Fed. Inter-Agency Sed. Conf., MP No. 970:512-550.
10. Price, W.A., 1947. Equilibrium of form and forces in tidal basins of the coast of Texas and Louisiana. Bull. Amer. Assoc. Petrol. Geologists, 31:1619-1663.
11. Price, W.A. and Kendrick, M.P., 1967. Dredging and siltation-cause and effect. Proc. Symposium on Dredging sponsored by Institution of Civil Engineers, London, p. 31-36.
12. Pritchard, D.W., 1955. Estuarine circulation patterns. Proc. Am. Soc. Civil Engineers, 81:1-11.
13. Rhodes, R.F., 1949. The hydraulics of a tidal river as illustrated by Savannah Harbor, Georgia. Spec. Rept. U.S. Army Corps of Engineers-Savannah District, 140p.
14. Simmons, H.B., 1965. Channel depth as a factor in estuarine sedimentation. U.S. Army Committee on Tidal Hydraulics, Tech. Bull. 8, 15p.
15. U.S. Army Engineer District, Norfolk, Corps of Engineers, 1974. Norfolk Harbor, Virginia, Report of survey investigation of Craney Island Disposal Area. Tech. Rept. with three Appendices, 65p.
16. U.S. Army Engineer District, Philadelphia, Corps of Engineers, 1973. Long range spoil disposal study, Part III, 140p.
17. U.S. Coast and Geodetic Survey Report, 1883. Appendix No. 8: 241-245.
18. Wicker, C.F. and Rosenzweig, O., 1950. Theories of tidal hydraulics. U.S. Army Committee on Tidal Hydraulics, Rept. 1:101-125.
19. Wright, L.D., Coleman, J.M. and Thom, B.G., 1973. Processes of channel development in a high-tide range environment. Cambridge Gulf-Ord River Delta, Western Australia. Jour. Geology, 81:15-41.

Two Waste Disposal Sites on the Continental Shelf off the Middle Atlantic States: Observations Made from Submersibles

D. W. Folger
H. D. Palmer
R. A. Slater

ABSTRACT

Nineteen dives were conducted in August 1974 and July 1975 to survey the Philadelphia sewage and DuPont acid dumpsites about 65 km southeast of the mouth of Delaware Bay. Though 1974 dives in the sewage dumpsite were carried out the day after 2 million gallons of sewage was released, little evidence of sewage remained in the water column or on the bottom within the area of the dumpsite. Bottom currents were typically weak (2-5 cm/s) and flowed in an easterly or northerly direction. The bottom was characterized by broad rounded ripples (30-100 cm wavelength; 2-5 cm wave height) or small hummocks and pits (15 cm diameter). Sand dollars, starfish, scallops, and crabs were most common on the bottom together with lesser numbers of hake, flounder, shrimp, skates, and snails. Oscillatory wave motion, tidal currents, and benthic organisms apparently were the main mechanisms redistributing sediment on the bottom during the period of the dives. In 1974, at the center of the acid waste site, where a 1-million-gallon release of acid (pH 0.06) had just taken place, visibility in the water column 6-9 m deep was reduced to zero. Below this opaque layer, suspended matter was composed mainly of large reddish-yellow flocks 1-3 cm long. The visibility reduction and the flocks were apparently due to precipitates from the acid waste that were concentrated at the top of the thermocline. Microtopography and epibenthic organisms in the acid site were similar to those in the sewage site, but the bottom was covered by a thin (<2 cm) yellowish to reddish, granular, flaky material. During dives conducted in 1975, bottom samples were collected by both the mother ship and the submersible. At five sites, in both dumpsites, gravel (including shell fragments) averaged 6.1%; sand, 88%; and silt plus clay, 2.1% of the sediment. Both coarsest and finest materials were accumulating in ripple troughs. Fe, Ti, and Cr in sediments were all higher at the acid site than in surrounding areas. Thus, sewage sludge apparently was being oxidized, eaten, or transported away from the dumpsite and accumulating elsewhere, whereas some acid waste was accumulating on the bottom in the disposal area.

INTRODUCTION

This study was conducted as part of a broader effort in progress by the U. S. Environmental Protection Agency (EPA) to document the effects of sewage and acid waste on the water column, bottom, and biology of the Philadelphia sewage sludge disposal site (lat. 38°20'N to 38°25'N; long. 74°10'W to 74°20'W) and the DuPont acid-waste disposal site (lat. 38°30'N to 38°35'N; long. 74°15'W to 74°25'W), approximately 65 km southeast of the mouth of the Delaware Bay (Fig. 1).

Setting and Previous Work

Dives were conducted on the middle Continental Shelf in water 39 to 56 m deep (Fig. 1). The Continental Shelf in this area is 130 km

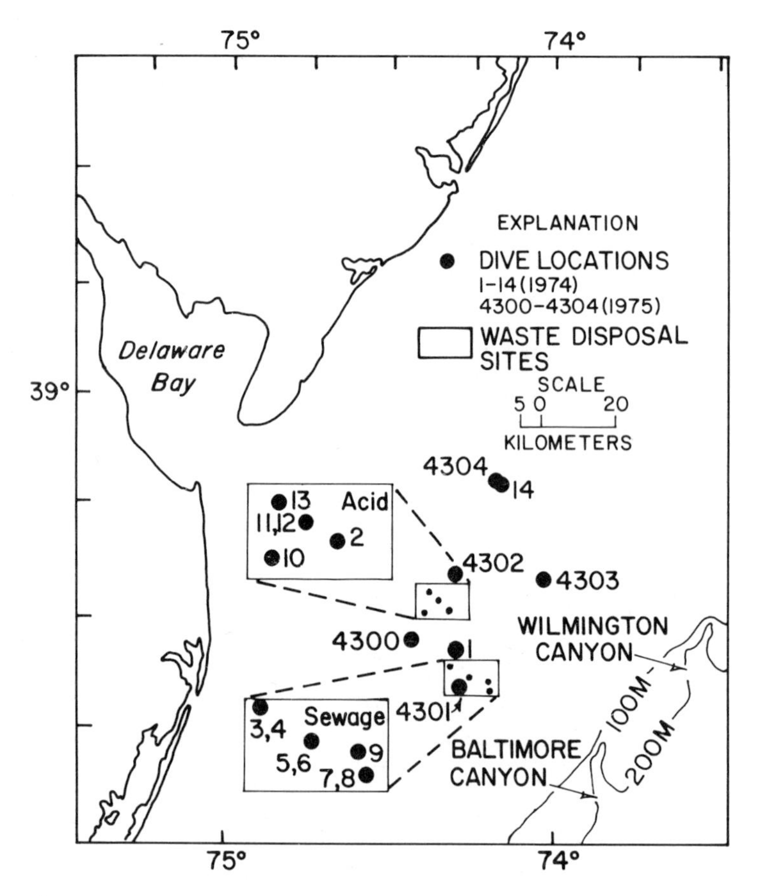

Figure 1
Map showing locations of sites where dives were conducted to assess physical and biological characteristics of the water column and bottom in and near areas where sewage and acid wastes have been released. In 1974, dives were numbered sequentially from 1 to 14; in 1975, several dives were carried out at each of the stations (4300-4304).

wide and comprises a gently dipping (1.5 m/km) plain. Relief is subdued and consists of broad low ridges and swales having wavelengths of 2 to 6 km and amplitudes of 2 to 14 m oriented subparallel to the shore (Uchupi, 1968; Duane and others, 1972; Swift and others, 1973; McKinney and others, 1974). The surface is mostly covered by a thin layer of fine to coarse sand interpreted to be, in part, relict from the last advance of sea level and, in part, derived from more modern deposits that have accumulated since the onset of Holocene time (Emery 1968a; Swift and others, 1972b; Frank and Friedman, 1973; Knebel and Spiker, 1977). The area is a dynamic one, affected by waves and currents associated with major storms and by moderate diurnal tides. The bottom supports an active fauna that has a biomass ranging from 10 to 1400 g/m^2 wet weight (D. Boesch, personal communication).

Morphologic and sedimentologic studies of the area have been extensive (see Emery, 1966; Schopf, 1968; Emery and Uchupi, 1972; Uchupi, 1968, 1970; Schlee and Pratt, 1972; Hollister, 1972; Milliman, 1972; and articles included and references cited in Swift and others, 1972a; McClennen, 1973; Knebel, 1975; and in Stanley and Swift, 1976). However, reports of direct observations of the sea floor have been sparse (McKinney and others, 1974; Edwards and Emery, 1968; Hard and Palmer, 1976).

METHODS

Data were collected during cruises in August 1974 (DSRV NEKTON GAMMA and R/V ATLANTIC TWIN) and in August 1975 (DSRV NEKTON BETA and R/V ADVANCE II. Stations were located by means of Loran A.

On board the submersibles, photographs and TV film were taken through the viewing ports. Grab samples were recovered by means of small closable cylinders. Temperature was recorded during most dive descents and ascents. Bottom-current speeds were estimated by repeatedly timing the passage of particulate matter across the submersible ports; direction was estimated more accurately by observing the movement of particulate matter in reference to the submersible's heading. Concentrations of epibenthic organisms were estimated by the biologists on the cruise. All observations were recorded on tape and transcribed for further study. To track the submersible's position, the mother ship followed a small buoy attached to the submersible by a 150-m polypropylene line.

On board the surface ship during the 1975 cruise, temperature profiles were acquired by means of a conventional bathythermograph; suspended matter was filtered (0.45 μm pore size) from 10-liter water samples; light transmission was measured by using a transmissometer, and bottom samples were collected by grab samplers. Elemental compositions were determined by emission spectroscopy. Textural analyses were made by rapid sediment analyzer and *Coulter counter. Organic carbon concentrations were determined by means of a *LECO WR12 carbon determinator.

The only previous inspection of the sea floor, aside from photos, in the dumpsites was a television survey conducted in 1973 (Palmer and Lear, 1973). Specific objectives of the 1974 reconnaissance cruise were 1) to look for evidence of sewage sludge and acid waste

*Any trade names in this publication are used for descriptive purposes only and do not constitute endorsement by the Geological Survey

in the water column and on the bottom; 2) to evaluate the effects of wastes in sedimentary sinks; 3) to measure near-bottom currents; 4) to describe and determine the origin of microtopography observed; and 5) to estimate the concentration and distribution of living organisms in the water column and on the bottom.

OBSERVATIONS - 1974

Sewage Sludge Site - Results

Release of sewage sludge (~4.5 million gallons/week) from the city of Philadelphia began in May 1973. The diving schedule was arranged so that while one release of 2 million gallons was being tracked at the surface by satellite, aircraft, and surface ships, the submersible would document its vertical distribution below the surface. Unfortunately, weather deteriorated the day of the release (winds reached 60 knots), and submersible observations after one dive (#1) north of the dumpsite could not be continued until the following day. Then, dives took place at sites along the track followed by the barge (~135°T). Two dives were made at each of six sites; during the first dive, a geologist was observer and during the second, a biologist. On each dive, NEKTON GAMMA ran on the bottom over a track about 100 m long. This allowed detailed observation of at least 1000 m^2 of bottom. One additional traverse (#9) about 1 km long was carried out parallel to and northeast of the barge track.

Water column: The top of the thermocline ranged from 10 to 15 m in depth and temperatures within it declined as much as 12°C down to 35 m. Bottom temperatures on deepest dives (56 m) were ~8°C. Visibility on three dives decreased in the thermocline at depths below ~20 m. Largest particles, of uncertain composition, seemed to be concentrated between 10 and 20 m, finer material being found above and below. Seston also comprised plankton, the largest of which (ctenophores, chaetognaths, and hydromedusae) were most concentrated in the lower part of the water column. Visibility on the bottom was most often 5 to 10 m.

Current flow near bottom, estimated during five dives, was in a northerly or easterly direction at speeds of 2 to 5 cm/s (Table 1).

Table 1
Bottom-current direction and speed estimated from the submersible in the sewage dumpsite, August 1974 (See Fig. 1 for dive locations).

Dive. No.	Flow Direction (Toward)	Speed (cm/s)
1	E	2 - 3
3	NW	3 - 5
5	-	2 - 4
7	NE	3 - 5

Thus, bottom currents were sluggish despite the intense storm that lasted most of the previous day.

Bottom: Broad rounded ripples (30 - 100 cm in wavelength; 2 - 5 cm in amplitude) or small hummocks and pits (15 cm in diamter) characterized the microtopography of the flat bottom (Figs. 2, 3). In the northwest corner and center of the area, hummocks and pits predominated but in the southeast corner, ripples were more common having crests oriented northeast, wavelengths of 30 to 100 cm, and wave heights of ~5 cm. No asymmetric ripples were observed; all were rounded (rounded trochoidal ripples of Komar and others, 1972), indicating that they were produced mainly by oscillatory wave motion. The predominance of hummocky topography, apparently due mostly to burrowing organisms such as crabs, hake, and scallops, suggests that the effects of physical processes in areas of rich benthic activity may be quickly obscured (Fig. 3).

The bottom was partially (60%) to almost completely covered by a thin (0 - 3 mm) veneer of gray-brown to dark-brown fine "flocky" or "tufted" material (Fig. 4). Commonly, it was thinnest on ripple and mound crests and thickest in troughs and depressions. The material remained suspended when agitated with the submersible manipulator arm, whereas the underlying very fine- to very coarse-grained tan-gray sand settled back to the bottom rapidly. Some of the flocky material present on shells of live scallops did not come off when the scallop moved; thus, it may have been, in part, organic growth.

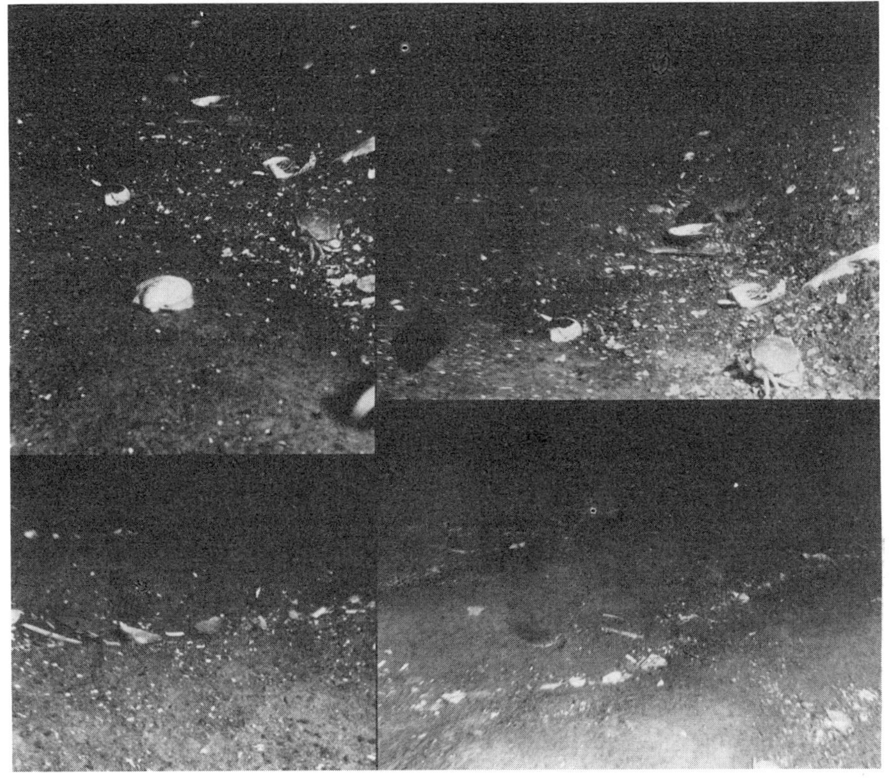

Figure 2
Photographs taken from the submersible port, showing broad (30 - 100 cm wavelengths) rounded symmetrical ripple marks that are common on the bottom throughout the area. Sand is most often exposed on crests, whereas flocky or tufted material, shells, and gravel are concentrated in troughs. Photograph at upper left was taken at Station 4301; upper right, at Station 4301; lower left, at Station 4300; and lower right, at Station 4303.

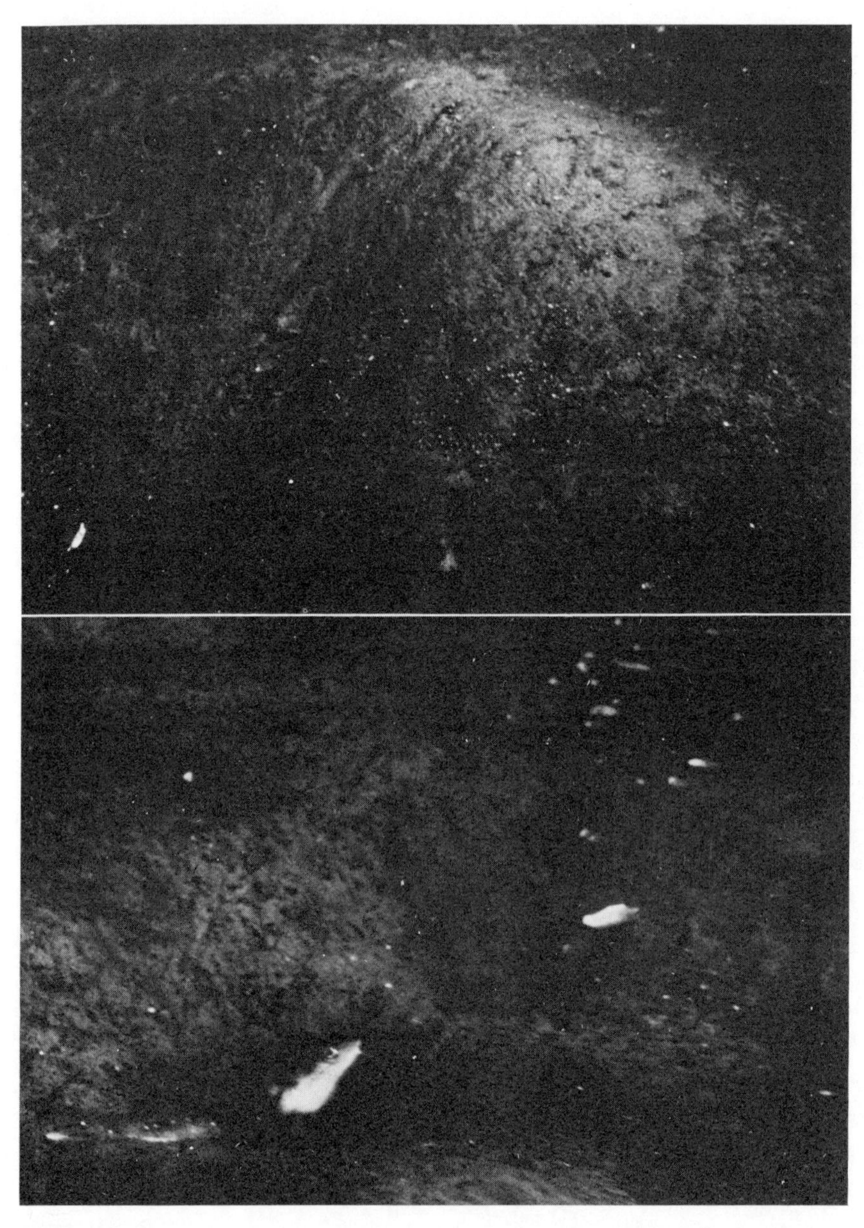

Figure 3
Photograph taken from the submersible, showing the mounds and depressions most often 15 - 30 cm in diameter and 5 - 10 cm deep or high that are common in areas where benthic organisms obliterate ripple marks. Upper photograph shows a particularly deep depression; lower photograph shows a hake working its way into the bottom and creating or deepening a depression.

Figure 4
Photograph from the submersible, showing sand dollars and the dark-brown-gray flocky or tufted layer overlying light patches of exposed tan to gray sand.

White shells and shell fragments of sea clams, razor clams, and sand dollars were common over most of the bottom (Fig. 5). Many whole shells lay concave-side up, but whether that position was due mostly to the absence of strong currents (Emery, 1968b; McKinney and others, 1974) or to live organisms such as crabs, which were observed turning the shells over, is not clear. Little scour was present around large shells or other objects on the bottom, yet many concave-up shells were partly filled with sediment. Probably bioturbation is mostly responsible for resuspension of bottom sediment during summer months.

Biology: The active bottom community was dominated by patches of sand dollars. An estimate of their greatest abundance in the northwest corner was 50 - 100/m^2. Starfish were most common in the northwest and central areas (~10/m^2). Sea scallops were present in concentrations of <1/m^2, although in some places they were more common. Many sand dollars lay on ripple crests and tops of mounds, whereas detritus such as broken shell fragments concentrated in ripple troughs and depressions (Fig. 6).

In summary, the sludge released on the previous day was either so well dispersed by the storm or simply so dilute that it was not discernible in the water or on the bottom as an obvious concentrated layer during the following day. Nor was the sludge clearly responsible for the flocky layer on the bottom. Consensus of experienced observers was that the layer could have resulted mostly from natural processes. It was thicker and more common in troughs and depressions. If similar material in the sludge were present, it probably also would accumulate there. However, the sludge may never have penetrated the thermocline and may have been transported away from the area by surface currents. Evidence that supports this conjecture was observed in the acid dumpsite.

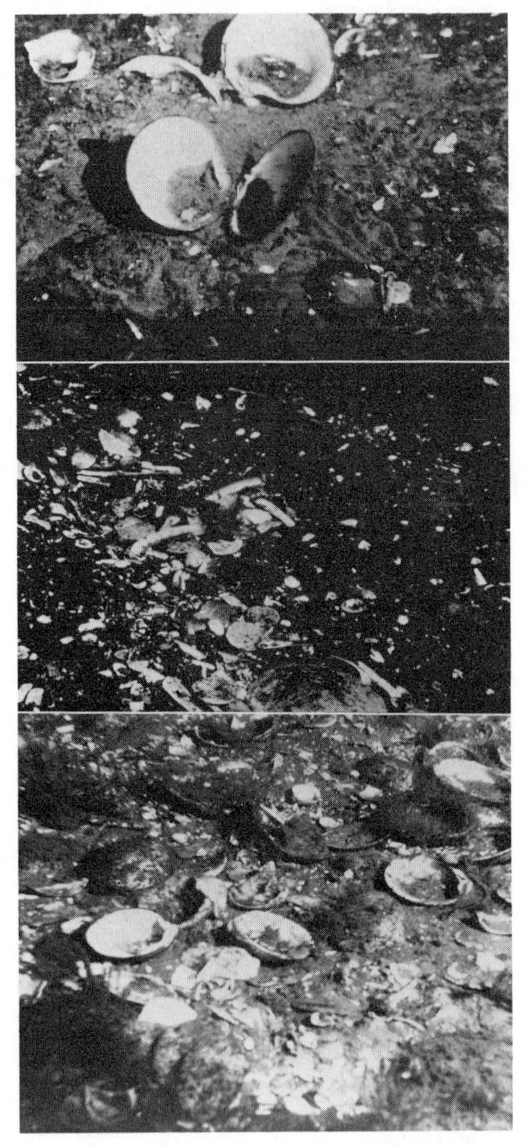

Figure 5
Photographs taken from the submersible, showing abundant white-shell debris consisting of whole shells and fragments of pelecypods such as sea clams, razor clams, and quahogs. Upper photograph taken on Dive 4; middle photograph taken on Dive 14; lower photograph taken at Station 4302.

Figure 6
Photograph from the submersible, showing abundant sand dollars that often pave ripple crests.

Acid Disposal Site - Results

Prior to the 1974 cruise, I. E. du Pont de Nemours and Company had been dumping a maximum of 20 million gallons of acid fluids (pH 0.06) per month at a rate of 1 million gallons per trip since July 1969 (personal communication, B. Reynolds). One such dump took place between 0135 and 0605, 9 August 1974, the day of the survey. At noon, several large areas of light-green water were still present near the center of the dumpsite, and several dives were undertaken to evaluate the vertical distribution of the acid.

Water column: The top of the thermocline (23° - 25°C) was between 8 and 10 m deep and its base (10° - 12°C) was between 30 - 40 m. Coldest bottom-water temperature at 45 m was about 9°C. The concentration of suspended matter observed in the area on dives 2, 10, and 13 was low and consisted mostly of some coarse "stringy" material near the top of the thermocline. However, near the center of the site in the area still affected by the acid waste, conditions in the water column were significantly different. On dive 11, turbidity was normal to a depth of about 6 m, but between 6 and 9 m, visibility was reduced to zero. Below this opague layer, from 9 to 15 m, suspended matter consisted mainly of large (1 - 3 cm long) reddish-yellow flocks (Fig. 7). The remainder of the water column contained fine particulate matter typical of that observed on previous dives. On the subsequent dive (#12), near the same location, the opaque layer was absent but the heavy flocks were at about the same level. The pH of the upper 15 m of the water column was reduced to about 7.5; thus, where the measurement was taken, the acid had been greatly diluted.

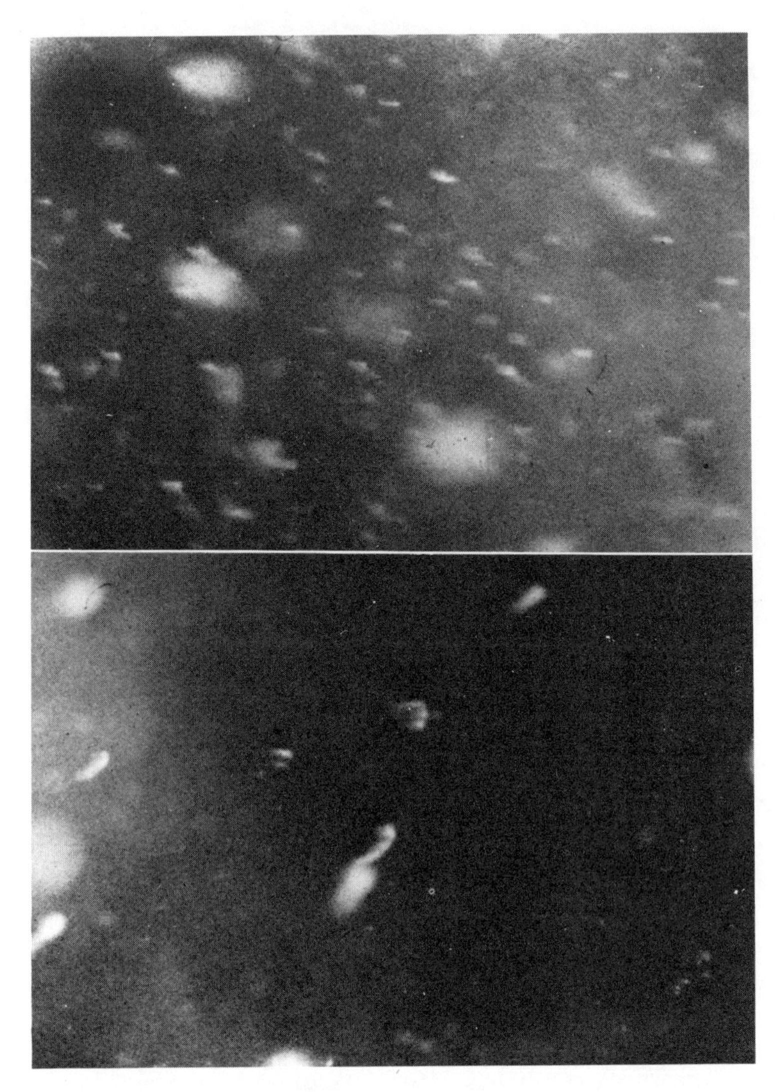

Figure 7
Photographs from the submersible, showing large flocks (1 - 3 cm long) of iron sulfate(?) precipitates at the center of the acid dumpsite 6 hours after a barge had released 1 million gallons of acid water.

Bottom: The microtopography differed little from that observed at the sewage sludge dumpsite. Both hummocks and ripples were present. On dives 2, 12, and 13, hummocks (10 cm diameter; 3 - 5 cm high [See Fig. 4]) were common, but no ripples were observed. On dives 10 and 11, symmetrical ripples were present that had wavelengths ranging from 15 to 80 cm and wave heights of 3 to 5 cm. Crests were mostly oriented north or northeast. A few sand waves having wavelengths of 3 to 5 m were also present at these sites. Bottom currents measured on two dives trended north at a speed of 4 to 5 cm/s.

Surficial material covering the bottom was different from that observed at the sewage site. Most of the bottom, particularly in ripple troughs, was covered by a thin (<2 mm) yellowish to reddish granular flaky material. Whether or not it was a stain resulting from the pollutants on the underlying fine- to medium-grained gray sand, or was a totally different material, was not resolved. It did not remain in suspension when stirred up with the submersible's manipulator arm.

Biology: As before, sand dollars covered large areas of the bottom, particularly on ripple crests and hummocks. At dive site 10, greatest concentrations reached $270/m^2$. Starfish were dominant ($6 - 12/m^2$) at dive site 2. Crabs were common at all sites. Flounder, hake, shrimp, skates, snails, scallops, and sponges, though occasionally abundant, were not as common as the sand dollars, crabs, and starfish. Fresh, open, articulated shells (clappers) appeared to be more abundant in the acid dumpsite than in the sewage dumpsite (Fig. 5).

In summary, the physiography and hydrology of the two areas differed little; abundant solids did not appear to be collecting at the bottom of either area. Effects on the benthic community, if any, were not obvious at a reconnaissance level; however, the reddish surficial layer appeared to be derived from or related to the effluent Clearly, the acid waste concentrated above and in the thermocline must have a localized drastic affect on plankton and nekton.

Control Site - Results

One dive, #14, was carried out 26 km north of the acid disposal site in order to document conditions where the effects of effluent should have been less. In the water column, the temperature structure was similar, but suspended particulates appeared to be less concentrated. Bottom-current flow was toward the southwest at about 5 cm/s. On the bottom, ripples were oriented northeast; they had wavelengths of 60 cm and wave heights of 7 to 10 cm. Little flocculant material was present, and the exposed very fine to very coarse sand was reddish brown. Among live organisms, sand dollars were dominant, in concentrations of as much as $180/m^2$ on ripple crests. Crabs, juvenile forms in particular, were probably the second most common organism. Shrimp were abundant; fewer hake, scallops, flounder, skates, starfish, and worms were present.

Thus, the main differences between the waste disposal sites and the control site were lower turbidity and less flocculant material on the bottom at the control site. During summer, when the thermocline is well developed and the settling of released effluents is obviously impeded, the distribution of such material in the water column by surface circulation is probably extensive; therefore, depending on water circulation, conditions at the control site might not be much different from those at the dumpsites. In winter, when settling is unimpeded, perhaps more acid waste and sludge would be present on the bottom in the dumpsites than at the control site.

DUMPSITE OBSERVATIONS - 1975

The five day cruise by R/V ADVANCE II and DSRV NEKTON BETA to the disposal sites in August 1975 was designed to gather more specific information on the character and composition of the sediments and the distribution of wastes. Several dives were carried out at each of the station locations 4300 - 4304 shown in Fig. 1. Because acid was being released when the ship arrived, no diving was carried out in the acid waste site. The senior author participated in only five dives at four sites, but some samples collected by the EPA team that made most of the observations during the cruise have been analyzed by the USGS, and the results are presented here. During several dives, 13 ripple crests and 12 ripple troughs were sampled from the submersible to ascertain whether texture varied significantly and to determine whether troughs actually serve as sinks for fine natural debris and possibly pollutants.

Water column: At the five stations occupied, surface-water temperatures ranged from 23° to 26°C; it diminished in the thermocline (~3 to ~30 m) to 8° - 9°C at the bottom. At the three eastern stations (4302, 4303, 4304), surface water was well mixed and isothermal to a depth of ~15 m. Currents were sluggish and variable near bottom at all five sites. Because no storms passed during the cruise, most water motion was probably due to tides. A slight north-south oscillation was noted at station 4300, whereas at 4303, north-flowing currents of 5 to 10 cm/s were actually scouring fine debris from the bottom. At other sites, speeds were <7 cm/s, and most often ~3 cm/s. At the control site (station 4304), water was flowing at only 3 cm/s to the southeast. From the submersible, water appeared to be most turbid above or in the thermocline and least turbid at the bottom where visibility was 10 to 15 m in ambient light. This observation was verified by transmissometer measurements (Fig. 8), which showed that most low values (high turbidity) were above and in the thermocline and high values (low turbidity) were near bottom. Suspended matter collected from the water column at stations 4301, 4302, and 4303 ranged from 0.10 to 0.50 mg/l in surface water; 0.25 to 0.50 mg/l in the thermocline; and 0.20 to 0.22 mg/l in bottom water (Table 2). Organic oxidizable components in the suspended matter were abundant in near-surface waters and declined near the bottom.

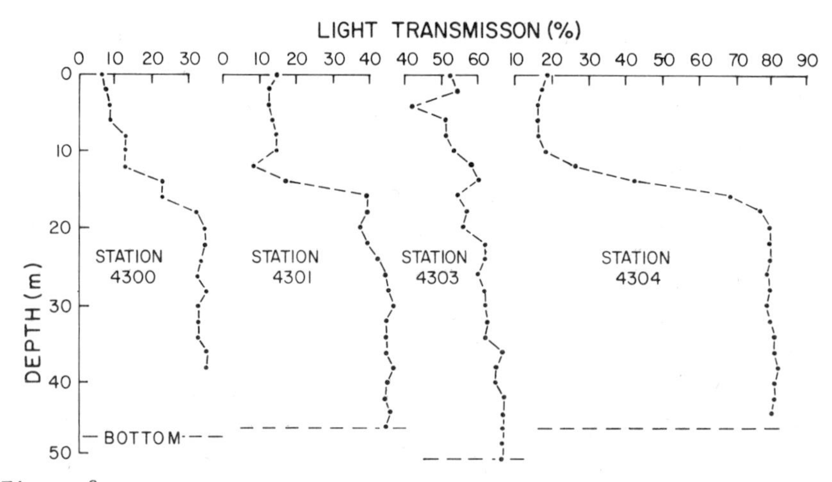

Figure 8
Transmissometer profiles, showing percent transmission increasing to the right, indicative of less suspended particulate matter in the water. Lowest values (high turbidity) are above or in thermocline.

Table 2
Concentrations of suspended sediment in the water column, ADVANCE II Cruise 75, Leg 4

Cruise	Stat #	Date 1975	Time	Lat. °N	Long. °W	Water Depth (m) Total/SPL	Filter* Type	Filter* No.	Water Vol. (L)	Insol Conc Mg/L	Insol %	Sol Conc Mg/L	Sol %	Total SM Mg/L
ADV 75-4	4300	8/17	2100	38°28.0'	74°25.0'	47/00	N	1801	5.50	0.19	75	0.06	25	0.26
"	"	"	"	"	"	47/11	M	4202	1.25	-	-	-	5?	0.52
"	"	"	"	"	"	47/11	N	1803	0.40	-	-	-	96?	0.70
ADV 75-4	4301	8/18	1932	38°21.4'	74°18.5'	46/00	N	1806	0.675	0.07	15	0.43	85	0.50
"	"	"	"	"	"	46/10	N	1827	0.85	0.25	50	0.25	50	0.50
"	"	"	"	"	"	46/46	N	1826	2.21	0.05	27	0.15	73	0.20
ADV 75-4	4302	8/19	2035	38°36.4'	74°17.7'	49/00	N	1831	5.25?	0.01	5	0.09	95	0.10
"	"	"	"	"	"	49/15	N	1807	0.60	0.00	0	0.40	100?	0.40
"	"	"	"	"	"	49/49	N	1832	2.20	0.20	78	0.05	22	0.25
"	"	"	"	"	"	49/49	M	4206	2.20	-	-	-	-	0.20
ADV 75-4	4303	8/20	2025	38°35.1'	74.03.5'	50/00	M	4204	1.80	0.19	73	0.07	27	0.26
"	"	"	"	"	"	50/00	N	1805	0.95	0.00	0	0.18	100?	0.18
"	"	"	"	"	"	50/12	N	1808	1.20	-	-	-	-	-
"	"	"	"	"	"	50/50	N	1809	1.30	0.17	79	0.05	21	0.22

*Filter type - M= millipore ®
N= nuclepore ®

Bottom: At most stations (4300, 4301, and 4304), ripples and sand waves were more abundant and larger than those observed the previous year. Many were ~60 cm in wavelength and 5 - 20 cm in amplitude (Fig. 2). The trend of crests was most often northeast (030°T) or north. At station 4301, some crests may have been oriented northwest, but the gyro compass on the submersible may have been in error. All ripples and waves were symmetrical, had rounded crests, and may have resulted, in part, from the passage of Hurricane Blanche through the area in July.

Flocculant dark-brown, fine-grained material covered much of the bottom. It was easily stirred up by the manipulator arm and was crisscrossed by many tracks and trails. At several stations, the material lay in broad stripes several meters wide that extended beyond the limit of visibility. These stripes could best be seen under ambient light when the submersible was 1 to 2 m above the bottom. Perhaps their geometry is related to Langmuir cells (Langmuir, 1925), to some other similar, slow-moving helical flow that affects mainly the finest particulate matter, or to short-period internal waves common in the area (Appel and others, 1975). Near the northeast edge of the dumpsite, the flocky layer was thicker and more widespread than that observed the previous year, but the granular reddish layer was not observed.

Grab Sample - Results

In all 47 grab samples collected at the five sites from the ship and the submersible, gravel (including shell fragments) ranged from 0 (by dry weight) to a maximum of 37.4%, with a mean of 6.1%; sand, from 61.5% to 99.0%, with a mean of 88.4%; and silt plus clay, from 0.2% to 41.4%, with a mean of 2.1%. Clay and silt were abundant (11.5% and 27.8%) in only one sample (Table 3). To obtain more specific information on local textural variations, some samples were taken from the submersible from adjacent ripples and troughs.

Gravel and silt plus clay concentrations (dry weight) in these samples collected from adjacent ripple crests and troughs are shown in Table 4. At eight of ten locations sampled, the mud (silt and clay) was more abundant in troughs than on crests but differences were small; and at seven of ten locations, gravel was also more abundant in troughs than on crests. Thus, the abundance of large rock and shell fragments is documented by the size analyses. Because of the wide range of values, however, it would be difficult to differentiate trough from crest samples collected blindly from the surface, solely on the basis of textural analyses.

Chemical analyses of grab samples show a significant buildup of some elements near the acid dumpsite but not at the sewage site (Table 5). For example, organic C, Fe, Ti, and Cr are low near stations 4300 and 4301 and increase by a factor of two or more at station 4302. Values are also high to the north and east at stations 4303 and 4304. Though few samples are involved, the data suggest that the sewage sludge was not accumulating in the sewage site when these samples were taken. The acid waste, however, may be reflected by the higher concentrations of Fe, Ti, and Cr. Fe_2O_3 and Cr concentrations near the acid site are anomalously high with respect to their regional distribution (M. Bothner, personal communication).

Table 3
Size analyses of grab samples collected from the surface ship and from the submersible.

Sta. #	Samp. #	Date	Lat.	Long.	Water Depth (m)	Med. Size (Phi)	Mean Size (Phi)	S.D. (Phi)	Gravel	Sand	Silt	Clay	Sand Fraction (%)				
													Very Coarse Sand	Coarse	Med. Sand	Fine Sand	Very Fine Sand
4300	A1	8/17/75	38°27.2'N	74°26.0'W	47	.39	.41	1.27	22.24	73.52	4.24	--	6.61	48.52	16.17	2.21	0.00
"	A2	"	"	"	47	.54	.53	1.06	16.53	83.27	0.21	--	4.16	45.80	26.65	6.66	0.00
"	A3	"	"	"	47	.51	.42	.90	14.73	84.95	0.32	--	2.54	62.01	18.69	1.70	0.00
"	B1	"	"	"	47	.32	.36	.60	1.66	96.56	0.81	--	17.38	73.39	5.79	0.00	0.00
"	B2	"	"	"	47	.21	.25	.84	9.45	90.30	0.25	--	25.29	48.77	12.65	2.71	0.00
"	C1	"	"	"	47	1.70	1.78	.60	0.24	98.20	1.56	--	0.00	0.00	71.69	24.55	1.96
"	C2	"	"	"	47	.86	1.02	.92	2.93	96.21	0.85	--	0.96	51.95	29.82	11.54	1.92
"	C3	"	"	"	47	2.40	2.45	.54	0.00	99.57	0.43	--	0.00	0.00	19.92	62.73	16.93
"	C4	"	"	"	47	.04	.15	.72	2.86	96.30	0.83	--	45.26	43.33	7.71	0.00	0.00
4301	A1	8/18/75	38°20.8'N	74°19.0'W	46	.76	.79	.74	4.30	94.94	0.76	--	0.00	63.61	28.49	2.85	0.00
"	A2	"	"	"	46	-.07	.06	.65	1.26	97.89	0.85	--	56.77	37.19	3.92	0.00	0.00
"	B1	"	"	"	46	.43	.58	.91	5.73	93.01	1.27	--	8.37	62.31	18.60	3.72	0.93
"	B2	"	"	"	46	-.03	-.01	.78	4.36	95.00	0.64	--	45.60	44.65	3.80	0.95	0.00
"	B3	"	"	"	46	.66	.77	.90	4.56	95.16	0.28	--	6.66	53.29	27.60	5.71	1.90
"	C	"	38°21.3'N	74°16.5'W	46	.85	.96	.74	1.47	97.97	0.56	--	0.00	56.82	34.28	6.86	0.00

Table 3, continued

Sta. #	Samp. #	Date	Lat.	Long.	Water Depth (m)	Med. Size (Phi)	Mean Size (Phi)	S.D. (Phi)	Percentage by Weight				Sand Fraction (%)				
									Gravel	Sand	Silt	Clay	Very Coarse Sand	Coarse Sand	Med. Sand	Fine Sand	Very Fine Sand
4302	A1	8/19/75	38°35.1'N	74°18.1'W	49	2.51	2.47	.70	1.41	96.85	1.73	--	0.00	0.00	14.53	68.76	13.56
"	A2	"	"	"	49	2.61	2.55	.65	1.29	97.96	0.75	--	0.00	0.00	9.76	71.51	16.65
"	A3	"	"	"	49	2.71	2.62	.80	2.42	97.09	0.48	--	0.00	0.00	3.88	66.03	27.18
"	B1	"	38°35.4'N	74°18.2'W	49	2.54	2.49	.57	0.00	99.17	0.83	--	0.00	0.00	21.81	60.47	16.86
	B2	"	"	74°18.0'W	49	2.48	2.42	.56	.41	98.06	1.54	--	0.00	0.00	20.59	72.57	4.90
"	C1	"	38°35.8'N	74°17.5'W	49	2.50	2.39	.63	1.23	98.27	0.49	--	0.00	0.00	19.65	75.67	2.95
"	C2	"	"	"	49	2.54	2.49	.52	0.10	98.47	1.43	--	0.00	0.00	19.69	69.91	8.86
"	C3	"	"	"	49	2.69	2.69	.57	0.69	98.63	2.47	--	0.00	0.00	2.90	77.45	16.46
"	D	"	38°36.0'N	74°17.0'W	49	1.99	1.97	.67	0.91	97.79	2.11	--	0.00	2.93	45.96	46.94	1.96
4303	A1	8/20/75	38°35.2'N	74° 2.3'W	50	2.53	2.44	.73	2.07	97.16	0.77	--	0.00	0.00	13.60	73.84	10.69
"	A2	"	"	"	50	2.08	2.06	.36	0.07	99.53	0.40	--	0.00	0.00	39.91	58.72	1.00
"	B	"	"	"	50	1.90	1.86	.84	2.58	94.43	3.09	--	0.00	0.00	50.05	43.44	1.88
"	C1	"	38°35.5'N	74° 3.5'W	50	2.32	2.23	.78	2.93	96.84	0.23	--	0.00	0.00	24.21	67.79	4.84
"	C2	"	"	74° 2.3'W	50	1.89	1.00	1.86	37.43	61.47	1.11	--	0.00	0.00	12.29	41.80	7.37
4304	D3	8/21/75	38°47.8'N	74°10.0'W	43	.88	.99	.71	0.48	98.58	0.94	--	0.00	56.20	38.44	3.94	0.00
"	E1	"	"	"	43	.10	.14	1.32	25.77	71.64	2.58	--	27.22	22.20	16.47	4.30	0.72
"	E2	"	"	"	43	1.88	1.75	.91	5.28	92.84	1.88	--	0.00	0.00	51.99	40.85	0.00

Table 3, continued

Sta. #	Samp. #	Date	Lat.	Long.	Water Depth (m)	Med. Size (Phi)	Mean Size (Phi)	S.D. (Phi)	Percentage by Weight				Sand Fraction (%)				
									Gravel	Sand	Silt	Clay	Very Coarse Sand	Coarse Sand	Med. Sand	Fine Sand	Very Fine Sand
4303	D	8/20/75	38°35.5'N	74° 3.5'W	50	-.05	.05	1.04	7.77	89.37	2.86	--	42.90	42.00	3.58	0.89	0.00
"	E1	"	"	"	50	1.81	1.08	1.89	35.17	62.17	2.66	--	0.00	1.86	14.92	34.19	11.19
"	E2	"	"	"	50	1.92	1.79	1.05	7.52	90.99	1.49	--	0.00	0.00	48.23	41.86	1.82
"	E3	"	"	"	50	2.39	2.28	.76	2.73	96.41	0.86	--	0.00	0.00	20.25	75.20	0.96
	F	"	"	"	50	1.56	3.81	3.80	0.94	59.78	11.46	27.82	0.00	11.36	44.84	3.60	0.60
4304	A1	8/21/75	38°47.6'N	74°10.1'W	43	.61	.77	.82	1.31	97.52	1.16	--	7.80	56.56	30.23	2.93	0.00
"	A2	"	"	"	43	.79	.88	.98	5.73	92.84	1.43	--	2.79	51.06	28.78	9.29	0.93
"	A3	"	"	"	43	1.09	1.17	.87	1.83	96.88	1.29	--	0.97	43.60	37.78	14.54	0.00
"	B1	"	38°47.8'N	74°10.0'W	43	1.69	1.49	1.47	15.18	81.54	3.29	--	0.00	9.78	35.88	26.09	9.78
"	B2	"	"	"	43	2.50	2.44	.59	1.39	98.16	0.45	--	0.00	0.00	7.85	86.38	3.93
"	C1	"	"	"	43	1.71	1.78	.58	0.23	98.68	1.09	--	0.00	0.00	71.05	25.65	1.97
"	C2	"	"	"	43	2.25	2.15	.90	3.58	92.53	2.78	--	0.00	0.00	29.61	61.07	1.85
"	C3	"	"	"	43	1.30	1.36	.70	0.78	97.74	1.48	--	0.00	36.12	51.80	9.77	0.98
"	D1	"	"	"	43	1.08	1.19	.86	2.99	96.33	0.68	--	0.00	43.34	40.46	12.53	1.92
"	D2	"	"	"	43	1.30	1.42	.86	1.25	95.49	3.26	--	0.00	31.51	51.56	10.50	1.90

Table 4
Texture, organic carbon, iron, titanium, and chromium concentrations in relation to microtopography. (Standard deviation of any single answer for Fe, Ti, and Cr should be taken as +50% and -33%.)

Sample	No.	Microtopography	Gravel (% by wt)	Silt & Clay (% by wt)	Org. C (% dry wt)	Fe (%)	Ti (%)	Cr (ppm)
4300	A-1	Trough	22.2	4.2	0.10	0.34	0.15	11.0
	A-2	Crest	16.5	0.2	0.06	0.48	0.23	5.7
	A-3	Crest	14.7	0.3	0.06	0.26	0.08	2.4
	B-1	Trough	1.7	0.8	0.12	0.26	0.08	4.2
	B-2	Crest	9.5	0.3				
4301	A-1	Trough	4.3	0.8	0.11	0.16	0.03	2.0
	A-2	Crest	1.3	0.9	0.06	0.14	0.02	4.6
	B-1	Crest	5.7	1.3				
	B-2	Trough	4.4	0.6				
	B-3	Crest	4.6	0.3				
4302	A-1	Trough	1.4	1.7	0.07	0.93	0.25	21.0
	A-2	Crest	1.3	0.8	0.15	1.40	0.37	18.0
	A-3	Crest	2.4	0.5	0.15	1.20	0.29	17.0
4303	A-1	Trough	2.1	0.8	0.15	1.30	0.26	14.0
	A-2	Crest	0.1	0.4	0.22	1.00	0.19	8.9
	E-1	Trough	35.2	2.7				
	E-2	Trough	7.5	1.5	0.14	0.82	0.13	18.0
	E-3	Crest	2.7	0.9	0.12	1.40	0.33	18.0
4304	A-1	Crest	1.3	1.2	0.08	0.24	0.04	4.2
	A-2	Trough	5.7	1.4	0.08	0.32	0.05	3.8
	A-3	Trough	1.8	1.3	0.09	0.44	0.08	5.0
	C-1	Crest	0.2	1.1	0.09	0.35	0.05	5.2
	C-2	Trough	3.6	2.8	0.09	0.44	0.05	3.6
	C-3	Trough	0.8	1.5	0.10	0.29	0.03	3.2
	E-1	Trough	25.8	2.6	0.08	0.32	0.04	3.7
	E-2	Crest	5.3	1.9	0.26	1.9(?)	0.19	22.0

Table 5
Chemical analyses of grab samples collected from the surface vessel and from submersible.
(Standard deviation of any single answer for Fe, Ti, and Cr should be taken as +50%/-33%.)

Sample	No.	% Fe	% Ti	% Mn	ppm Cr	% Fe_2O_3	% TiO_2	% MnO	Org. C
4300	A-1	0.34	0.15	0.019	11	0.49	0.25	0.025	0.10
	A-2	0.48	0.23	0.015	5.7	0.69	0.38	0.019	0.06
	A-3	0.26	0.081	0.012	2.4	0.37	0.14	0.016	0.06
	B-1	0.26	0.085	0.0067	4.2	0.37	0.14	0.0087	0.12
	C	0.15	0.012	0.012	1.6	0.21	0.020	0.016	0.06
4301	A-1	0.16	0.033	0.0070	2.0	0.23	0.055	0.0090	0.11
	A-2	0.14	0.022	0.0059	4.6	0.20	0.037	0.0076	0.06
	C	0.29	0.48	0.015	4.7	0.42	0.80	0.019	0.07
4302	A-1	0.93	0.25	0.039	21	1.3	0.42	0.050	0.07
	A-2	1.4	0.37	0.040	18	2.0	0.62	0.052	0.15
	A-3	1.2	0.29	0.034	17	1.7	0.48	0.044	0.15
	B-1	1.2	0.29	0.045	14	1.7	0.48	0.058	0.14
	B-2	1.3	0.23	0.049	13	1.9	0.38	0.063	0.16
	C-1	2.0	0.36	0.072	22	2.9	0.60	0.093	0.16
	C-2	2.2	0.58	0.060	24	3.2	0.97	0.078	0.15
	C-3	1.4	0.39	0.045	15	2.0	0.65	0.058	0.17
	D	1.7	0.46	0.043	32	2.4	0.77	0.056	0.17
4303	A-1	1.3	0.26	0.038	14	1.9	0.43	0.049	0.15
	A-2	1.0	0.19	0.030	8.9	1.4	0.32	0.039	0.22

Sample	No.	% Fe	% Ti	% Mn	ppm Cr	% Fe_2O_3	% TiO_2	% MnO	Org. C
4303	B	1.1	0.22	0.041	13	1.6	0.37	0.053	0.16
	C-1	1.1	0.18	0.032	16	1.6	0.30	0.041	0.19
	D	1.1	0.23	0.031	15	1.6	0.38	0.040	0.25
	E-2	0.82	0.13	0.028	18	1.2	0.22	0.036	0.14
	E-3	1.4	0.33	0.043	18	2.0	0.55	0.056	0.12
	F-1	1.1	0.21	0.035	11	1.6	0.35	0.045	0.22
	F-2	0.88	0.15	0.039	10	1.3	0.25	0.050	0.20
4304	A-1	0.24	0.042	0.0074	4.2	0.34	0.070	0.0096	0.08
	A-2	0.32	0.050	0.0092	3.8	0.46	0.083	0.012	0.08
	A-3	0.44	0.083	0.013	5.0	0.63	0.14	0.017	0.09
	B-1	5.1	0.27	0.12	47	7.3	0.45	0.16	0.26
	C-1	0.35	0.049	0.014	5.2	0.50	0.082	0.018	0.09
	C-2	0.44	0.053	0.019	3.6	0.63	0.088	0.025	0.09
	C-3	0.29	0.031	0.016	3.2	0.42	0.052	0.021	0.1
	D-1	0.30	0.050	0.012	4.9	0.43	0.083	0.016	0.09
	D-2	0.44	0.059	0.015	5.8	0.63	0.098	0.019	0.09
	D-3	0.28	0.048	0.016	3.7	0.40	0.080	0.021	0.12
	E-1	0.32	0.044	0.0075	3.7	0.46	0.073	0.0097	0.08
	E-2A	4.4	0.32	0.077	62	6.3	0.53	0.099	0.26
	E-2B	1.9	0.19	0.046	22	2.7	0.32	0.059	0.35

The same elements were analyzed in eight paired ripple-crest--trough samples. At five of eight stations, organic C concentrations were higher in troughs than on crests (Table 4), but differences were small and the number of observations inadequate to show a statistical trend. The coarse detritus that also accumulates in troughs is commonly associated with little organic matter, which would cause the percentage of total organic C in the whole sample to be lower. Variations in concentrations of Fe, Ti, and Cr were, for most crest-trough pairs, insufficient, in view of the analytical error, to draw any conclusion about preferential accumulation (Table 4).

In summary, the bottom microtopography and texture in 1975 reflected a higher energy regime at the bottom than that during the previous year. This increase may have been caused by the passage nearby of a July hurricane. The few chemical analyses carried out suggest that the sewage sludge is being transported away from the dumpsite or is being dissipated and destroyed before it accumulates on the bottom. The distribution of elements associated with acid waste, however, suggests that the waste is reaching bottom in the area of the dumpsite and may be spreading mainly to the east and north.

CONCLUSIONS AND RECOMMENDATIONS

1. During summer, the well-developed thermocline in the area provides a density barrier that inhibits settling of some particulate wastes and may result in their widespread dispersal throughout the area.
2. Turbidity appears to be low during summer, but particulate matter is probably sufficiently abundant to agglomerate some pollutants.
3. Bottom-water circulation is slow (<10 cm/s) when no storms are traversing the area. However, when coupled with activity of benthic organisms that stir up bottom sediment and oscillatory flow due to waves associated with distant storms, tidally driven currents probably still disperse considerable sediment.
4. The flocculant, dark-brown fine-textured layer that covers much of the bottom may be a sink for certain pollutants and should be investigated further.
5. Microtopography is mainly characterized by rounded symmetrical ripples probably formed by waves. In many areas, these ripples were obliterated by bioturbation that replaced them with either featureless or hummocky topography.
6. Organic pollutants in and near the Philadelphia sewage dumpsite did not appear to be building up significantly on the bottom; however, evidence of acid-waste accumulation was present. This reconnaissance survey clearly shows that further studies are needed to document the distribution and concentration of both pollutants in the water column and on the bottom.

ACKNOWLEDGEMENTS

The cruises to the disposal sites were supported by the EPA. We wish to thank the cruise participants whose observations contributed to the report. For the 1974 cruise, personnel included B. Reynolds and G. Morrison of EPA, H. Palmer and J. Forns of Westinghouse Corp., F. Childress of U. S. National Oceanic and Atmospheric Administration (NOAA), and B. Oostdam of the Marine Sciences Consortium. G. Shiller was pilot of the submersible. In August 1975, the cruise was manned by a large EPA team headed by J. Pesch and D. Lear; the U.S. Geological Survey group was led by D. W. Folger and comprised P. Cousins, S. Rindge, S. Purdy, C. Morse, and D. Edwards. Submersible pilots included R. A. Slater, J. O'Donnell, and R. Czahara. S. Wood, S. Rindge, and S. Wieber carried out grain-size analyses; L. Poppe, X-ray analyses; and P. Cousins and S. Wieber processed photographs. Fe, Ti, Mn, and Cr analyses were carried out by J. L. Harris at the USGS in Reston, Virginia, and C analyses were carried out by C. Parmenter. M. Bothner assisted in the interpretation of chemical data. S. Rindge and S. Wood were of particular assistance assembling data; J. Moller, P. Forrestel, L. Sylwester, and M. Roy drafted the illustrations; E. Winget and S. Merchant prepared the manuscript.

REFERENCES

Appel, J. R., H. von Byrne, J. R. Proni, and R. L. Charwell, 1975, Observations of oceanic internal and surface waves from the Earth Resources Technology Satellite. Jour. Geophys. Research, 80(6): 845-881

Duane, D. B., M. E. Field, E. P. Meisburger, D. J. P. Swift, and S. J. Williams, 1972, Linear shoals on the Atlantic inner Continental Shelf, Florida to Long Island. In Swift, D. J. P., D. B. Duane, and O. H. Pilkey, eds., Shelf Sediment Transport: Process and Pattern, Dowden, Hutchinson, and Ross, Inc., Stroudsburg, Pa., 447-498.

Edwards, R. L., and K. O. Emery, 1968, The view from a storied sub, the "Alvin" of Norfolk, Va. Comm. Fisheries Review, 30: 48-55.

Emery, K. O., 1966, Atlantic Continental Shelf and Slope of the United States, geologic background. U. S. Geol. Survey Prof. Paper 529-A: 1-23.

Emery, K. O., 1968a, Relict sediments on continental shelves of the world. Am. Assoc. Petroleum Geologists Bull., 63: 1105-1108.

Emery, K. O., 1968b, Positions of empty pelecypod valves on the Continental Shelf. Jour. Sed. Petrol., 38: 1264-1269.

Emery, K. O., and Elazar Uchupi, 1972, Western North Atlantic Ocean: topography, rocks, structure, water, life, and sediments. Am. Assoc. Petroleum Geologists Mem., 17: 1-532.

Frank, W. M., and G. M. Friedman, 1973, Continental Shelf sediments off New Jersey. Jour. Sed. Petrol., 43: 224-237.

Hard, C. G., and H. D. Palmer, 1976, Sedimentation and ocean engineering: ocean dumping. In Stanley, D. J., and D. J. P. Swift, eds., Marine sediment transport and environmental management, J. Wiley and Sons, New York: 557-577.

Hollister, C. D., 1972, Atlantic Continental Shelf and Slope of the United States--texture of surface sediments: New Jersey to southern Florida. U. S. Geol. Survey Prof. Paper 529-M.

Knebel, H. J., 1975, Significance of textural variations, Baltimore Canyon Trough area. Jour. Sed. Petrol., 45: 873-882.

Knebel, H. J., and Elliot Spiker, 1977, Thickness and age of the surficial sand sheet, Baltimore Canyon Trough area. Am. Assoc. Petroleum Geologists Bull., 61(6): 861-871.

Komar, P. D., R. H. Neudeck, L. D. Kulm, 1972, Observations and significance of deep-water oscillatory ripple marks on the Oregon continental shelf. In Swift, D. J. P., D. B. Duane, and O. H. Pilkey, eds., Shelf Sediment Transport: Process and Pattern, Dowden, Hutchinson and Ross, Inc., Stroudsburg, Pa., 601-619.

Langmuir, C. F., 1925, Surface motion of water induced by the wind. Science, 86: 119-123.

McLennen, C. F., 1973, New Jersey Continental Shelf near-bottom current meter records and recent sediment activity. Jour. Sed. Petrol., 43: 371-380.

Milliman, J. D., 1972, Atlantic Continental Shelf and Slope of the United States--Petrology of the sand fraction of sediments, northern New Jersey to southern Florida. U. S. Geol. Survey Prof. Paper 529-J, 40 p.

McKinney, T. F., W. L. Stubblefield, D. J. P. Swift, 1974, Large scale current lineations on the central New Jersey shelf: investigations by sidescan sonar. Marine Geology, 17: 79-102.

Palmer, H. D., and D. W. Lear, 1973, Environmental survey of an interim dumpsite, Middle Atlantic Bight. U. S. Env. Protection Agency Rpt. 903-001-A, 134 p.

Schopf, T. J. M., 1968, Atlantic Continental Shelf and Slope of the United States--Nineteenth century exploration. U. S. Geol. Survey Prof. Paper 529-F, 12 p.

Schlee, J. S., and R. M. Pratt, 1970, Atlantic Continental Shelf and Slope of the United States--gravels of the northeastern part. U. S. Geol. Survey Prof. Paper 529-H, 39 p.

Stanley, D. J., and D. J. P. Swift, 1976, Marine sediment transport and environmental management. J. Wiley and Sons, New York, NY, 602 p.

Swift, D. J. P., D. B. Duane, T. F. McKinney, 1973, Ridge and swale topography of the Middle Atlantic Bight, North America: Secular response to the Holocene hydraulic regime. Marine Geology, 15: 227-247.

Swift, D. J. P., D. B. Duane, O. H. Pilkey, 1972a, Shelf sediment transport process and pattern. Dowden, Hutchinson and Ross, Inc., Stroudsburg, Pa., 656 p.

Swift, D. J. P., J. W. Kofoed, F. P. Saulsbury, Philip Sears, 1972b, Holocene evolution of the shelf surface, central and southern Atlantic shelf of North America. In Swift, D. J. P., D. B. Duane, O. H. Pilkey, eds., Shelf Sediment Transport: Process and Pattern, Dowden, Hutchinson and Ross, Inc., Stroudsburg, Pa., 499-574.

Uchupi, Elazar, 1968, The Atlantic Continental Shelf and Slope of the United States--Physiography. U. S. Geol. Survey Prof. Paper 529-C, 30 p.

Uchupi, Elazar, 1970, Atlantic Continental Shelf and Slope of the United States--Shallow structure. U. S. Geol. Survey Prof. Paper 529-I, 44 p.

Mathematical Modeling Predictions of the Geological Effects of Sewage Sludge Dumping on the Continental Shelf

David O. Cook

SUMMARY

A far-field (beyond the initial mixing zone) water quality math model has been adapted to predict geological implications of ocean dumping. The example simulated is ongoing discharge of digested sewage sludge by the City of Philadelphia at a disposal site located at mid-shelf in the Middle Atlantic Bight. After calibration of the model using hydrographic field data, predictions were made of the horizontal distributions of suspended solids and deposition following an individual dumping incident. The distributions shift south-southwest of the discharge line in response to non-tidal currents and disperse over a larger area with time. Following a 1.3×10^5 kg discharge of sludge solids, the maximum predicted far-field suspended solids concentration is 2.06 mg/l (after initial mixing), and the maximum depositional rate is $30.9 \times 10^{-5} g/cm^2$. The predictions are highly dependent upon assumed settling characteristics of the sludge.

The hydrodynamic simulation of the study area appears to be reasonably accurate. The geological predictions, in particular those for deposition, represent a first order estimate because of uncertainties in sludge solid settling behavior and pre/post depositional sludge fate. Regarding this fate, the effects of sludge recycling by zooplankton in the water column and detritus feeders on the sea floor are unknown. Also, the presence of physical structures in bottom photographs suggests that settled sludge solids would be subject to periodic resuspension. Until these uncertainties are resolved and quantified, the primary use of predicted depositional patterns is to indicate those areas of the sea floor where the benthic community may potentially be affected by sludge disposal.

INTRODUCTION

In recent years, waste disposal on the continental shelf has been recognized as a potential threat to the environmental quality of shelf waters and the adjacent sea floor. Efforts to evaluate the impact of such disposal operations have fallen into two categories: field studies of water chemistry and bottom sediment characteristics in existing dumping grounds, and mathematical simulation of dumped material dispersion. Mathematical models have been developed for and applied to the dumping of dredged materials from a barge (Koh and Chang, 1973). However, much of the industrial and municipal waste being disposed offshore consists of a dominant fluid phase and a low density particulate phase. Sewage sludge, as an example, contains less than 5 to 10 percent solids, and the specific gravity of these particles averages 1.5. These wastes are not readily amenable to models of dredged materials because of their different physical characteristics. Also, models such as Koh and Chang's focus on dispersion of dumped material in the near field region prior to completion of initial mixing. Low-density sludge particles, however, can be expected to remain in suspension for long periods and settle to the sea floor far from the release site.

The purpose of this paper is to describe the extension of a math model utilized for predicting far field water quality impacts from sludge dumping to also predict zones and quantities of sludge solids reaching the sea floor beyond the immediate dump site. This study was performed to evaluate both the potential value and the shortcomings of mathematically simulating the sedimentary behavior of sludge solids.

The study derives from a program which evaluated dumping of digested sewage sludge by the City of Philadelphia at the EPA Region III Interim Ocean Disposal Site. This site (Figure 1) is located in the Middle Atlantic Bight approximately 65 kilometers off the coasts of Delaware and Maryland where waters are 40 to 50 meters deep. Ocean currents constitute the primary sludge dispersal mechanism in this mid-shelf environment and, therefore, must be understood before dispersion processes can be reliably assessed. Therefore, the first phase of the program entailed the deployment of current meters and tide gauges to augment existing circulation data. The second phase employed mathematical models to estimate the rate of sludge transport and dispersion from the discharge region. Those parameters of the sludge investigated included coliform bacteria, biochemical oxygen demand (BOD), dissolved oxygen (DO), heavy metals, and suspended solids.

This paper dwells upon the suspended solids predictions and extrapolation of solids accumulation on the bottom. The site physical characteristics, models, and sediment-related predictions are discussed in order, followed by an evaluation of the validity of these predictions.

PHYSICAL CHARACTERISTICS OF THE SITE

Physical characteristics of the EPA Region III Interim Ocean Disposal Site vicinity are known from other studies and field measurements conducted in support of the modeling program. These field measurements, made in the fall of 1975, included recording of currents and tide as well as bottom photography. Current data was obtained at the locations shown in Figure 1 using ENDECO Model 105 meters. Shallow to mid-depth currents from station 4, which typifies the area, are plotted as a function of time on Figure 2 and in progressive vector fashion on Figure 3.

Data from these measurements highlight the tidal nature of the currents. Tidally-influenced currents are partially rotary, flooding in a westerly direction and ebbing towards the south-southeast. The

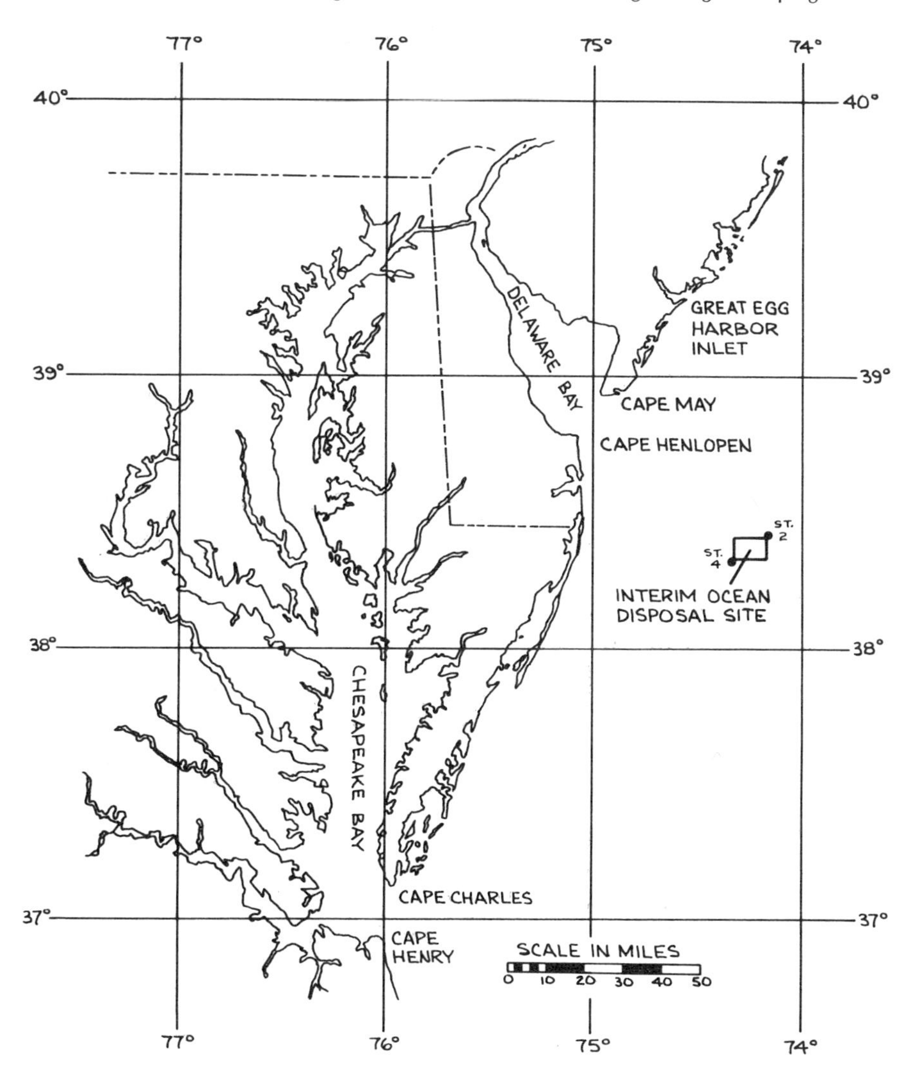

Figure 1
Location map.

strongest currents observed were 21 cm/sec whereas average currents ranged from 3 to 12 cm/sec. Progressive vector plots reveal a net south-southwesterly current drift in the study area. This approximates a flow parallel to the Maryland-Delaware shoreline. Although the magnitude of the drift is storm-dependent, typical net speeds were 8 km/day at station 5 and 15 km/day at station 2 located farther offshore. Previous measurements (McClennen, 1975; Beardsley and others, 1976) have demonstrated the existence of a southwest non-tidal drift in the mid-shelf region of the Middle Atlantic Bight in all seasons. Bottom water drift takes place at much slower speeds, and the site lies just within the 55 to 65 m isobaths which, according to

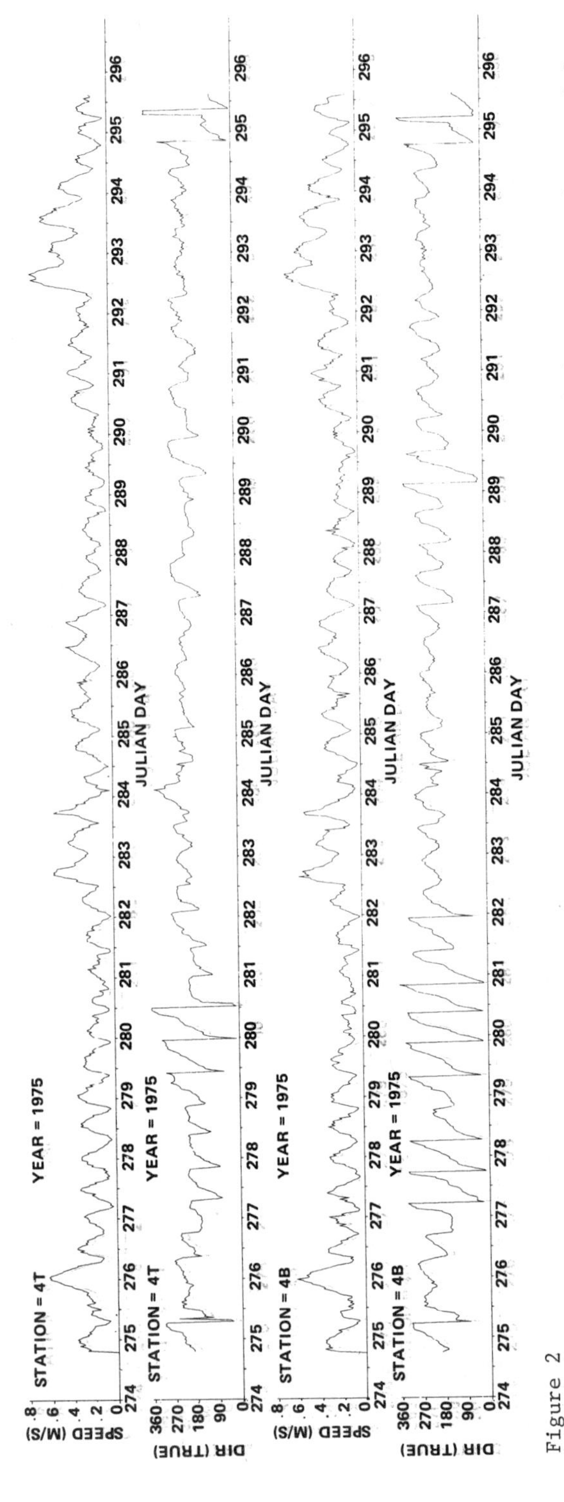

Figure 2
Current speed and direction vs time at station 4, depths of 7.6 m (4T) and 15.2 m (4B). The records extend from Oct. 1 to 23 (Julian days 274 to 296) 1975.

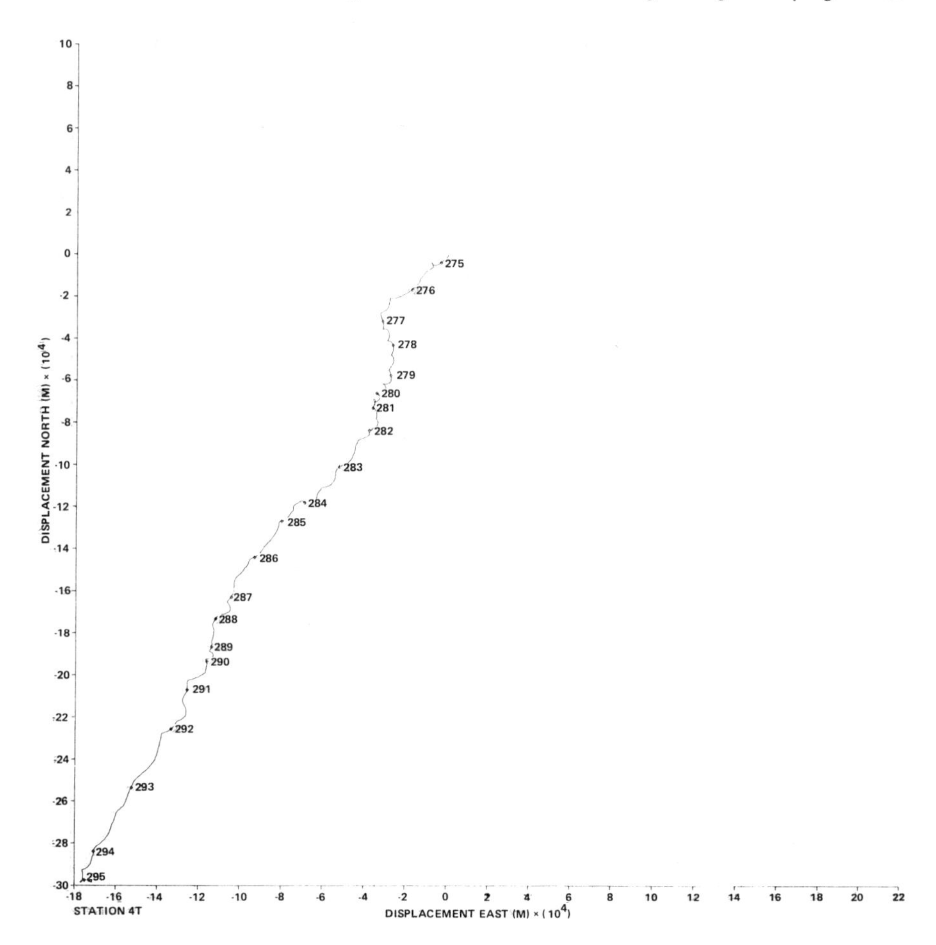

Figure 3
Progressive current vector plot for station 4, depth 7.6 m, Oct. 1 to 23 (Julian days 274-296) 1975.

Bumpus (1965), represent the division between onshore bottom drift on the inner shelf and offshore drift on the outer shelf.

The hydrography of shelf water on the Middle Atlantic Bight is summarized by Bumpus (1973). The mid-shelf region in the vicinity of the Interim Ocean Disposal Site experiences annual temperatures ranging from 5°C in late winter to 25°C in late summer. The water column is well-mixed in winter, and a seasonal thermocline which normally develops by May persists through October. During summer months, surface waters in the upper 10 to 15 m of the water column are 10 to 15°C warmer than bottom waters below 30 m. Salinity varies annually from 32 to 33.5 ‰, highest in winter and lowest in late spring following spring runoff from land. The dumping site is sufficiently distant from shore to minimize vertical salinity gradients, and temperature is the primary contributor to density stratification.

From a geological standpoint, data pertaining specifically to the Interim Ocean Disposal Site have been collected by Palmer and Lear

(1973) and Folger (1978). Bottom sediments consist of well sorted medium to fine sands, whose micro-relief is characterized both by physical and biological structures. For further information on disposal site sediments, the reader is referred to the paper by Folger which appears in this volume.

MATHEMATICAL MODELS

Excess sludge parameter concentrations in the far-field region (beyond the initial mixing zone) were predicted using a far-field advection-diffusion model in which concentration distributions are determined by advection (currents), turbulent diffusion, and natural decay processes. This Waste Dispersion Model (WDM) is a two-dimensional computer model which predicts the concentration of the discharged material over the study area. Inputs to the far-field dispersion model include the spatially- and temporally-varying advective currents. A far-field circulation model is used to generate these currents. This tandem set of models is described in the following sections, the Circulation Model first, then the Waste Dispersion Model.

Circulation Model

The Circulation Model predicts currents throughout a vertically well-mixed, two-dimensional body of water from a knowledge of the tide heights and/or currents at the open boundaries of the study region. The model is "dynamic" in that time-varying currents can be predicted from time-varying boundary currents or tide heights. Once calibrated, the model can be used to predict the circulation patterns for a wide variety of conditions.

The Circulation Model consists of a digital computer algorithm which yields a numerical solution to the vertically-averaged hydrodynamic equations of motion. The equations of motion describe water currents driven by horizontal pressure gradients which are produced by tidally-induced changes in surface elevation.

The components include a "continuity equation":

$$\frac{\partial \eta}{\partial t} + \frac{\partial(h + \eta)}{\partial x} + \frac{\partial(h + \eta)}{\partial y} = 0$$

and the momentum equations:

$$\frac{\partial u}{\partial t} + u\frac{\partial u}{\partial x} + v\frac{\partial u}{\partial y} + g\frac{\partial \eta}{\partial x} + g\frac{u(u^2 + v^2)^{1/2}}{C^2(h + \eta)} = 0$$

$$\frac{\partial v}{\partial t} + u\frac{\partial v}{\partial x} + v\frac{\partial v}{\partial y} + g\frac{\partial \eta}{\partial y} + g\frac{v(u^2 + v^2)^{1/2}}{C^2(h + \eta)} = 0$$

where:

u = vertically-averaged velocity in x direction
v = vertically-averaged velocity in y direction
η = incremental tide height about mean value

h = water depth to reference plane (MLW)
C = Chezy coefficient
g = acceleration of gravity
t = time.

These equations were solved numerically by the multioperational method (Leendertse, 1967) which possesses the good stability attributes of a purely implicit scheme and the computational efficiency of an explicit scheme.

Waste Dispersion Model

A computerized Waste Dispersion Model (WDM) is used to estimate the far-field excess concentration of waste constituents resulting from the release of sludge from a barge in the offshore area. This model, previously developed by Raytheon (Callahan and Wickramaratne, 1974) predicts the vertically averaged excess concentrations of coliform, BOD, DO, suspended solids and heavy metals throughout a horizontal two-dimensional plane. It is assumed that the waste remains in a fixed surface layer after near-field mixing has taken place.

The ambient transport processes of advection and turbulent diffusion are included in the model, along with the internal decay mechanisms. The spatial grid over the study area used in the circulation model is also used for the WDM, and the dynamic current distributions predicted by the circulation model are used as inputs to the WDM model. The WDM translates the steady effluent discharge into time-varying excess constituent concentration patterns throughout the study area.

The WDM model is a digital computer model of the two-dimensional, time-varying advection and diffusion characteristics of a vertically well-mixed water body (Leendertse, 1970).

The basic equation which describes the dependence of the concentration (C) on the distance variables (x, y), time (t), currents (U, V), constituent decay coefficient (k), diffusion coefficient (E_x, E_y) and discharge rate (S) is:

$$\frac{\partial(Ch)}{\partial t} = -\frac{\partial(UCh)}{\partial x} - \frac{\partial(VCh)}{\partial y} + E_x \frac{\partial^2(Ch)}{\partial x^2} + E_y \frac{\partial^2(Ch)}{\partial y^2} - kC + S$$

The first five terms on the right-hand side of this equation represent, respectively, the x- and y- advection, x- and y- turbulent diffusion, and decay processes. The sixth term represents the effluent source rates. This is the basic conservation equation for a single constituent with each term representing the rate change of constituent concentration due to the particular processes.

The numerical methods for solving the equation above are developed by discretizing the time and space variables to form a finite difference equation. A "forward difference" is used for time and "central differences" are used for the diffusion terms. The finite differences for the advective terms depend on the sign of the currents,

U and V, and are chosed to satisfy conservation of concentration in the discrete spatial volumes.

For a particular grid location m

$$\frac{\partial(UCh)}{\partial x} = \frac{U_{m-1}\, h'_{m-1}\, C'_{m-1} - U_m\, h'_m\, C'_m}{\Delta x}$$

where C'_m and C'_{m-1} are determined by

$$C'_m = \begin{cases} C_m \text{ if } U_m > 0 \\ C_{m+1} \text{ if } U_m < 0 \end{cases}$$

An analogous definition gives advection in the y-direction. The resulting explicit equation is programmed for solution on a computer. Boundary conditions are built into the program by classifying each spatial grid volume and adjusting the equation such that no constituent is advected or diffused across a land boundary. Once a constituent crosses a non-land boundary, it does not re-enter the model region.

MODELING RESULTS

Simulated Conditions

In order to apply the Waste Dispersion Model to the Interim Ocean Disposal Site, the grid system illustrated in Figure 4 was established. Two grid sizes were utilized: a fine 0.6 x 0.6 km system in the disposal area proper, and a coarse 1.85 x 1.85 km system in the surrounding far-field region. A total area of 46 x 46 km was incorporated in the model, and it was anticipated that dilution of most sludge parameters to background levels would be accomplished within these confines.

The Dispersion Model was used to simulate a release of 7.6 x 10^6 l of digested sewage sludge from a barge along the 10.2 km-long line oriented NW-SE within the Interim Ocean Disposal Site as shown on Figure 4. This volume of sludge is currently being dumped every four days in the study area by the City of Philadelphia. The sludge was assumed to be initially distributed in the twelve fine-grain model grid squares illustrated in the figure. The initial near-field dilution was conservatively estimated at 825:1 by assuming that wastes are distributed throughout a mixing zone 100 m wide and 20 m deep extending along the 10.2 km discharge line. After the near-field dilution has taken place, the waste was assumed to be mixed uniformly throughout the upper layer of the water column as defined by the thermocline, which was found to be approximately 16 m deep in the early fall (September-October) of 1975.

The model computes excess concentrations (introduced by sludge disposal and superimposed on ambient concentrations) of suspended solids, BOD, DO, coliform bacteria, and various trace metals at 3.1 hour intervals. Changes in concentration with time are caused by advection, diffusion, and, in the case of chemical parameters, decay. In view of the sedimentary orientation of this paper, only the results

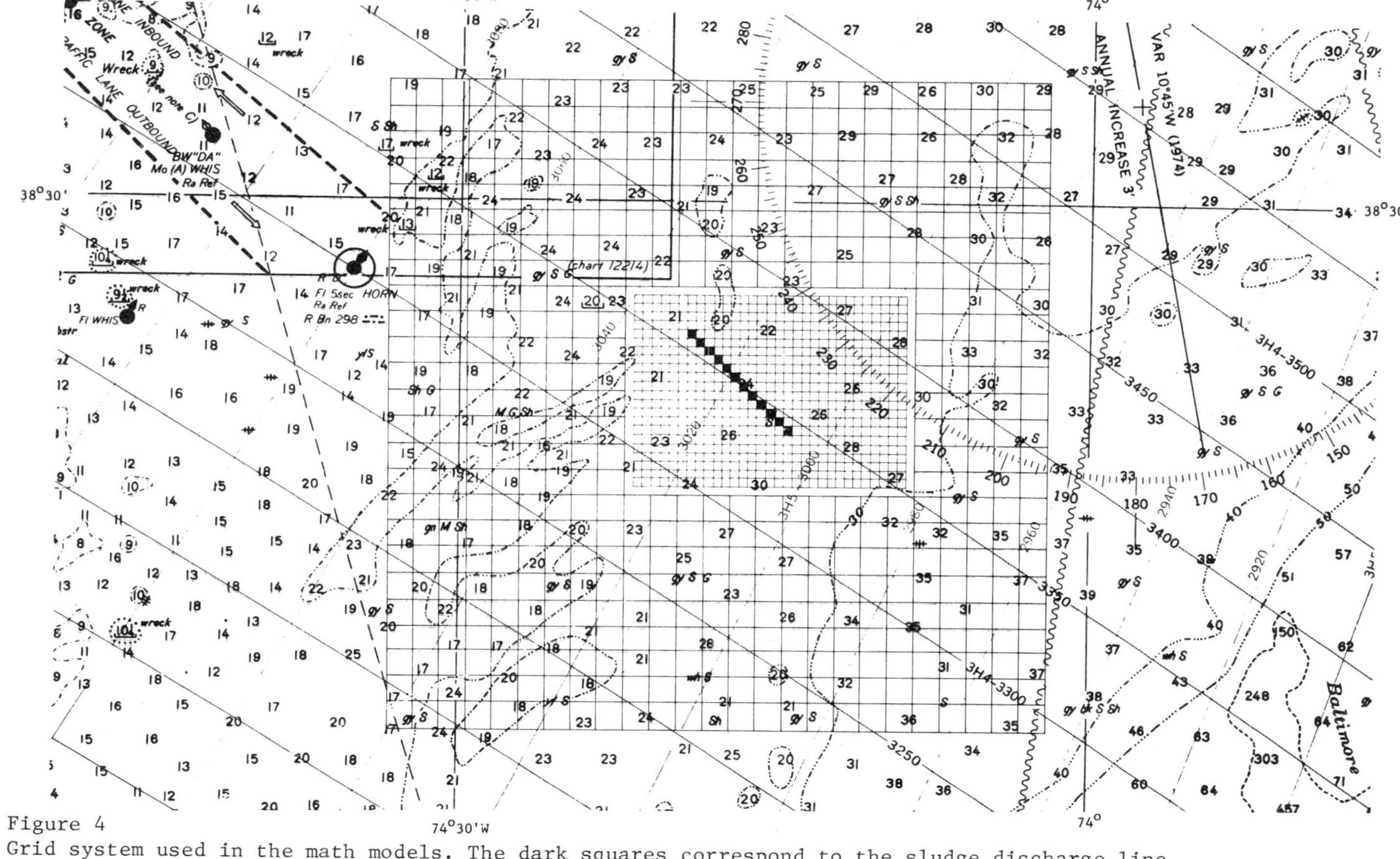

Figure 4
Grid system used in the math models. The dark squares correspond to the sludge discharge line.

pertaining to suspended solids and their deposition are addressed herein.

In order to interpret the model output, appropriate source levels of the substance must be known. The initial concentration of solids in the sludge was assumed to be 17,000 mg/l on the basis of sewage sludge measurements performed by the City of Philadelphia.

The physical characteristics of sludge solids which control settling characteristics, a necessary input to the suspended solids predictions, are at present poorly known. Callaway and others (unpublished manuscript) estimated sludge particle settling rates from nephelometry profiles taken in conjunction with dumping of New York City sewage sludge in the New York Bight. Settling rates estimated from these field trials range from 0.001 to 1 cm/sec, and both of these values have been used in the model to bound the depositional predictive outputs.

Suspended Solids

This paper deals primarily with estimates of sludge deposition, but it is also instructive to examine model predictions of suspended solids. Two conservative assumptions were made in computing suspended solids concentrations: (1) the particles are confined to the upper 16 m of the water column by the thermocline, and (2) no loss of solids occurs by settling through the water column. Thus, the predictions can be considered representative of "worst case" conditions with regard to environmental peturbation.

Figures 5 and 6 show predicted horizontal distributions of suspended solids in the upper layer 12.5 and 100 hrs (1 and 8 tidal cycles) after a sludge release has occurred. By examining the distributions a unit number of tidal cycles after release, rotary tidal current effects are eliminated to reveal net dispersion. The three suspended solids concentrations displayed, 2.06, 0.21, and 0.02 mg/l, correspond to 10, 1, and 0.1 percent dilutions of initial mixing zone concentrations.

After 1 tidal cycle, the suspended solids are distributed in an elongate concentric pattern with the long axis parallel to the discharge line. The solids have dispersed over an area of 153 km^2, and the overall distribution has been shifted southwest of the discharge line by non-tidal currents. Maximum suspended solids concentrations are 2.06 mg/l.

Continuing dispersion has, after 8 tidal cycles, caused the distributional pattern to expand markedly, depart almost completely the disposal site proper, and extend southward of the study area. The solids are distributed over 426 km^2 of the region modeled, and probably twice this value in total. Maximum suspended solids concentrations are now 0.21 mg/l.

It is instructive to compare these conservative predicted concentrations to suspended matter observed in the Mid-Atlantic Bight by Meade and others (1975) in the fall of 1969. Measured "fair weather" suspended solids concentrations decreased from 1.0 mg/l near shore to 0.1 mg/l at the shelf edge; concentrations of approximately 0.5 mg/l occurred near the mid-shelf disposal site. These values increased substantially when storms caused temporary resuspension of bottom sediment. Thus, the predicted concentrations of sludge are in the

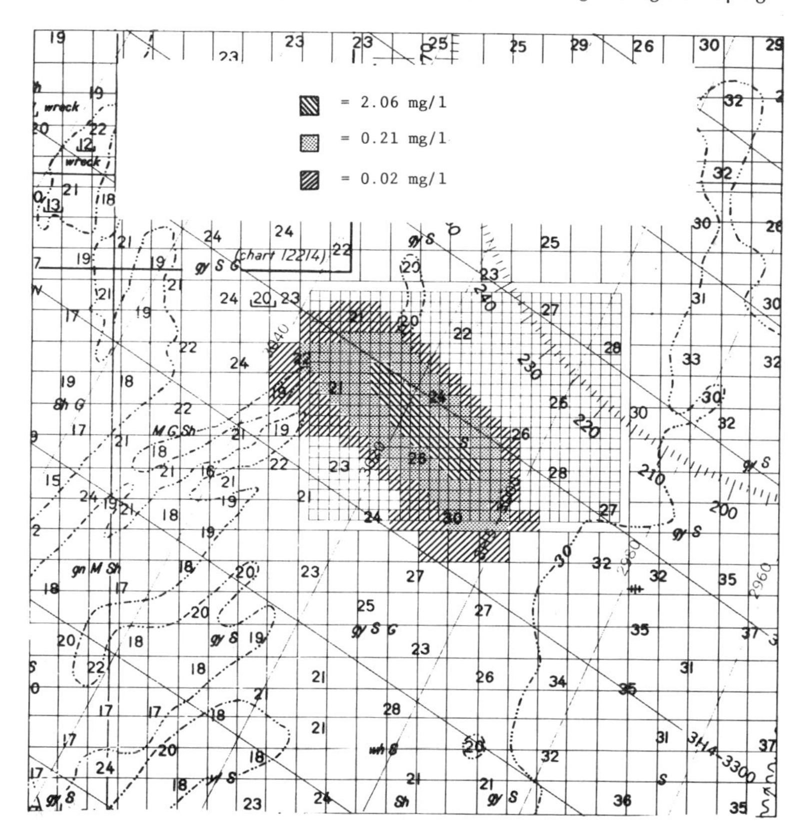

Figure 5
Predicted suspended sludge solid distribution 12.5 hr after discharge.

same order of magnitude as those for ambient "fair weather" suspended matter.

Also pertinent is the relationship of suspended solids transport predictions to the sludge dumping schedule of the City of Philadelphia. Dumps of sludge volumes similar to that modeled are made approximately every four days. The 8 tidal cycle (4 day) distributional pattern shows no suspended sludge solids remaining in the discharge zone, implying that additive effects from repetetive dumping are minimal as regards solids in the water column.

Sludge Deposition

Distributions of sludge particles settling to the sea floor, as predicted by the model, are shown on Figures 7 to 10. The first two distributions are based upon a uniform particle settling rate of 0.001 cm/sec, and the other two assume the settling rate to be 1.0 cm/sec. For each settling rate, the patterns of sludge deposition have been computed for 12.5 and 100.0 hours (1 and 8 tidal cycles) after the discharge has occurred.

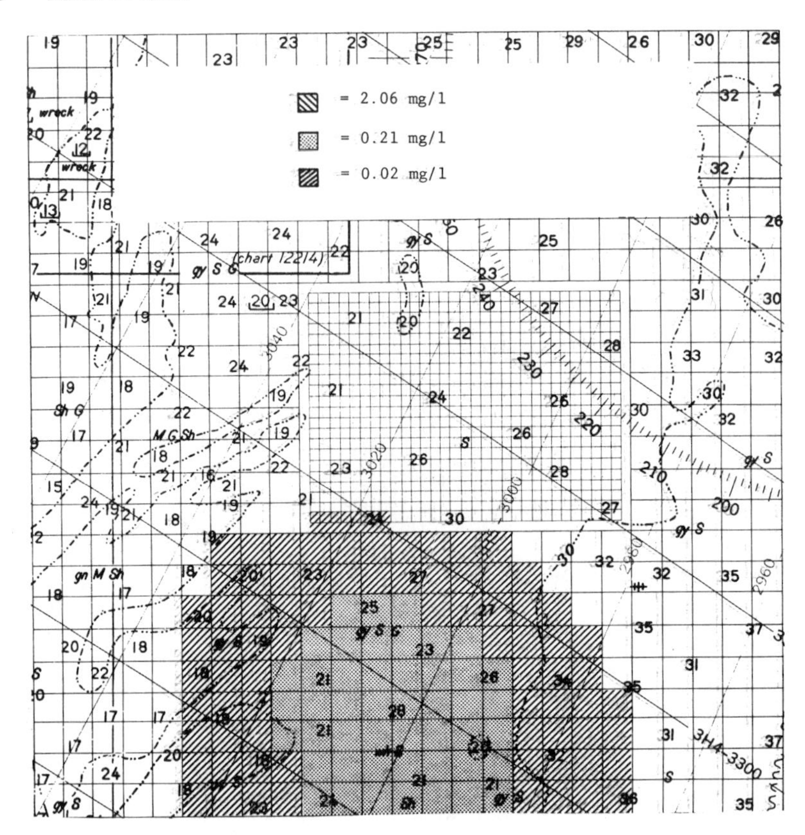

Figure 6
Predicted suspended sludge solid distribution 100 hr after discharge.

Predictions for the 0.001 cm/sec settling rate, appearing on Figures 7 and 8, indicate meager deposition on the order of 10^{-7} g/cm^2. After 100 hours, only .043 percent of the solids dumped have reached the bottom. The depositional pattern extends south-southwestward from the discharge zone in response to non-tidal currents, and given additional time, the pattern would extend well beyond the study area. Computer time constraints did not allow the model to be exercised beyond the 100-hour predictions.

Deposition takes place much more rapidly with the 1 cm/sec settling rate as illustrated on Figures 9 and 10. The same south-southwest displacement from the discharge line develops, but after 100 hours most of the deposition is confined to the dumping site proper. Quantities of sludge deposited are in this case 10^{-5} g/cm^2, and the percentages of total discharged solids reaching the sea floor after 12.5 and 100 hours are, respectively, 25.6 and 76.7.

The total areas impacted by deposition for each settling rate could not be directly calculated because the computer simulation was limited to 100 hours. With the 1.0 cm/sec settling rate, extrapolation of results suggests that all solids will be deposited within a 65 km^2 area 170 hours after the discharge. Particles settling at 0.001 cm/

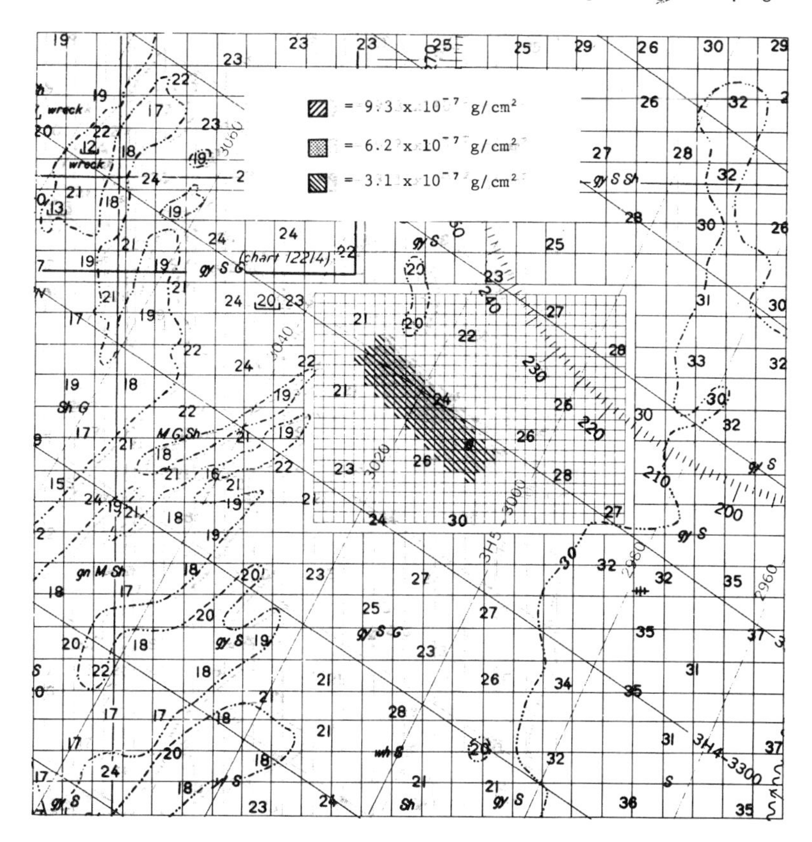

Figure 7
Predicted depositional pattern of sludge solids settling at 0.001 cm/sec 12.5 hr after discharge.

sec will, estimating from available predictions, remain in the water column for weeks and deposit over many thousand square kilometers.

These depositional predictions are based upon a single dumping episode, whereas discharges of this magnitude are made by the City of Philadelphia approximately every 4 days. To place this anthropogenic sediment source in a geological context, the annual weight of sludge solids dumped at the Interim Ocean Disposal Site, 6 x 10^4 kg, is equal to 1.7 percent of the annual suspended load of the Delaware River at Trenton and 0.7 percent of the annual Susqehanna River suspended load at Harrisburg (Curtis and others, 1973). Estimates of sludge accumulation at the sea floor caused by repetitive dumping can be developed easily if resuspension and biological consumption are ignored. Hypothetical sludge spheres with a settling rate of 1 cm/sec and a specific gravity of 1.5 have a diameter of 0.28 mm. Given the 30.9 x 10^{-5} g/cm^2 accumulation rate per single dump and a dumping frequency of 4 days, that portion of the bottom subjected to this accumulation rate would be covered by a 1 grain thick layer of 0.28 mm sludge spheres in approximately 9 months.

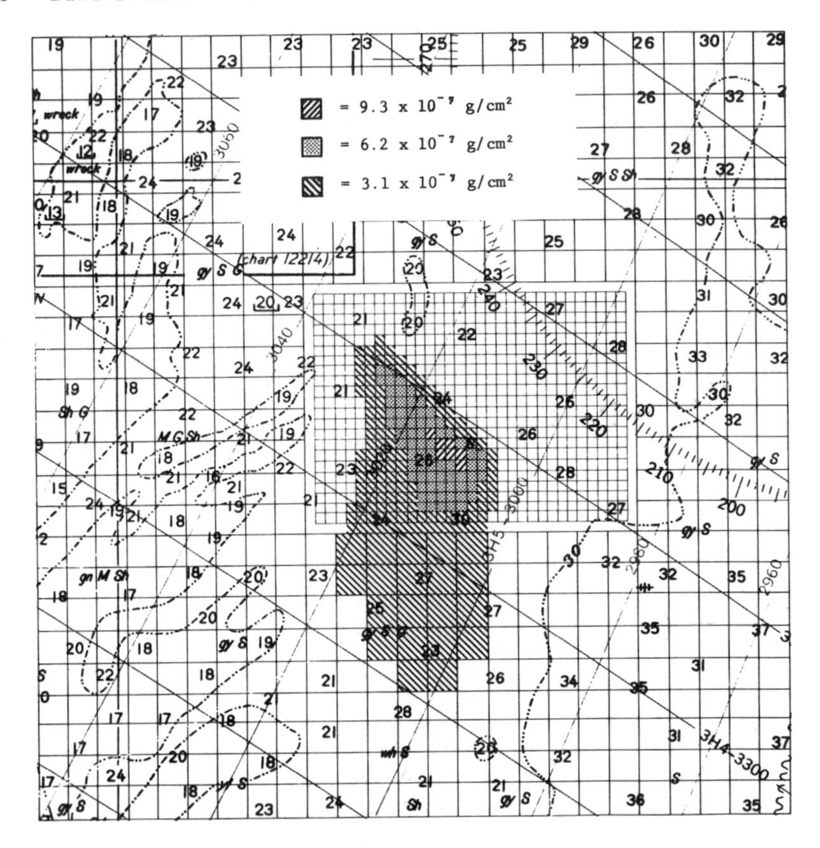

Figure 8
Predicted depositional pattern of sludge solids settling at 0.001 cm/sec 100 hr after discharge.

DISCUSSION

Having demonstrated application of the Waste Dispersion Model to predict geological implications of sewage sludge dumping on the continental shelf, the validity of these predictions must be examined. The accuracy of any mathematical simulation of environmental processes is directly related to the validity of input assumptions. The assumptions incorporated in this modeling effort fall into three categories: shelf hydrography, sludge characteristics, and pre/post depositional sludge fate. An evaluation of these areas is presented below.

Shelf Hydrography

The model was adjusted to simulate early fall hydrographic conditions on the Middle Atlantic Bight mid-shelf area. Measured tidal heights were used to drive the circulation model, and predicted currents corresponded closely to the southwesterly drift known to characterize this mid-shelf region most of the year. Diffusion coefficients were not measured at the study site, but values which were

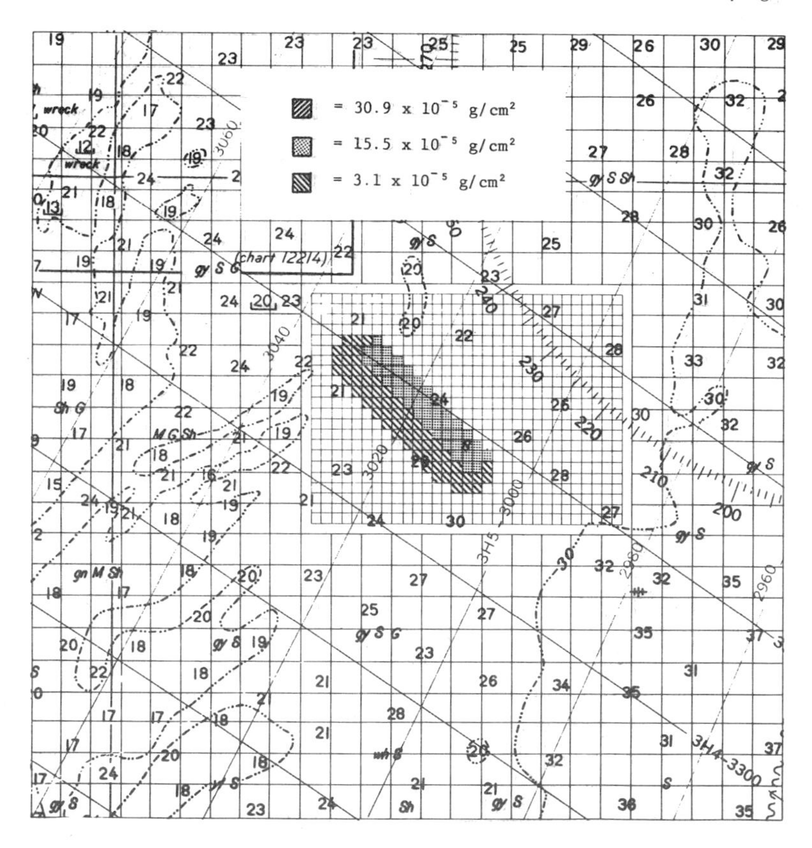

Figure 9
Predicted depositional pattern of sludge solids settling at 1.0 cm/sec 12 hr after discharge.

extracted from the literature had been validated in similar environmental settings. The seasonal thermocline is known to confine dumped wastes to the upper layer of the water column, leading to significantly higher concentrations than would occur if the shelf water was in its well-mixed mode. In summary, the model successfully simulated the typical flow regime and the worst-case stratification at the site. Other conditions could undoubtedly have been simulated with equal success by inputting appropriately representative field data. From a modeling standpoint, it was fortunate that the Middle Atlantic Bight hydrography has been reasonably well defined by previous studies so that an extensive field program was not necessary to define typical or worst case conditions at the site.

Sludge Characteristics

Although the chemical character of digested sewage sludge has been well-documented, the physical characteristics of sludge solids are poorly known. Two bounding settling rates were used in the model, but in reality the solids would exhibit a spectrum of settling rates.

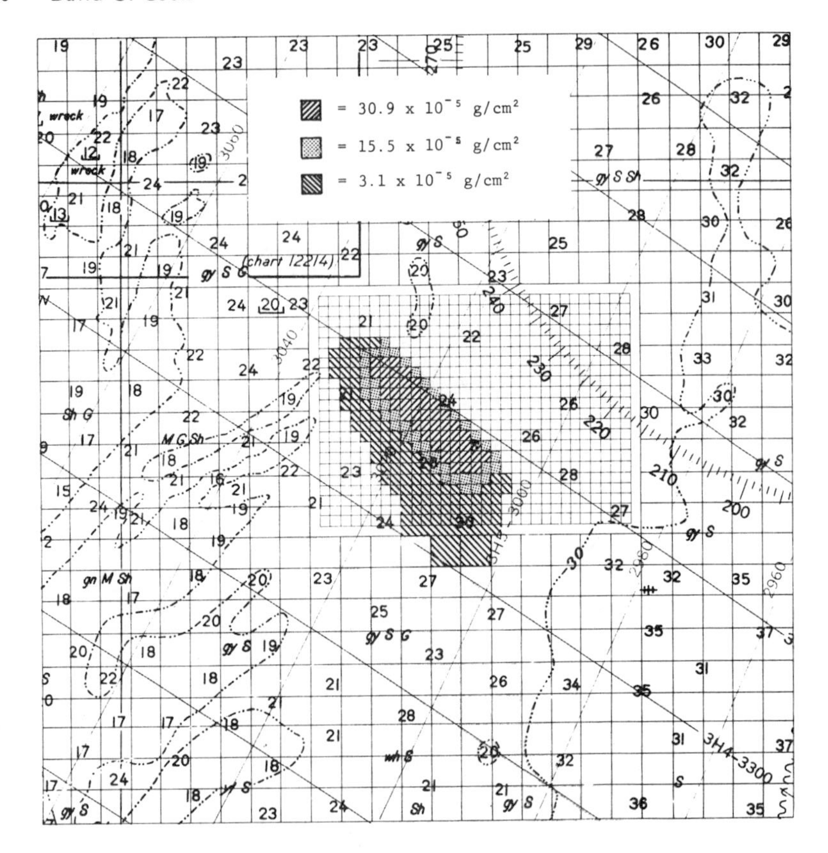

Figure 10
Predicted depositional pattern of sludge solids settling at 1.0 cm/sec 100 hr after discharge.

The settling characteristics of finer solids would also vary markedly with seasonal factors such as water temperature and stratification. The potential for flocculation is another unknown. While this information gap may not heavily impact the suspended solids predictions, depositional patterns can only be bounded by the model at the present time. It is obvious that the physical nature of sewage sludge is an area of needed research.

Pre/Post Depositional Sludge Fate

The most tenuous assumptions incorporated in the depositional estimates concern the absence of biological recycling of sludge solids and resuspension after deposition. It is known that certain zooplankton and microzooplankton feed on organic particles in the size range exhibited by sludge solids. This could conceivably cause significant recycling of the sludge as fecal pellets, particularly the finer sludge solids which have a long residence time in the water column. Once the sludge reaches to sea floor, it is subject to similar recycling by benthic detritus feeders. While the potential for such biological effects on suspended and deposited sludge is recognized, the meager data available on feeding rates does not permit quantification of these effects. Research is required here also before the depositional process can be simulated with confidence.

Sludge solids reaching the sea floor and escaping detritus feeders are subject to resuspension by currents. For the coarser of the hypothetical sludge particles used in the model, the threshold velocity is approximately 20 cm/sec; the finer particles could require higher velocities if flocculation has not occurred and the work of Hjulstrom (1939) obtains. Normal tidal currents at the site do not exceed 20 cm/sec. However, substantially stronger alongcoast flows can result under storm conditions (Beardsley and Butman, 1974), and storm waves are capable of stirring the sand bottom at 40 to 50 m depths. Evidence for resuspension of settled sludge is available in the form of bottom photographs from the Interim Ocean Disposal Site vicinity. Examples of bottom photos taken in Fall 1975 as part of a biological study are shown on Figure 11. Both current-generated ripple marks and biological structures (mounds, burrows, trails) are present on the predominately sand bottom. Some organic matter which is either sludge-related or of biologic origin appears to have accumulated in ripple troughs, but in this area where the City of Philadelphia has dumped sewage sludge since 1973, no carpet of sludge exists. This conclusion is supported by the submersible observations of Folger (1978) at the disposal site which revealed no obvious sludge deposition on the sea floor in August 1974 and July 1975. Undoubtedly the bottom here undergoes infrequent vigorous storm-related mixing which resuspends fines, separated by long periods of quiessence when biological reworking is dominant and fines temporarily accumulate in microtopographic lows. In view of the resuspension process, the sludge depositional patterns predicted by the model may be most relevant to delineating zones where sludge solids may become available to benthic detritus feeders during quiessent periods.

ACKNOWLEDGMENTS

The math modeling and field work described in this paper were performed by Raytheon Company, Oceanographic & Environmental Services under contract to the City of Philadelphia Water Department. The field measurements were conducted using Raytheon's research vessel SUB SIG I, and contributions of Robert Lobecker, C. Frederick Willett, and Paul Higley to the field program are gratefully acknowledged. Overall technical guidance was provided to the study by Dr. Geraldine Cox, Raytheon's program manager, and by Mr. Steven Townsend of the Philadelphia Water Department.

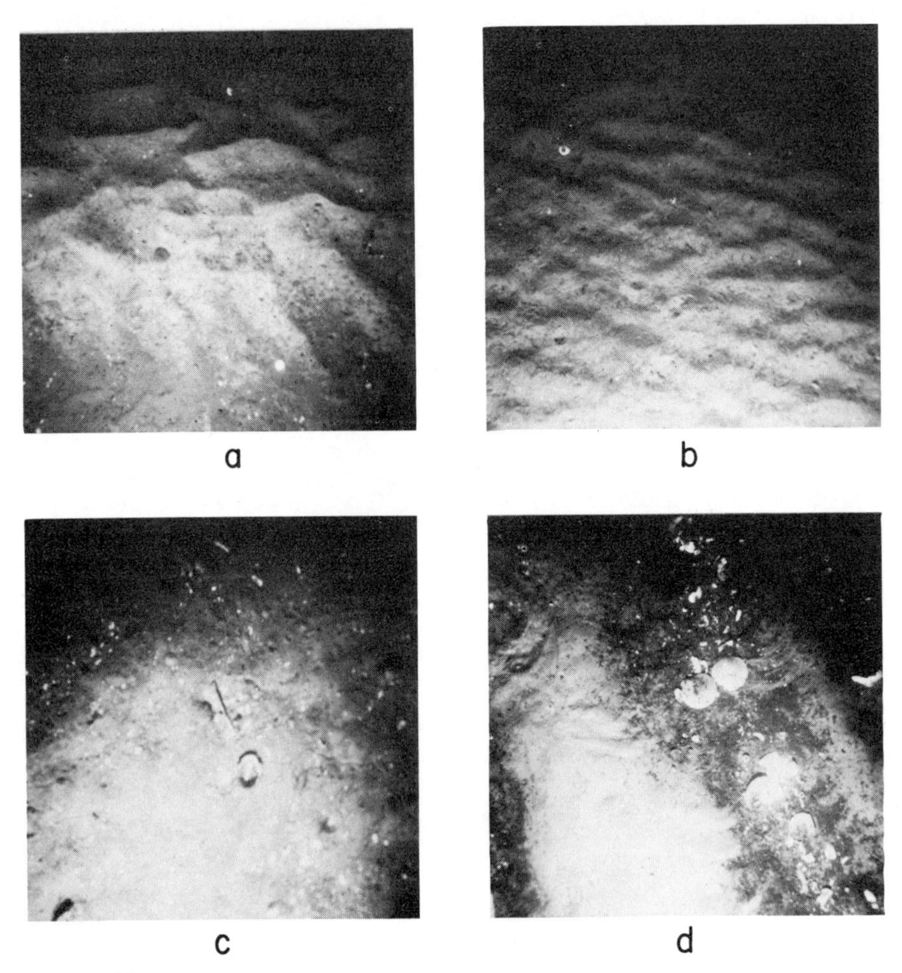

Figure 11
Bottom photographs (1.5 x 1.5 m) taken in the interim ocean disposal site vicinity during fall 1975. a) relatively fresh ripple marks; b) ripples partially degraded by biological activity; c) biological mounds and burrows; d) large ripple marks with clam shells and organic debris (not necessarily sludge) in troughs.

REFERENCES

Beardsley, R.C. and Butman, B., 1974. Circulation on the New England continental shelf: Response to strong winter storms. Geophys. Res. Lett., 1:181-184.

Beardsley, R.C., Boicourt, W.W., and Hansen, D.V., 1976. Physical oceanography of the Middle Atlantic Bight, in Middle Atlantic Continental Shelf and New York Bight, vol 2. M.G. Gross ed. Am. Soc. Limnol. Oceanogr. Special Symposium. pp 20-34.

Bumpus, D.F., 1973. Physical oceanography, in Coastal and off-shore environmental inventory, Cape Hatteras to Nantucket Shoals; S. Saila ed. Marine Publication Series No. 2, Univ. of Rhode Island, Kingston, R.I.

Bumpus, D.F., 1965. Residual drift along the bottom on the continental shelf in the Middle Atlantic Bight area. Limnol. and Oceanogr. Suppl. to 10:R50-R-53.

Callahan, J.D. and Wickramaratne, P.J., 1974. A finite difference effluent dispersion model. Unpublished manuscript, Raytheon Oceanographic & Environmental Services, Portsmouth, R.I.

Callaway, R.J., Teeter, A.M., Browne, D.W., and Ditsworth, G.R., no date. Preliminary analysis of the dispersion of sewage sludge discharged from vessels to New York Bight Waters. Unpublished manuscript, U.S. Environmental Protection Agency, Corvallis, Ore.

Curtis, W.F., Culbertson, J.K., and Chase, E.B., 1973. Fluvial-sediment discharge to the oceans from the conterminous United States. U.S. Geol. Survey Circular 670, 17 pp.

Folger, D.W., Palmer, H.D., and Slater, R.A., 1978. Two waste disposal sites on the continental shelf off the Middle Atlantic states--observations made from submersibles. In this volume.

Hjulstrom, F., 1939. Transportation of detritus by moving water, in recent Marine Sediments, P.D. Trask ed., Am. Assoc. Petrol. Geol., pp 5-31.

Koh, R.C.Y. and Chang, Y.C., 1973. Mathematical Model for barged ocean disposal of wastes. U.S. Environmental Protection Agency Rep. No. EPA 660/2-73-029.

Leendertse, J.J., 1967. Aspects of a computational model for long period water wave propagation. Rand Corp. Report No. RM-5294-Pr. Santa Monica, CA.

Leendertse, J.J., 1970. A water quality simulation model for well-mixed estuaries and coastal seas: volume I, principles of computation. Rand Corp., Report No. RM-6230-RC. Santa Monica, CA.

McLennen, C.E., 1973. New Jersey continental shelf near bottom current meter records and recent sediment activity. Jour. Sed. Petrology 43:371-380.

Meade, R.H., Sachs, P.L., Manheim, F.T., Hathaway, J.C., and Spencer, D.W., 1975. Sources of suspended matter in waters of the Middle Atlantic Bight: Jour. Sed. Petrology 45:171-188.

Palmer, H.D. and Lear, D.W., 1973. Environmental survey of an interim ocean disposal site. Westinghouse Corp. Report on contract 68010481. Annapolis, MD.

Swift, D.J.P., Duane, D., and McKinney, T.F., 1973. Ridge and swale topography of the Middle Atlantic Bight, North America: secular response to the Holocene hydraulic regime. Mar. Geol. 15: 227-247.

Distribution of Suspended Particulate Matter near Sewage Outfalls in Santa Monica Bay, California

Ronald L. Kolpack

ABSTRACT

Between 1961 and 1976 more than 12 x 10^8 l/day of primary and secondary-treated sewage effluent and 19 x 10^6 l/day of sludge were discharged into Santa Monica Bay in southern California. During 1971 the amount of total suspended solids discharged from the effluent and sludge outfalls was about 6.1 x 10^4 metric tons. No significant accumulation of particulate material around the discharge sites has been detected in precision-depth and high-resolution seismic-reflection profiles.

Distribution and transport paths of sewage particles within the water column on the Santa Monica Shelf were studied with more than 500 light-transmission profiles made on eight cruises between September, 1970 and November, 1972. Additional measurements of temperature, salinity and bottom currents were used to interpret the distribution of suspended matter in the water column.

The sewage plume in Santa Monica Bay rises rapidly from an initial discharge depth of about 60 meters to about 20 meters below the water surface. On occasion the sewage effluent can be detected at the water surface. Wave action in the water column tends to disperse the sewage plume over a large part of the Santa Monica Shelf. The dominant direction of initial transport during most of the year is toward the shoreline southeast of the discharge site. Subsequently, at several reasonably consistent locations within the bay the net transport of the sewage material, as well as material derived from nearshore mixing, is directed toward the west and transported offshore in well-defined zones which are approximately normal to the coastline. The amount of suspended material transported to the offshore area via Santa Monica and Redondo canyons appears to be rather negligible during normal conditions. On the other hand, a significant portion of the suspended material in Santa Monica Bay is transported within the near-surface water column toward the offshore basins. This pattern of particle transport also takes place at the water surface and at the sediment-water interface, although there is somewhat more variation near these surfaces.

INTRODUCTION

Disposal of large quantities of municipal wastes is a common environmental problem in many large urban areas. In southern California this problem has been handled by discharging sewage via ocean outfalls. Santa Monica Bay has received large amounts of waste substances. During the past 15 years, for example, about 12×10^8 l/day of primary and secondary-treated sewage effluent and an additional 19×10^6 l/day of sludge have been discharged from the Hyperion outfalls near the head of Santa Monica Canyon (Fig. 1).

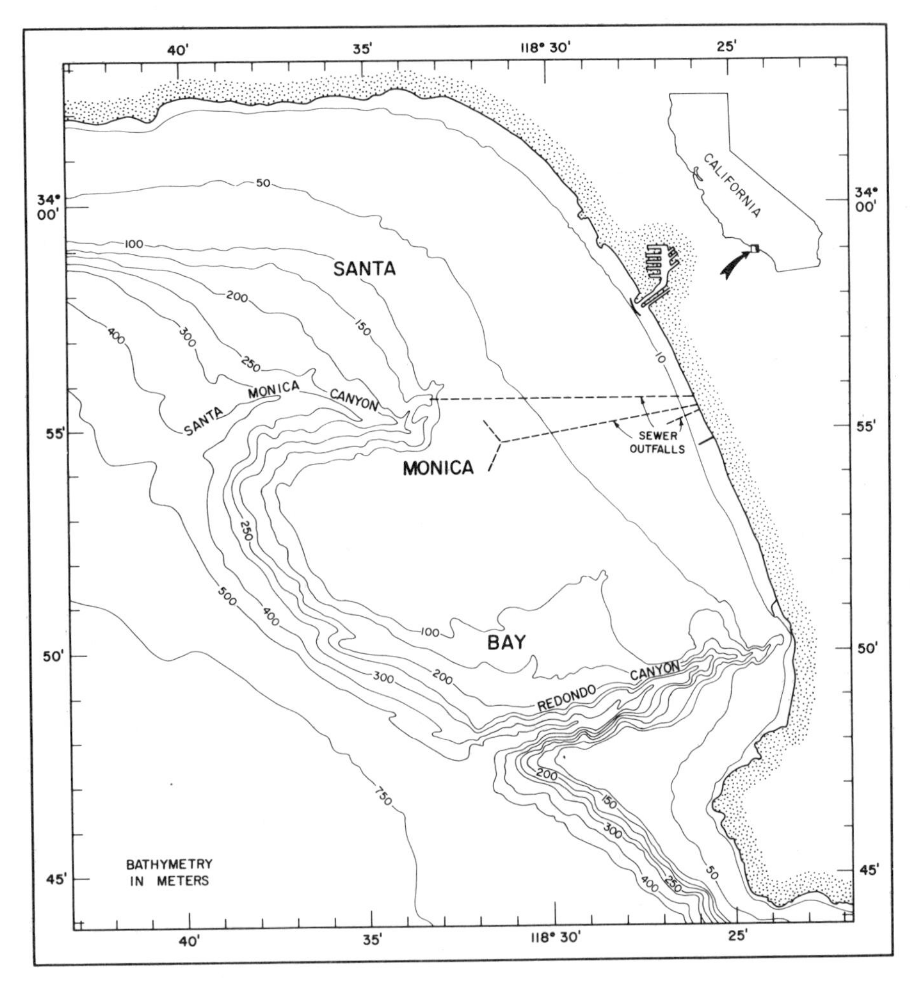

Figure 1
Location map and generalized bathymetry of Santa Monica Bay. The sludge disposal line, shown by the upper dashed line, terminates at the head of Santa Monica Canyon and the middle dashed line depicts the effluent outfall line. The original outfall, shown by the short dashed line, is used once a week for a brief period of time.

Effluents discharged into Santa Monica Bay contain a variety of materials such as oils and greases, trace metals and pesticides, in addition to large quantities of suspended solids (SCCWRP, 1973). Some of these substances, such as pesticides and certain heavy metals, are known to have adverse effects on some marine organisms when the concentration in organisms reaches a threshold level (Merlini, 1971; Føyn, 1971; National Academy Sciences, 1975). Furthermore, organisms such as some birds which derive their food from the sea have exhibited adverse effects that are attributable to the discharge of organic and inorganic wastes from human activity (Risebrough, 1971; Ketchum, 1972). Heavy metals and pesticides are often moved with suspended matter from sewage effluent owing to the large surface area to volume ratio of the particles (Gross, 1972). The large amount of organic material is also of concern because its degradation consumes large amounts of oxygen. If the organic material accumulates on the bottom at a rate that exceeds the local degradation capacity, anoxic conditions may result which restricts or alters the benthic fauna (Richards, 1971).

Reconnaissance high-resolution seismic profiling surveys in the area did not reveal what could be regarded as a significant accumulation (> 0.5m) of material in the vicinity of the outfalls, and inspection of bottom samples likewise did not reveal large areas of anoxic sediments. The fate of this relatively steady discharge of particulate material within a small portion of the inner shelf area of Santa Monica Bay for a long time became a matter for scientific study. Consequently, an investigation utilizing light transmission measurements was carried out to determine the behavior of the sewage plume in the water column and the pathways of transport leading to the areas of deposition. This study is an attempt to acquire a better understanding of the dispersal pathways for the sewage effluent discharged in Santa Monica Bay so that additional research efforts might be focused on distinct aspects of the system.

STUDY AREA

Santa Monica Bay is located along a northwest-southeast trending coastline in southern California. It is bordered by land on the eastern and northern margins and protected from ocean swell from the northwest by the northern Channel Islands and from the south by Santa Catalina Island. Sediment from the mainland comes from several small streams that drain the Santa Monica Mountains along the northern side of the bay, a few flood control channels, and erosion of coastline deposits. A former source of sediment supply from the Los Angeles River was diverted to the San Pedro Bay upon completion of a diversion channel in 1884 (Troxell and others, 1942). Thus, the major source of material to the bay at the present time is the municipal waste from the Hyperion sewage treatment plant.

Sewage is discharged through two outfalls. One outfall, for the disposal of sludge material, terminates at the head of the Santa Monica Canyon. The second outfall, for the disposal of primary and secondary-treated effluent, terminates on the Santa Monica Shelf in a Y configuration diffuser, eight kilometers offshore. The discharge from this outfall is from a series of diffuser outlets along the 2400 meters length of the two arms of the Y (Pomeroy, 1960).

Santa Monica Shelf is cut by two large submarine canyons. Santa Monica Canyon, in the central portion of the area, terminates about ten kilometers offshore. North of this area the shelf is about 8-10 kilometers wide and parallels the mainland coastline until it narrows at the northwestern part of the bay where it is incised by a canyon off Point Dume. At the southern end of Santa Monica Bay the shelf is incised by Redondo Canyon, which terminates within a few hundred meters of the coastline. The widest part of the Santa Monica Shelf is located between Redondo and Santa Monica canyons (Fig. 1). Surficial sediments in portions of this area include gravel, relict deposits from periods of lower sea level, authigenic sediments and coarse-grained detrital sediments (Emery, 1960; Kolpack, unpublished data).

PREVIOUS WORK

Before 1950 most of the studies in Santa Monica Bay involved work on some aspect of geology such as sediment distribution, near-shore processes, structure and submarine topography (Shepard and Macdonald, 1938; Shepard and Emery, 1941; Kerr, 1938). Following this period, however, plans for construction of the two new Hyperion outfalls stimulated additional interest in the area and a series of geological, biological and oceanographical studies were completed (Stevenson and others, 1956; Allan Hancock Foundation, 1965; Emery, 1952; Hartman, 1956; Tibby, 1960; Hartman, 1960; Bandy and others, 1965). Studies of suspended sediments were made by Rodolfo (1964), Wildharber (1966), Beer (1969), Drake and Gorsline (1973) and Drake (1974). The emphasis of the latter studies was primarily on the distribution and transport of particles in the vicinity of Redondo Canyon at the south end of Santa Monica Bay. Except for Drake and Gorsline (1973) and Drake (1974), the other workers studied the suspended material by filtering water samples.

Previous oceanographic studies related to the Hyperion project (for example, Stevenson and others, 1956; Allan Hancock Foundation, 1965; Tibby, 1960) provided information about water clarity and movement of water at the water surface and in the upper 20-25 meters of the water column. A major reason for this emphasis, with regard to water clarity, was the limitation imposed by the instruments available for that work.

METHODS

Field work for this study was carried out in conjunction with other projects. Generally all stations during a cruise were occupied without interruption once the measurements started to obtain quasi-synoptic measurements. Since the objective of this work was to determine distributions and transport paths of material discharged from the Hyperion outfalls, emphasis was placed on obtaining continuous vertical profiles of light transmission, using a continuously recording beam transmissometer with a maximum depth capability of 10^3 meters. The instrument was designed by the Visibility Laboratory of Scripps Institution of Oceanography (Petzold and Austin, 1968; Tyler and others, 1974).

The transmissometer utilizes a 20 mm diameter current-regulated

20-W white light source with a folded-path length of 1 meter to measure light attenuation. Power is supplied to the instrument by conductor cable from a 110 V source aboard ship. Return signals for light transmission and temperature measurements, with respect to depth, are also supplied by conductor cable from the beam transmissometer to an on-board recorder so that real-time data can be inspected and preserved for further processing and analysis.

No attempt was made to relate light transmission values to particle concentrations. Consequently, the light transmission values presented in this study should be regarded as measurements of the amount of inorganic and organic particulates as well as dissolved substances in the water. This is especially true of the area in the immediate vicinity of the outfalls. In the southern portion of Santa Monica Bay, however, studies by Drake and Gorsline (1973) and Drake (1974) show a consistent correlation between light transmission values and total suspended particle concentrations. Their data indicate that in the vicinity of Redondo Canyon 10% T/M (Transmission/meter) is equivalent to approximately 5 mg/l total suspended particles and 80% T/M is equivalent to between 0.2 and 0.3 mg/l total suspended particles.

Current measurements and salinity-temperature-depth (STD) profiles were made during the study. A Hydro-Products Model 502 current meter was mounted in a tripod with the rotor and vane one meter above the bottom. Salinity and temperature profiles were made by Plessey 9040 and 9060 STD instruments. Three-dimensional isometric projection plots were generated with the SYMVU computer graphics program developed by the Laboratory for Computer Graphics and Spatial Analysis at Harvard University and run on an IBM 370/158 computer.

RESULTS

Light transmission profiles were measured at stations in two closely spaced grids around the Hyperion outfalls in order to determine the behavior of the sewage plume in the water column immediately after its release from the diffuser ports. One series of 56 stations was occupied in a radial pattern of eight traverses radiating outward from the southern arm of the Y diffuser of the effluent outfall. Following this, a second series of 48 stations was occupied in a circular pattern around the outfalls (Fig. 2). Each series of light transmission profiles was completed within a 12 hour period and all environmental conditions were reasonably constant during the entire sampling period.

In most instances, current meter data from stations occupied on the inner shelf in southern California show a good correlation between tidal period and the direction of water current flow. However, during the time the stations mentioned previously were occupied, data from a current meter positioned one meter off the bottom at a depth of 30 meters (Fig. 2) showed an almost steady current flow toward the southeast. During the 24 hour period preceding the time the stations shown in Fig. 2 were occupied, the current direction was exceptionally stable and varied less than 10° from 110°. At the time the circular grid of stations was occupied, the bottom current flow was somewhat less stable and ranged from 85-130° with the average direction of current flow also at 110° from true north.

Light transmission values from the circular grid of profile stations were contoured in unfolded cross sections to outline the

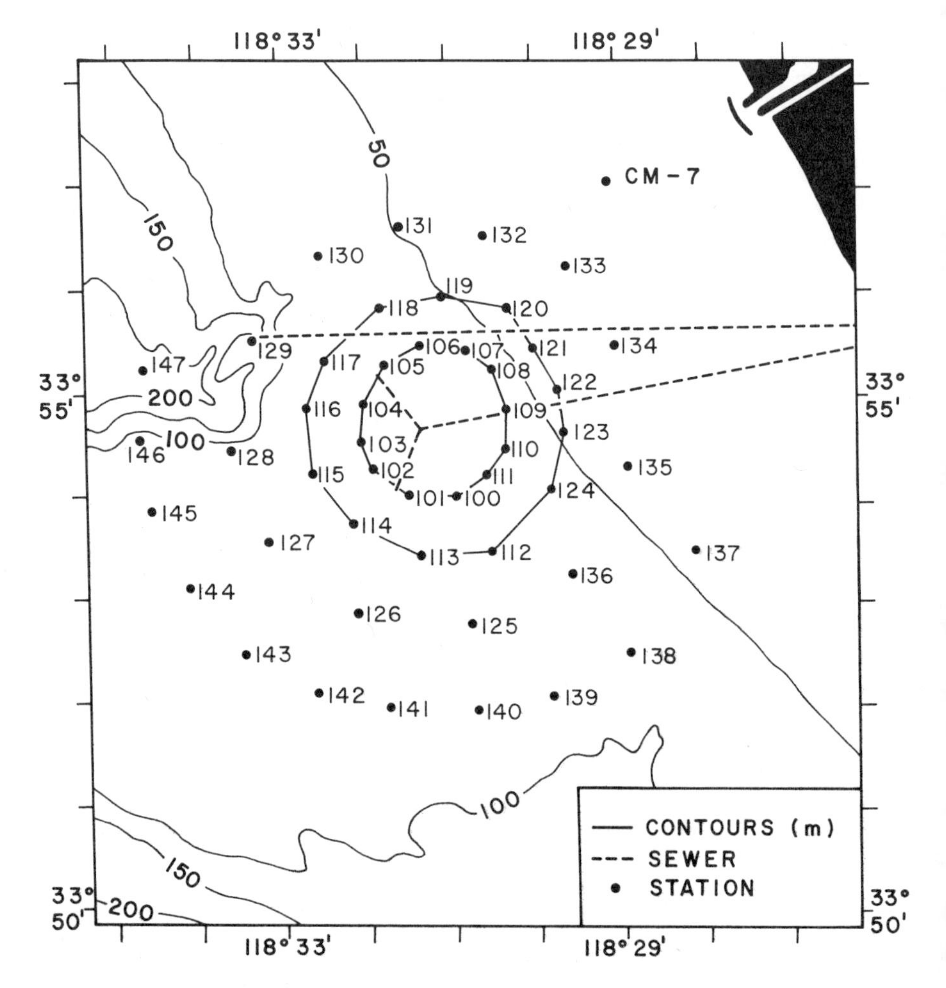

Figure 2
Station locations for a 12 hour survey of light transmission values in the vicinity of Hyperion outfalls (November 3-4, 1972). The first two digits (18) of all station numbers were omitted in this diagram for the purpose of simplification. A current meter was positioned one meter off the bottom in 30 meters of water at station CM-7 during the survey.

main sewage plume within the water column. Profiles from the innermost ring of stations (Fig. 3) show that the plume rises rapidly from a depth of approximately 60 meters to a depth of about 30 meters. For instance, at station AHF 18105, the station nearest the end of the northern Y of the effluent outfall, the lowest light transmission values are at a depth of 30-35 meters. The top of the plume at this station is at a depth of about 20 meters; whereas, the water at a depth greater than 40 meters is clearer and the bottom of the main plume can be designated to be at this depth. This characteristic behavior is a result of the buoyancy effect produced by the influx of

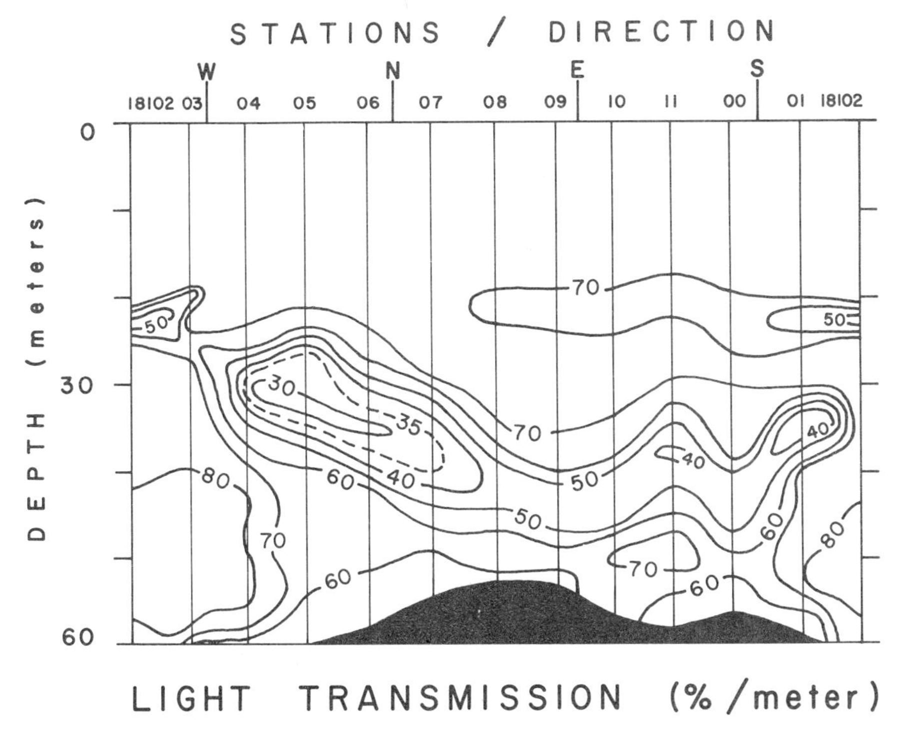

Figure 3
Cross section of light transmission values from innermost circle of stations shown in Fig. 2. The northern end of the Y diffuser is near station 18105 and the southern end is near station 18101.

the warmer fresh water used to dilute and transport the sewage effluent to its discharge point. This cross section also shows that much of the material discharged from the effluent outfall was transported toward the northwest during the study period. Profiles from the second circle of stations (Fig. 4) more clearly define this trend and in addition show that some materials discharged from the sludge outfall at the head of Santa Monica Canyon also move toward shore. The latter feature is outlined by a turbid lobe around a depth of 50 meters at station 18117. It is also interesting to note that a lobe of clear water overrides the plume from the sludge outfall and appears to penetrate the plume from the effluent discharge, causing a bifurcation of the latter plume. This phenomenon is more clearly illustrated in the light transmission profiles (Fig. 5) which show the development and decay of several lobes of turbid water with increasing distance from the effluent outfall.

Temperature and salinity measurements show that water density is an important factor in determining the distribution of sewage material within the water column. For example, a water mass with a temperature of 10-11°C and a salinity of 33.70-33.80 o/oo appears to move into the shelf area from the vicinity of Santa Monica Canyon. Since surveys in the area before the outfalls were operative (Stevenson and others, 1956) did not record a similar water mass, it seems reasonable to conclude that the elevated temperature of this water mass is a result of the discharge from the sludge outfall at the head of Santa Monica Canyon. Although the main portion of the sewage plume exhibits numerous instabilities in water density, there is a tendency for layering in the water column which can be attributed primarily to variations in salinity around the outfall area.

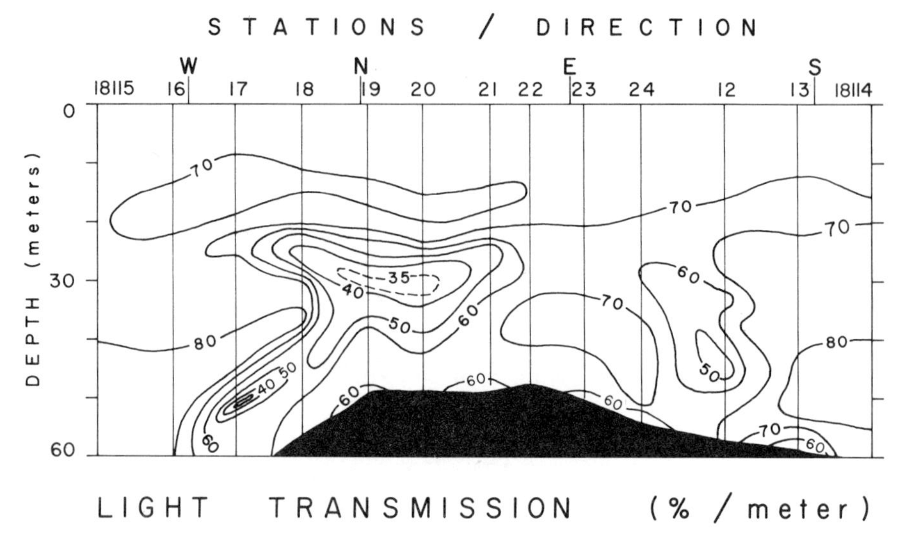

Figure 4
Cross section of light transmission values from second circle of stations shown in Fig. 2. The main effluent field is located at a depth of about 30 meters near stations 18119-20. The low light transmission values at a depth of about 50 meters near station 18117 are from the sludge discharge.

Many lenses of turbid water are associated with variations in water temperature, as reported in other studies (for example, Drake and others, 1972). However, in this situation the vertical structure of the turbid lobes appears to be a result of a combination of effluent particle density, dilution of the ocean water by the fresh water in the effluent discharge, and thermal discontinuities. Since the sewage effluent is introduced at the bottom of the water column and subsequently rises, the density control on stratification of the particulate material is occasionally at the top of the turbid lobe rather than at the bottom, as is the normal case for particles which settle after being introduced at or near the water surface. Separation of the main turbid lobe into several lobes with increasing distance from the source indicates that there is a significant segregation of material based on particle density. It is reasonable to expect such a segregation because the particles that are discharged range from low-density material commonly referred to as "floatables" to high-density particles with specific gravities more than twice that of sea water.

Underwater photographs taken near other southern California outfalls show considerable turbulence at the point of sewage discharge (Anonymous, 1973; Palmer, 1977). This turbulence assists in keeping the particles in suspension for a short time, but presumably becomes less effective as the sewage field is modified by advection and diffusion. Consequently, low-density particles continue to rise toward the water surface and high-density particles tend to settle out. This segregation of the sewage field in response to the vertical stratification in the water column is outlined by the sequence of light transmission profiles shown in Fig. 5.

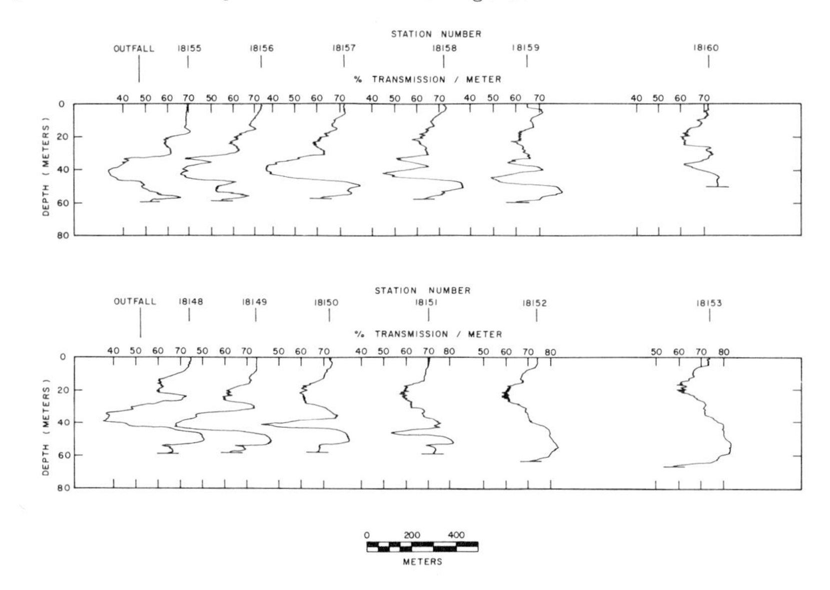

Figure 5
Light transmission profiles with respect to increasing distance from the southern arm of the Y diffuser. Upper series of stations in southwest (210°) direction and the lower series in southwest (240°) direction. Each series of stations occupied within two hour time interval on November 4, 1972. Horizontal line at bottom of each profile represents bottom contact.

Dispersion of the sewage field is of paramount importance for evaluating areas of possible environmental impact. Therefore, the available light transmission measurements from the area immediately around the discharge, as well as the surrounding shelf area in Santa Monica Bay, were plotted in a variety of forms to show the distribution of the sewage plume. During the survey on November 3-4, 1972, which was focused on the area in the immediate vicinity of the effluent discharge, the lowest light transmission values in the water column were located around the northern Y of the outfall. Contours of the minimum light transmission values for all stations during the survey show that the plume moved shoreward and tended to branch near the shoreward station control (Fig. 6). In the vertical dimension,

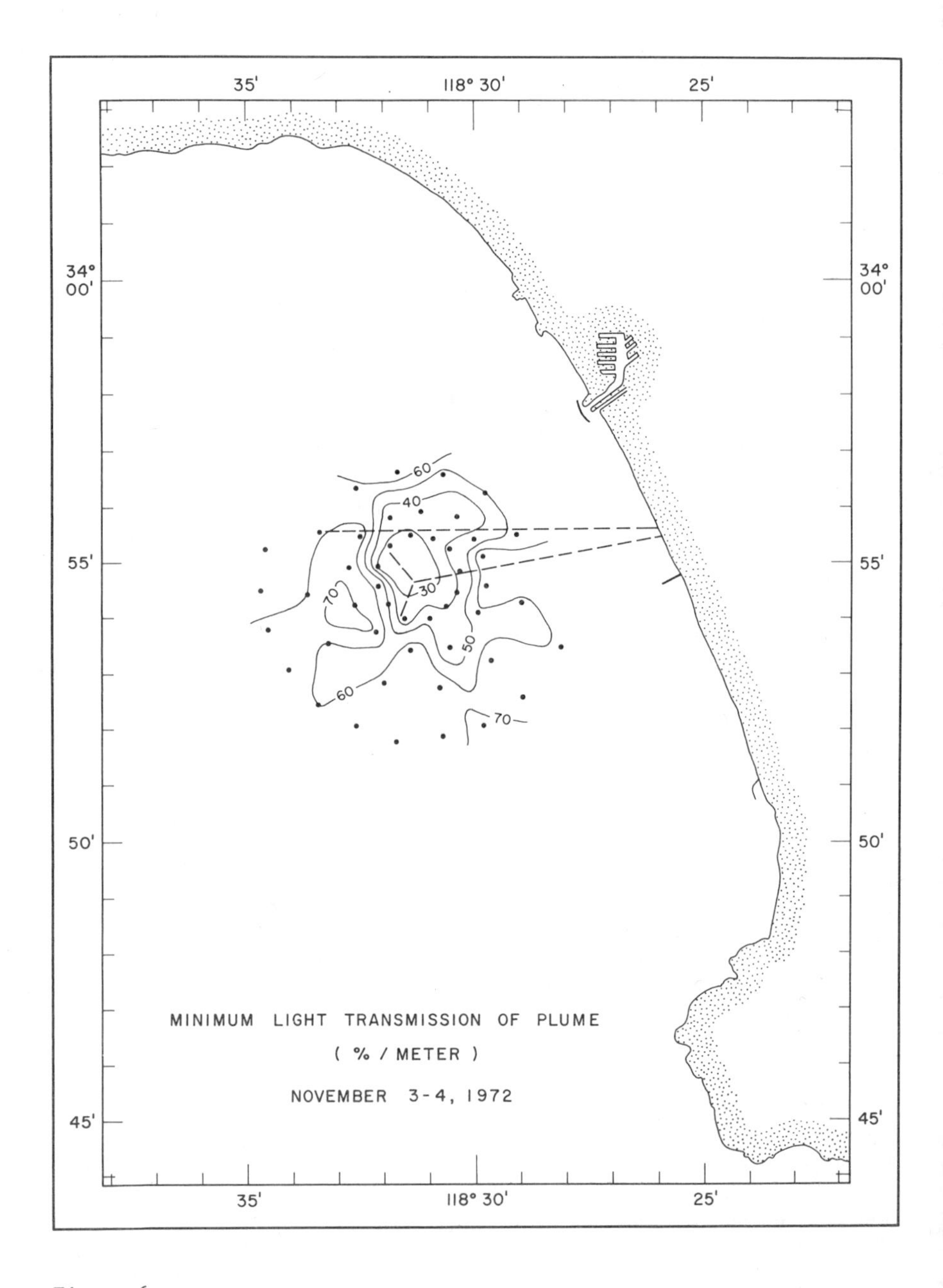

Figure 6
Distribution of light transmission values from the main sewage plume. Contours represent minimum light transmission values in the plume.

the plume was at its greatest depth just east (landward) of the discharge area and then tended to rise in the water column with increasing distance away from the deepest portion of the plume (Fig. 7). However, the ascent of the main portion of the plume was arrested at a depth of about 20 meters below the water surface. Three-dimensional plots of water turbidity and the depth to the middle of the plume (Fig. 8) also suggest that after the plume becomes stabilized at a distance ranging from 1.0 to 2.5 kilometers from the discharge, the vertical location of the suspended portion of the sewage effluent is subject to modification by wave action in the water column. This observation will receive greater attention in a subsequent portion of this discussion when the characteristics of the plume over a larger area, such as the survey during November 1-4, 1971, are evaluated.

The distribution of light transmission values for the surface water during this cruise (Fig. 9) indicates that particle concentrations were low except for the nearshore zone. The feature of primary interest during this time interval is the offshore movement of a tongue of turbid surface water in the vicinity of Redondo Canyon. This indicates that material brought into suspension by wave action in the surf zone can escape from that system to be transported into the water overlying the shelf.

It is worthwhile to note that the term "surface water" as used in this paper actually represents a depth of about 15 centimeters below the surface. This convention was adopted because the configuration of the transmissometer housing creates some turbulence and entrapment of bubbles which interfere with light transmission when the instrument is at or very near the water surface. Consequently, the surface microlayer where much of the floatable material from the effluent discharge is concentrated is not accounted for in this study. The actual light transmission values at the water surface are presumably somewhat lower than reported herein.

The behavior of the sewage plume with increasing distance from the discharge area suggests that some particles gradually settle out to be deposited at the sediment surface. Inasmuch as some of this material may be resuspended and transported to a new depositional site, some attention was devoted to the light transmission values near the sediment-water interface. A depth interval of 0.5 meters above the bottom was selected as representative of bottom light transmission values in order to avoid erroneous values caused by turbulence as the transmissometer approached and contacted the bottom.

The light transmission values at the bottom during November 1-4, 1971 (Fig. 10) are in general lower than in surface waters. The distribution of turbidity suggests that the shelf sediment dynamics in Santa Monica Bay have some influence on the eventual distribution of particles derived from the sewage discharge and from the nearshore zone. Several features are of particular interest. The largest such feature is a lobe of more turbid water which originates in the nearshore vicinity of a yacht harbor and drainage discharge at Marina del Rey (33°58'N;118°27'W). This lobe of turbid water moves toward the southwest across much of the shelf and then changes abruptly to a westward direction. On either side of the area where the change in direction occurs there is a shoreward penetration of clearer water that is located in the vicinity of the break in shelf created by Redondo Canyon. However, the southernmost line of stations from this

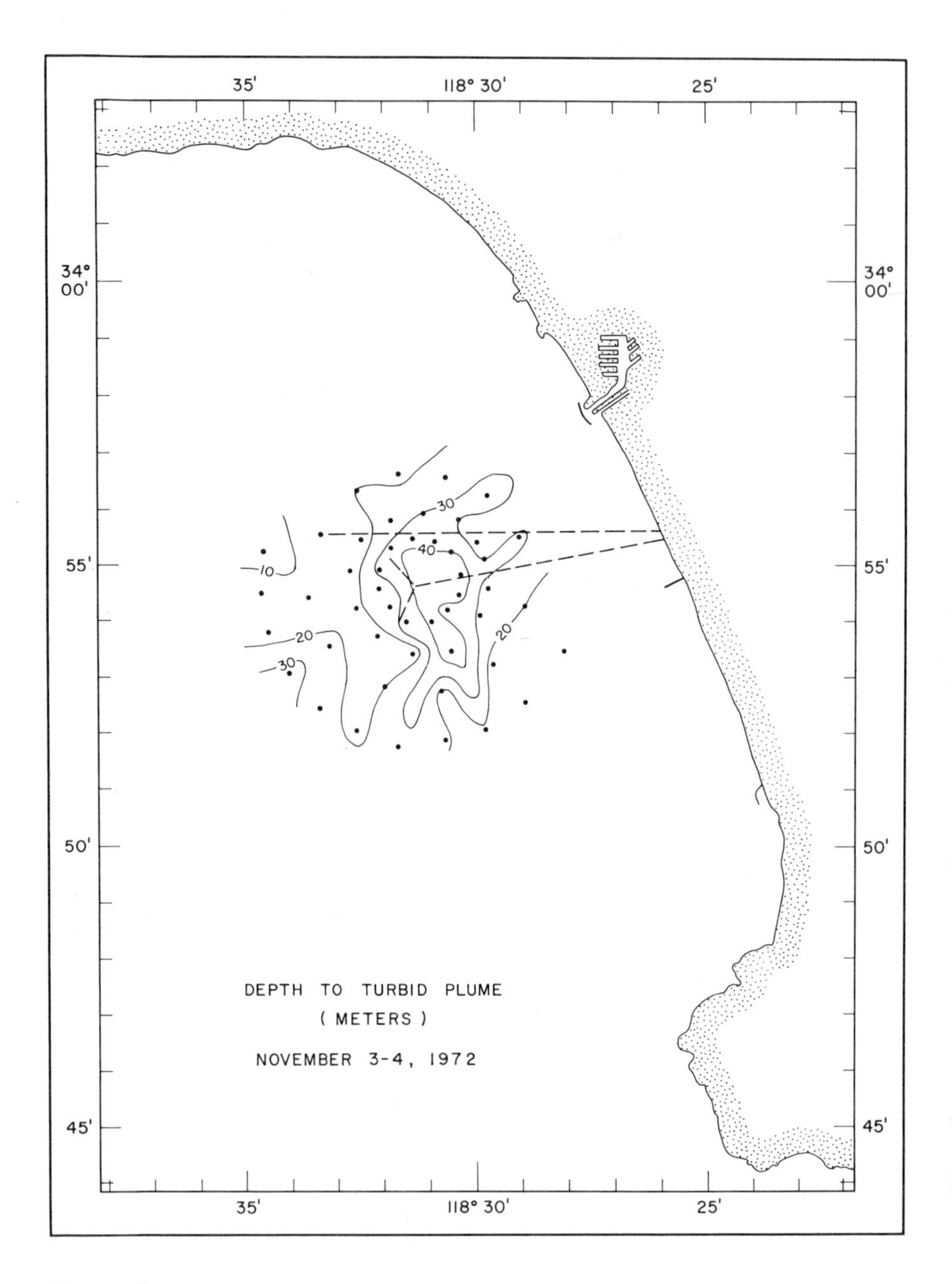

Figure 7
Depth from water surface to minimum light transmission portion of sewage plume. Solid circles represent station control.

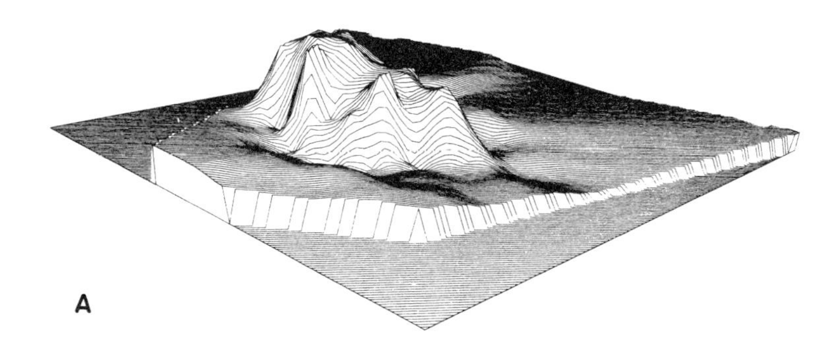

TURBIDITY

NOVEMBER 3-4, 1972

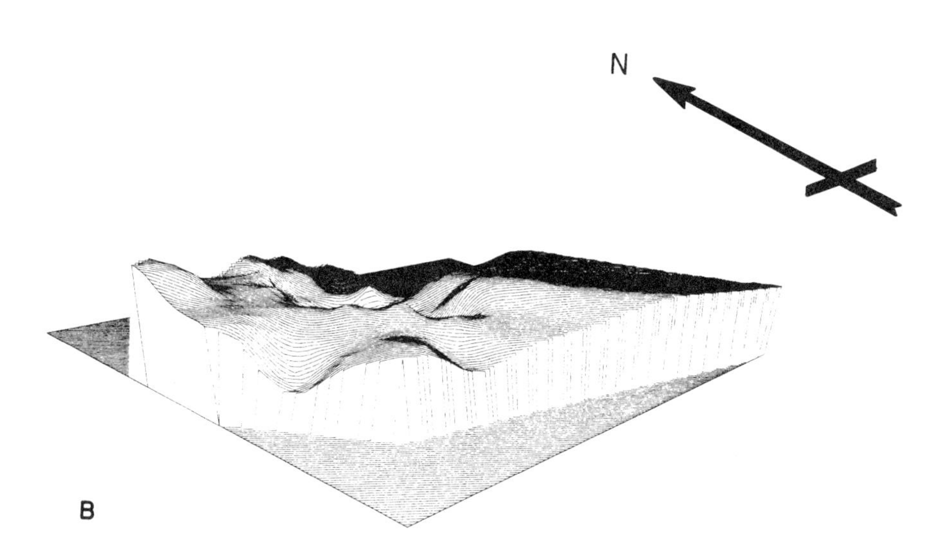

DEPTH TO PLUME

NOVEMBER 3-4, 1972

Figure 8
Three-dimensional depiction of middle of sewage plume from station control shown in Fig. 2. View is from the southwest at a viewing angle of 30° from the horizontal. High turbidity (a) represents low light transmission values (100 -% T/M). The diagram showing depth to middle of plume (b) illustrates pattern in which plume ascends with increasing distance from source.

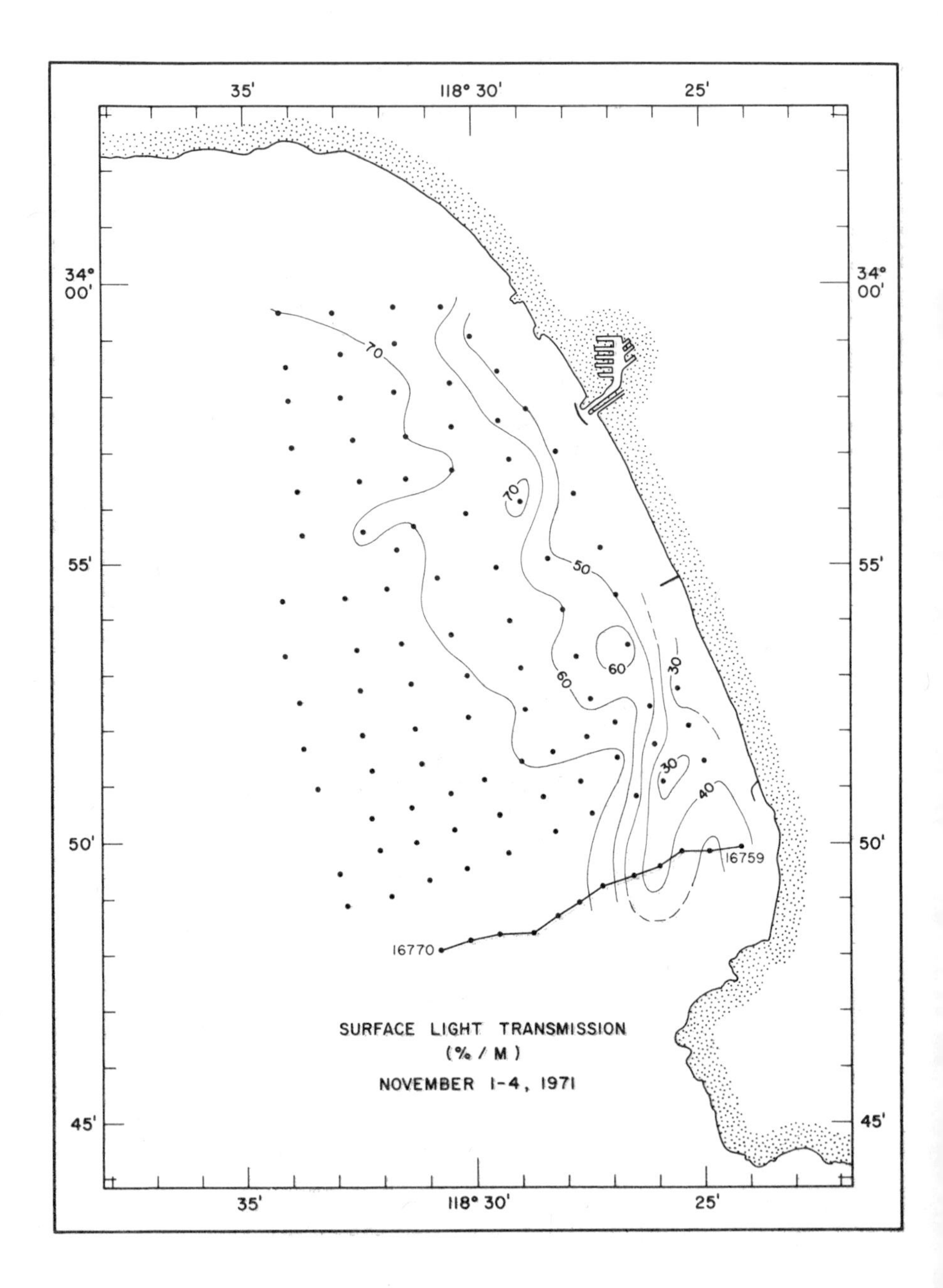

Figure 9
Surface water transparency (% T/M) during November 1-4, 1971.

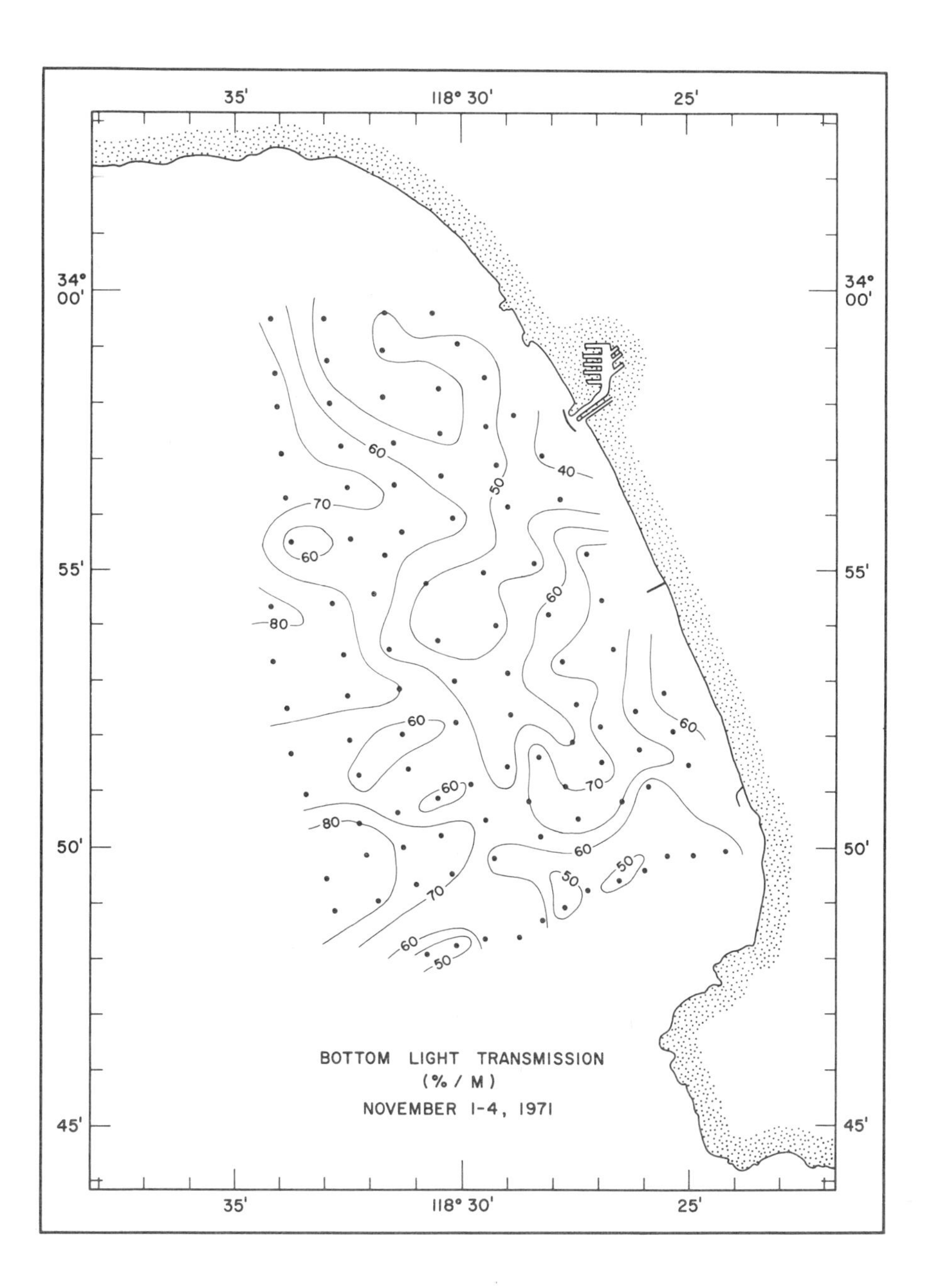

Figure 10
Water transparency (% T/M) within 0.5 meters of the bottom during November 1-4, 1971.

survey is in the axis of Redondo Canyon and the lower light transmission values at these stations provide limited evidence for movement of particles near the floor of that canyon. A cross section view of light transmission values from the stations occupied in the axis of Redondo Canyon (Fig. 11) reveals that the more turbid water in the canyon is a near-bottom feature. The question of whether the movement of the near-bottom turbid layer is up or down canyon can not be resolved from this evidence alone. There is some evidence, especially at a depth of about 400-420 meters at station 16765, that particles derived from the canyon floor are mixed upward into the water column and subsequently transported in an upcanyon direction.

The light transmission profiles show a significant portion of the material discharged from the effluent outfall to remain suspended within the water column. This could be traced over the entire shelf area that was investigated. Therefore, the minimum light transmission values within the main part of the sewage plume were contoured (Fig. 12) to evaluate the dispersion and transport of particles from the sewage discharges. This task was subject to a considerable amount of interpretation because there were several sources for the material in the area. In some cases it was difficult to distinguish which turbid layer actually represented the main portion of the plume. Nevertheless, the distribution of minimum light transmission values within the water column provides some insight into the areal movement of the sewage effluent.

The elleptical pattern of low values near the terminus of the sludge outfall indicates that there is either a shoreward transport of at least a portion of the suspended plume which originates from the sludge discharge or there is a northwestward transport of the sewage plume from the effluent discharge. If the plume outlined is interpreted as being from the sludge discharge, then the pattern is consistent with the salinity and temperature data which indicate a shoreward penetration of water that is derived from Santa Monica Canyon. In addition, there are two lobes of turbid water which originate in the nearshore area and are transported seaward. The northernmost lobe (at approximately 33°55'N) is derived from an intermittent discharge of sewage from the short outfall shown on the location map (Fig. 1). The southern lobe of turbid water appears to be an offshore transport of water from the nearshore zone. The latter feature is modified by a lobe of clearer water that penetrates from the seaward direction.

In addition to the somewhat complicated pattern of minimum light transmission values which suggests an offshore-onshore alignment of the sewage plume, there was a suggestion of a similar pattern in the depth of the plume. When the depth to the mid-point of the plume was contoured (Fig. 13), a striking pattern showing alignment normal to the coastline was clearly evident. This situation is indicative of a wave-induced modification of the turbid plume in the water column.

In order to better visualize these patterns, the data were processed into three-dimensional views (Fig. 14) which vividly outline the features just discussed. One troublesome problem, however, is the fact that the pattern for the depth to the plume is so consistent for such a long (four day) sampling interval over a relatively large area. The maximum turbidity associated with the sewage outfall, at a distance of more than 2 kilometers from the discharge, is near the thermocline. This portion of a stratified water column is also the

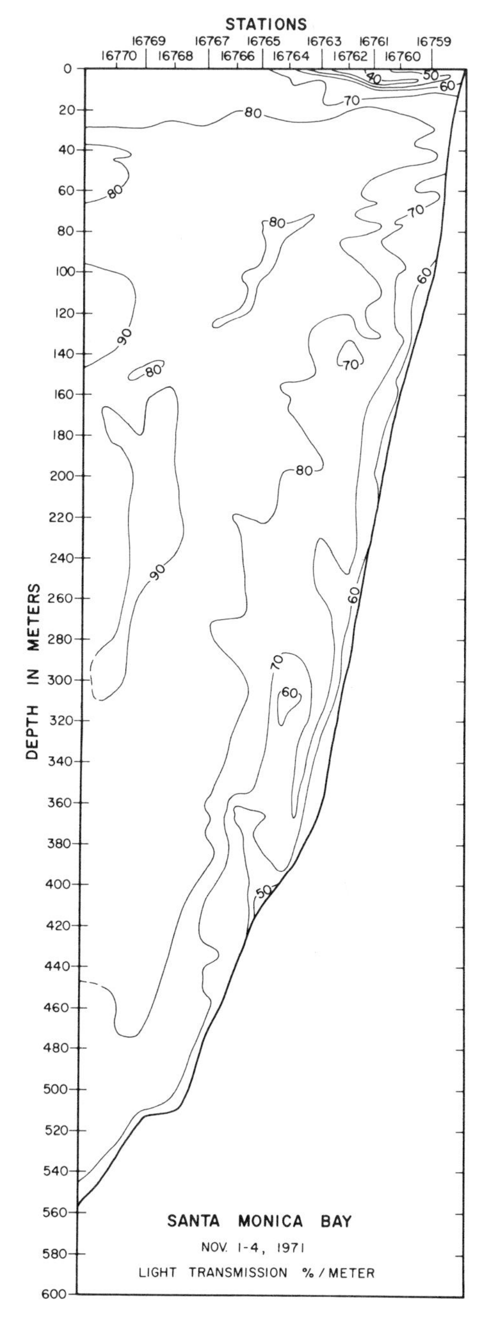

Figure 11
Cross section of light transmission in axis of Redondo Canyon. Refer to Fig. 9 for station locations.

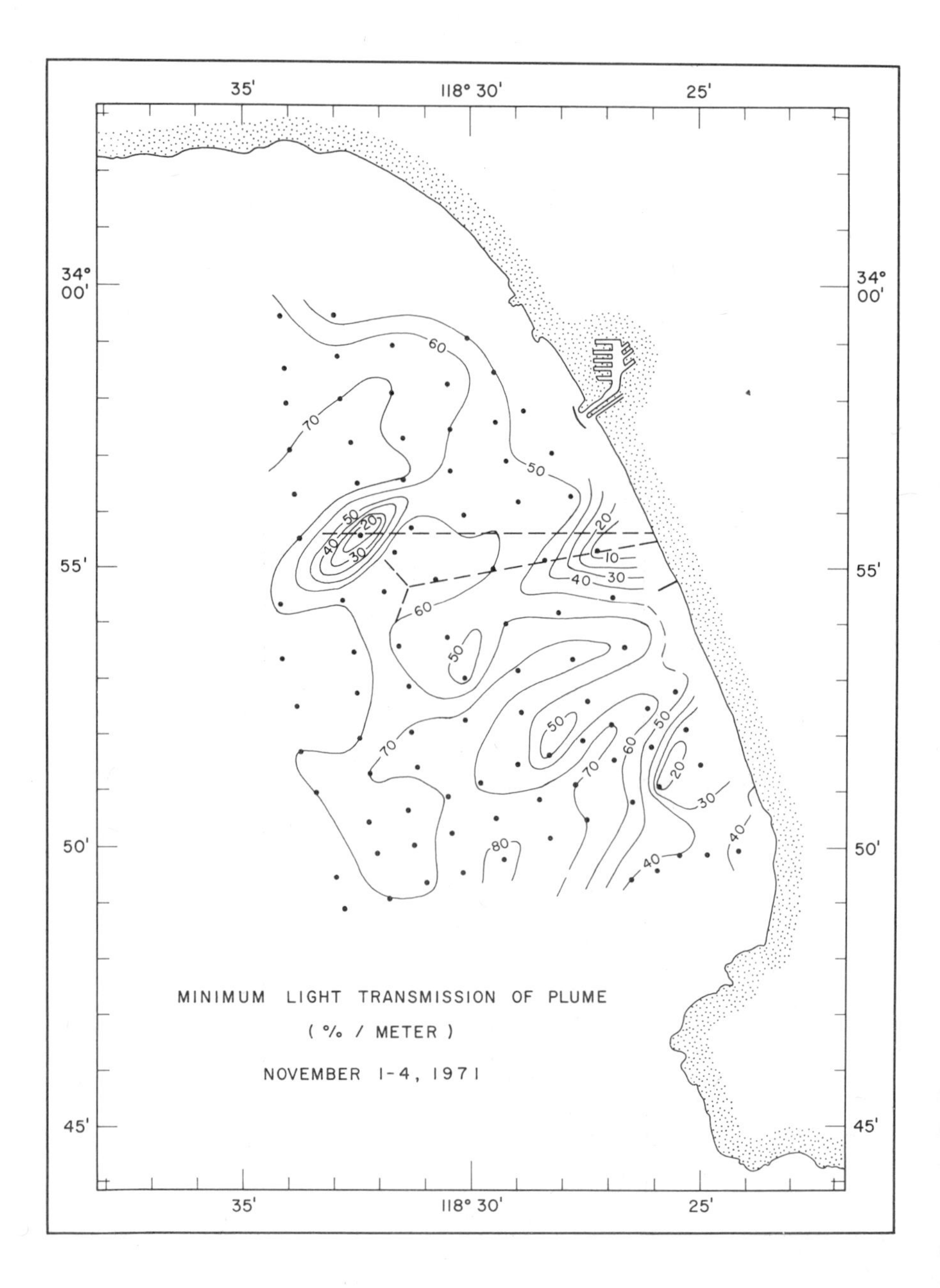

Figure 12
Distribution of minimum light transmission values within principal sewage plume. Low values along the coastline probably represent seaward transport of particles from the littoral transport system.

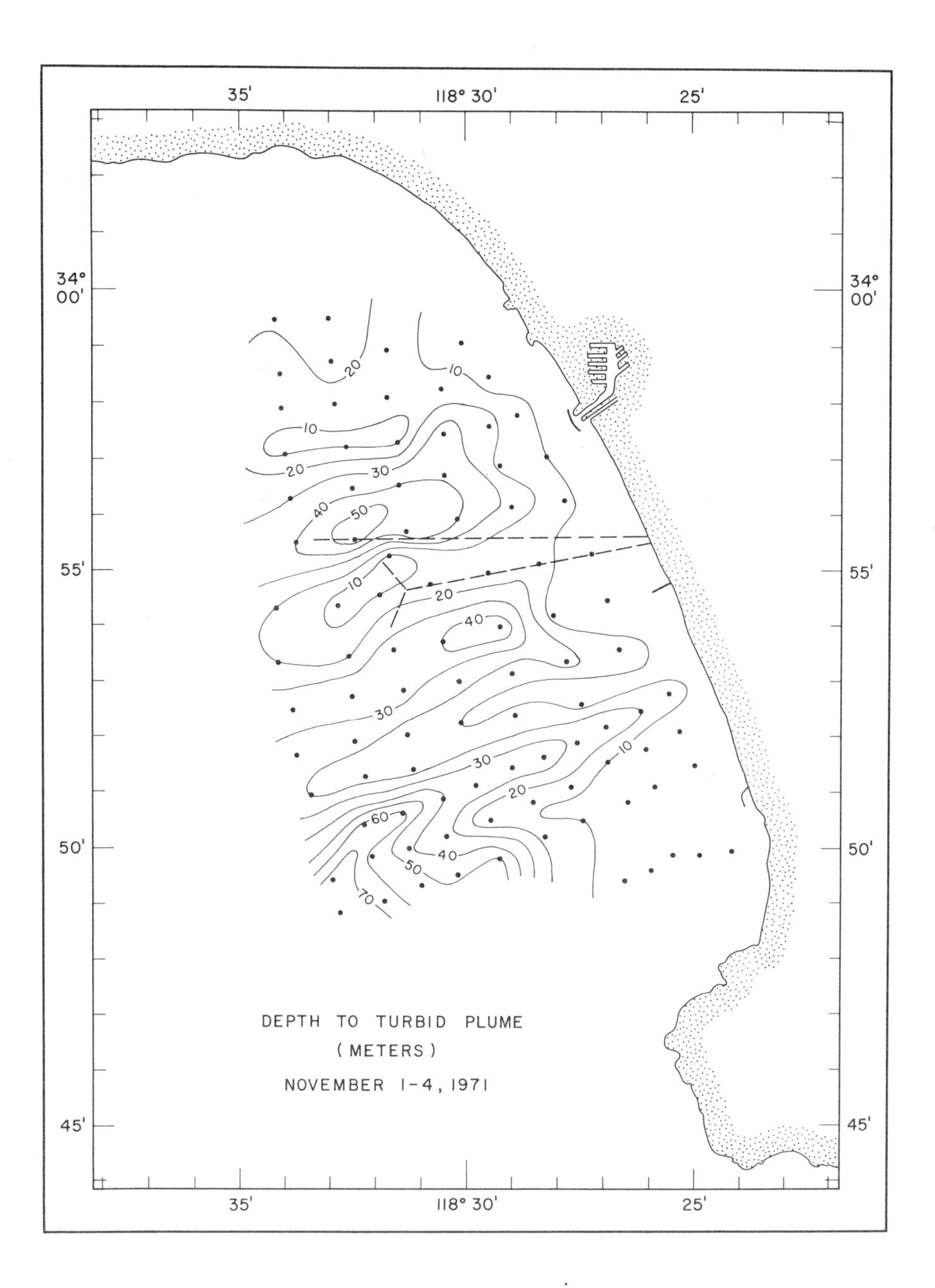

Figure 13
Depth from water surface to minimum light transmission portion of sewage plume during November 1-4, 1971.

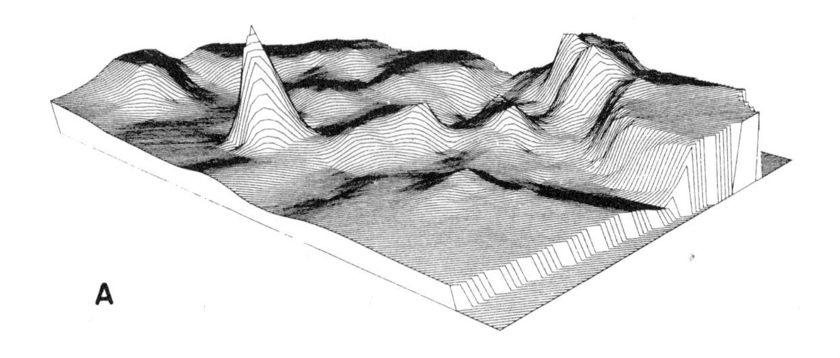

TURBIDITY

NOVEMBER 1-4, 1971

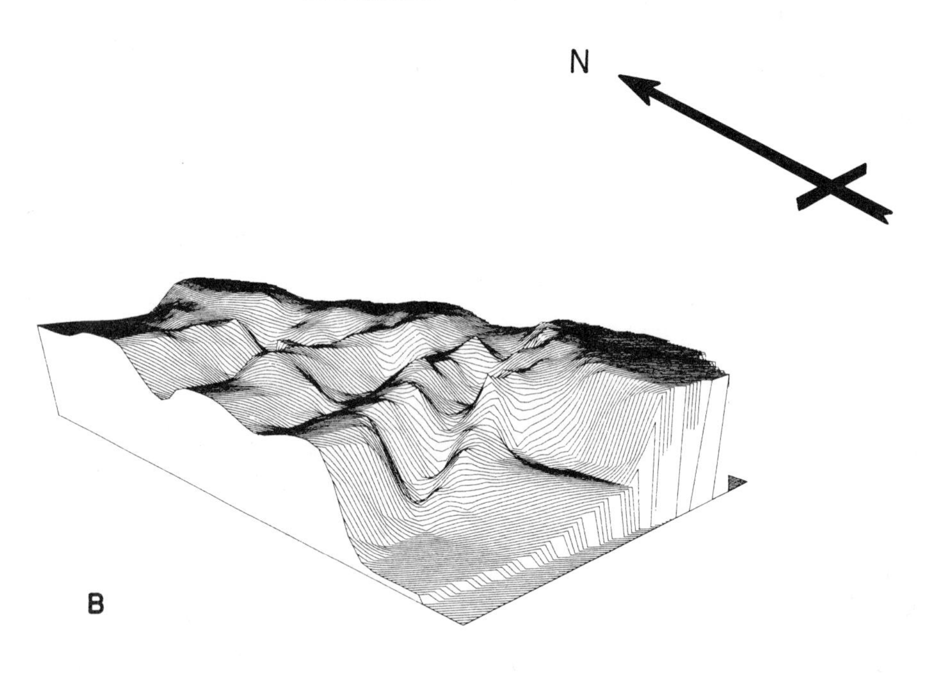

DEPTH TO PLUME

NOVEMBER 1-4, 1971

Figure 14
Three-dimensional views of (a) turbidity (100-% T/M) of sewage plume and (b) depth from water surface to middle of plume. Main discharge is shown by peak in upper left portion of top diagram. View is from the southwest at an angle of 30° from the horizontal.

location of the maximum amplitude of internal waves (LaFond, 1962). The spacing and large amplitude of the wave forms normal to the coast depicted in Fig. 14 suggest that long period waves are propagated to the north or south in the water column as a result of the offshore topography and configuration of the Santa Monica Bay coastline. It is possible that Santa Monica Bay may effectively produce a trapping of these waves in the manner outlined by Munk and others (1956), Munk (1962), and Leblond and Mysak (1977). Alternative interpretations include standing tidal and internal waves within Santa Monica Bay. In addition to the east-west alignment just discussed, there is a secondary north-south alignment of wave crests in the vicinity of the shelf break. These forms are not particularly well defined in Fig. 14 but are more evident in the data from other cruises and will be discussed later.

The most comprehensive coverage of the area was obtained during a cruise on May 19-28, 1971 when 126 stations were occupied. The distribution of surface water light transmission values (Fig. 15) demonstrated the control that density stratification has on determining the position of the plume within the water column. During May the density stratification in the water column was less pronounced than, for example, during November and a greater portion of the sewage field reached the water surface in Santa Monica Bay. This reduced the light transmission values at the water surface to about 50% of the values measured during November. There is also a surface expression of the plume immediately over the effluent discharge. This is somewhat unusual for the Hyperion outfall because the normal density stratification and currents are commonly strong enough to prevent a significant portion of the plume from rising to the surface directly over the discharge area. In the southern part of the study area, in the vicinity of Redondo Canyon, there is an indication of an east-west alignment of more turbid surface water with intervening lobes of clearer water derived from the offshore limits of the area sampled during the cruise.

At the sediment-water interface two lobes of turbid water extend from the nearshore zone seaward to about the middle of the shelf (Fig. 16). With the exception of these two seaward intrusions of turbid water, the bottom water is significantly clearer than the surface water. A cross section of light transmission values through the central portion of the area and parallel to the coastline (Fig. 17) outlines the movement of the sewage plume toward the water surface. In this case, the plume rises toward the southeast from the discharge site, breaches the thermocline, and is then dispersed in the upper 15 meters of the water column. Consequently, any particles which might tend to settle out from the zone of well mixed surface water are inhibited by the density barrier at the thermocline. Movement of turbid water off the shelf and into Santa Monica and Redondo canyons is limited to an interval of about 10 meters off the bottom and is a relatively minor avenue of transport during the early summer season. There is more suspended material near the bottom in Santa Monica Canyon than in Redondo Canyon, which is attributed to the discharge of sewage sludge at the head of Santa Monica Canyon. Measurements made down the axis of both canyons at other times of the year (Figs. 18 and 19) indicate that on the average, there is only a limited amount of suspended material available for transport to the offshore area via the submarine canyon systems in Santa Monica Bay. However, other studies, primarily of sediments, show that the

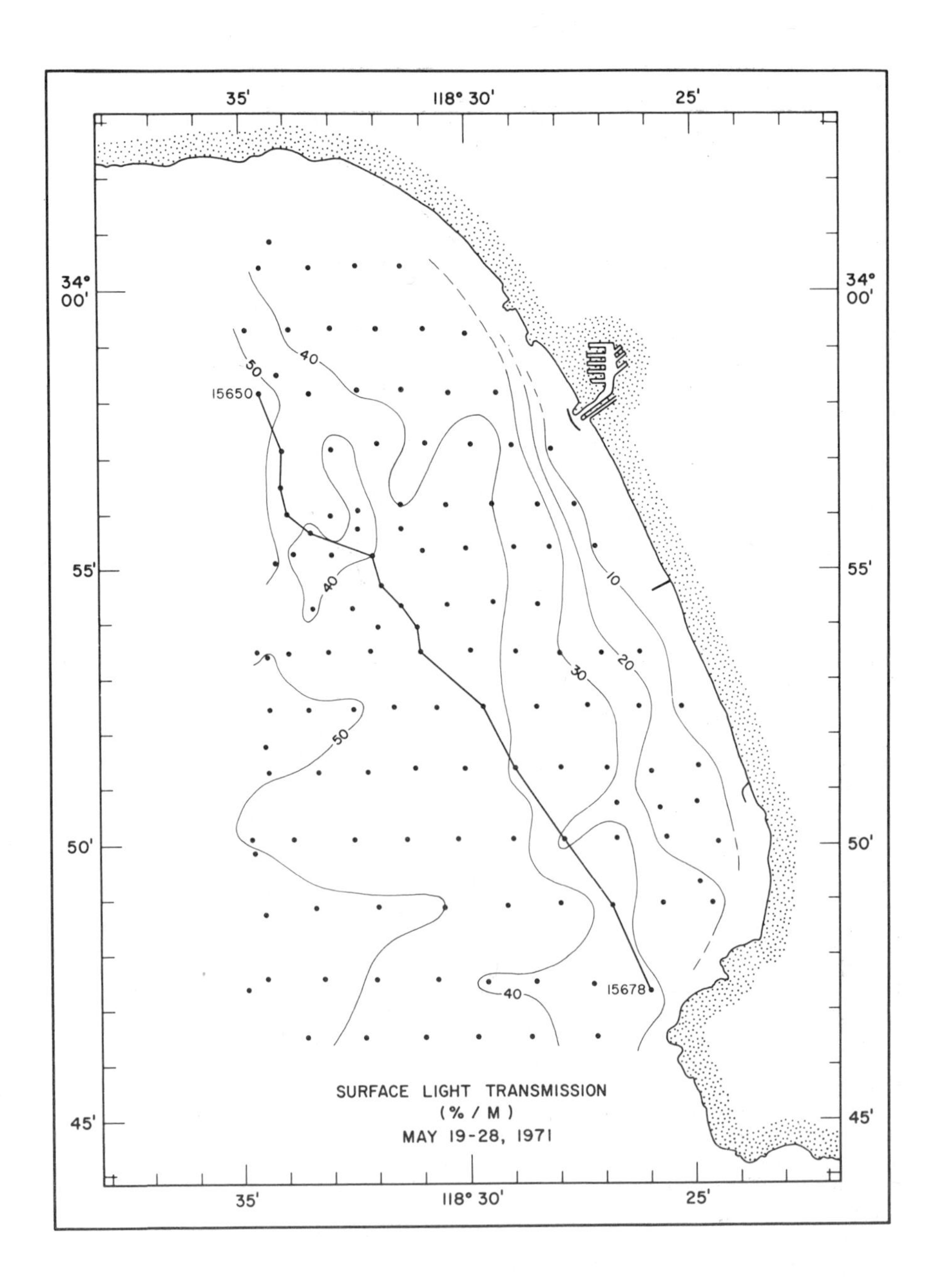

Figure 15
Distribution of light transmission values for water surface during May 19-28, 1971.

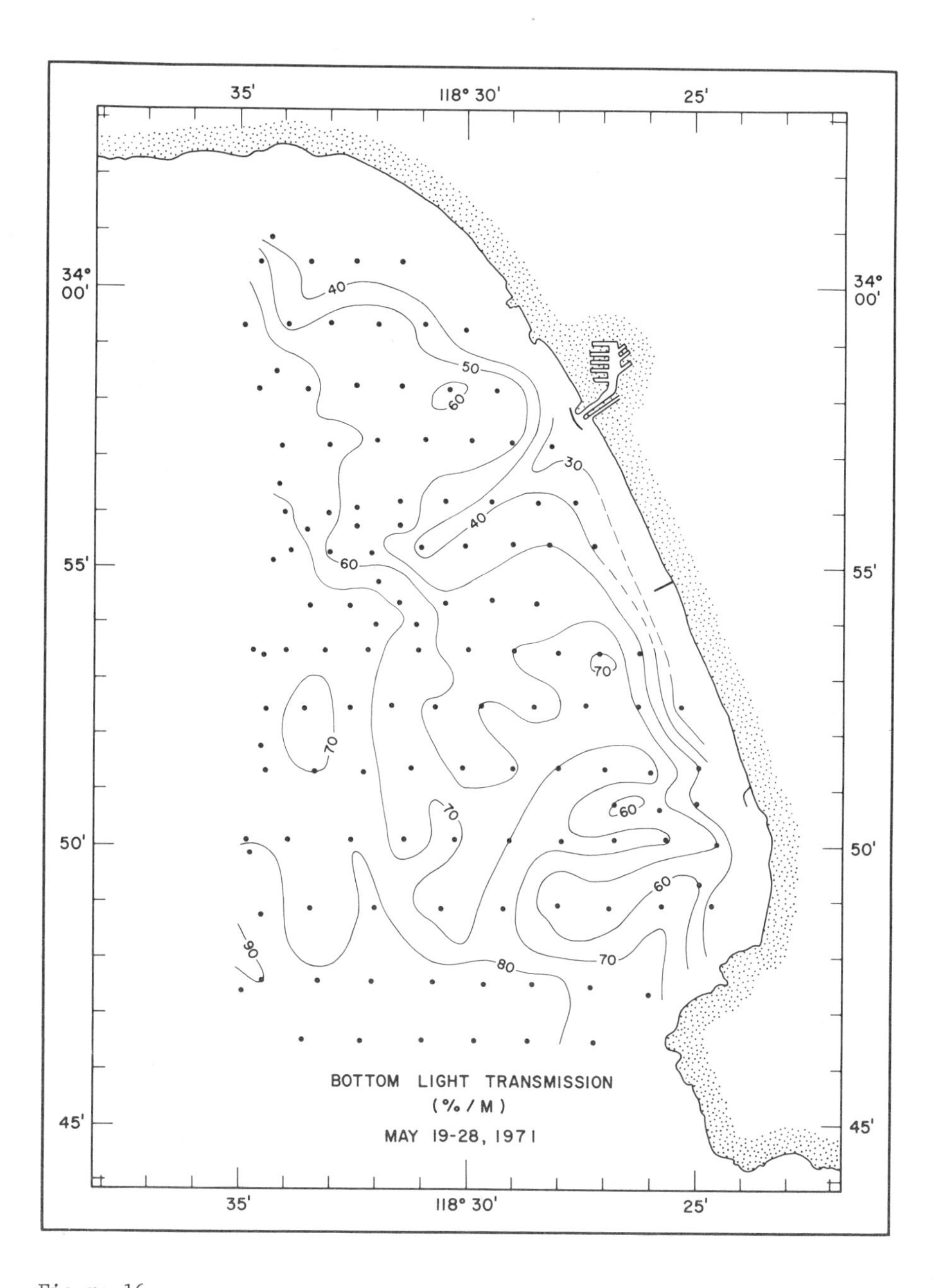

Figure 16
Distribution of light transmission values within 0.5 meters of bottom during May 19-28, 1971.

canyons act as conduits to transport coarse-grained sediments from the nearshore area to the basin (Gorsline and Emery, 1959; Haner, 1969). Presumably the majority of this material is transported episodically and it would indeed be fortuitous to measure the concomitant high turbidity values associated with a density type of flow containing the wide range of particle sizes which piston and box core samples indicate are involved in one of these events. Light transmission values obtained from traverses on both sides of the canyons also indicate that most of the suspended particles derived from nearshore mixing are confined to the water column above the thermocline. For example, in the vicinity of Redondo Canyon, material is obviously transported offshore from the headlands of Palos Verdes peninsula (Fig. 20-B) in the upper 10 meters of the water column. There is no appreciable amount of material in suspension near the bottom. However, the undulating character of the light transmission and temperature contours suggest that internal wave activity can resuspend fine-grained material near the bottom at or near a distinct change in bottom slope.

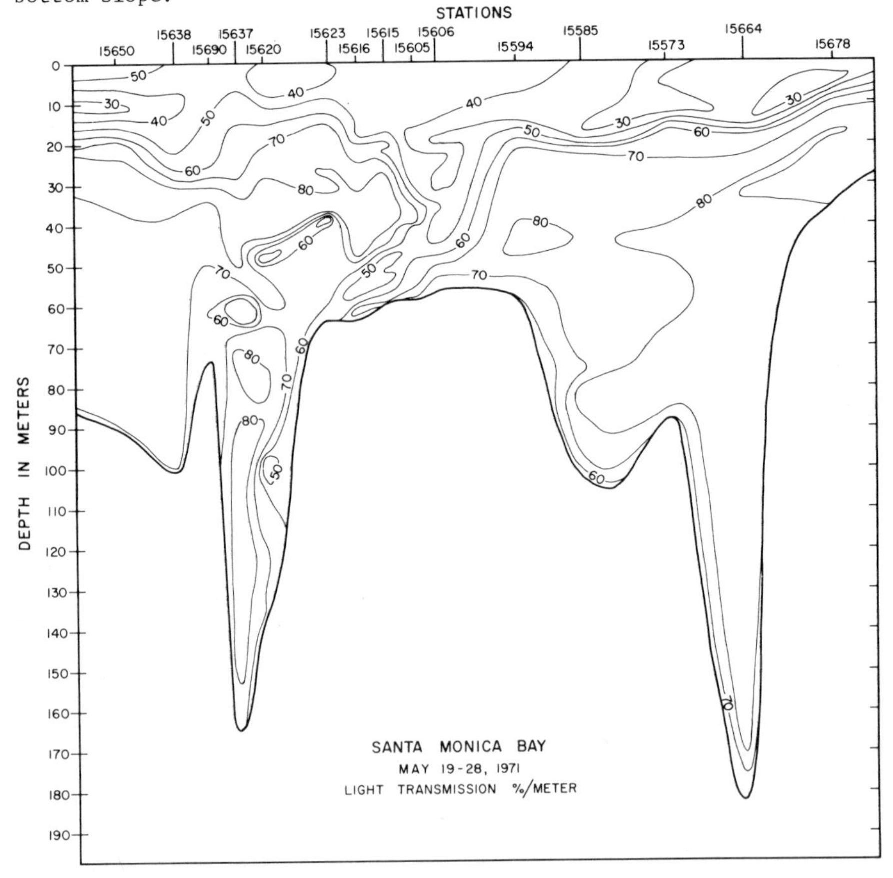

Figure 17
North-south cross section of light transmission values in Santa Monica Bay. Redondo Canyon is at the right and Santa Monica Canyon is at the left. Refer to Fig. 15 for station locations.

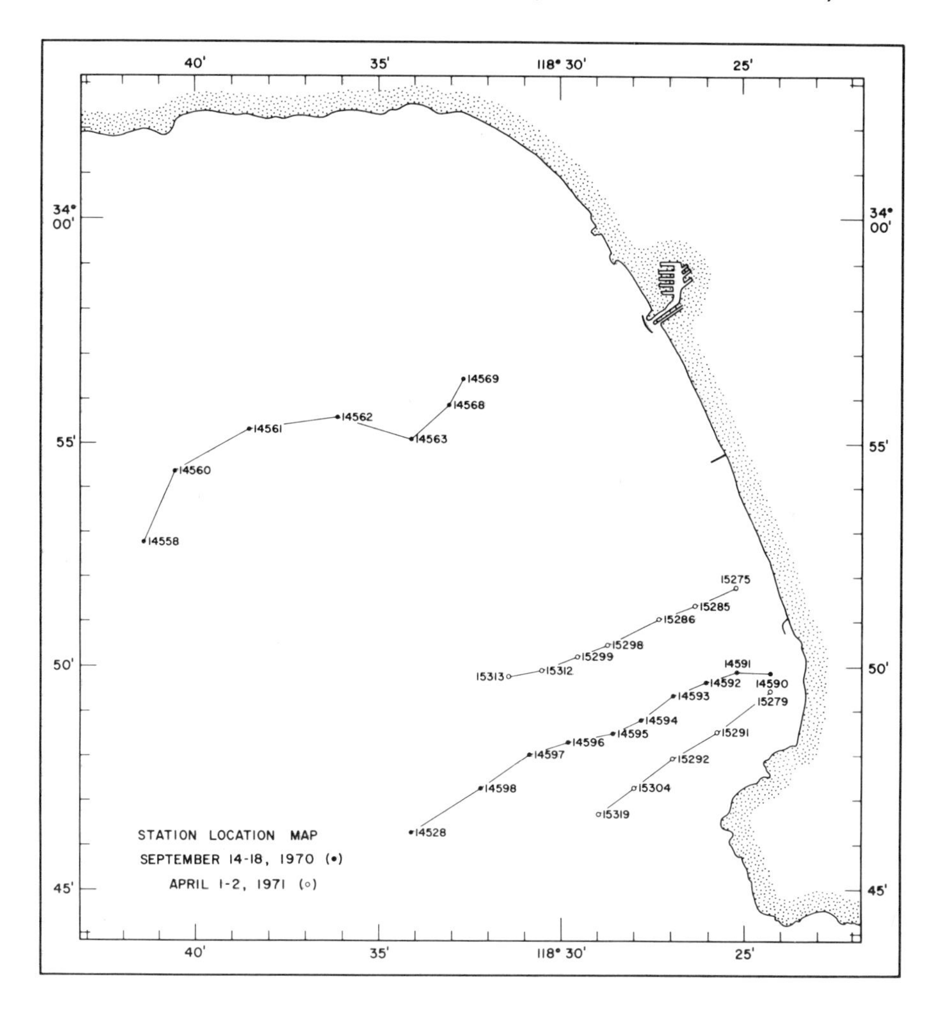

Figure 18
Station locations for September 14-18, 1970 (solid circles) and April 1-2, 1971 (open circles) cruises. The September stations were occupied in the axis of Santa Monica and Redondo canyons.

Internal waves were present during most of the cruises. For instance, the cross sections of light transmission values and the three-dimensional view of the depth to the plume during November, 1971 (Fig. 14-B) indicate internal wave activity. However, in July, 1971 the internal wave activity during a survey of the central portion of the shelf (Fig. 21) was more pronounced, as is evident in the three-dimensional view of the depth to the sewage plume (Fig. 22). Cross sections of selected traverses through the area during July, 1971 summarize the typical position of the sewage field within the water column and also provide excellent examples of long-period wave activity in the area. A north-south cross section (Fig. 23) through the main portion of the sewage field from the effluent

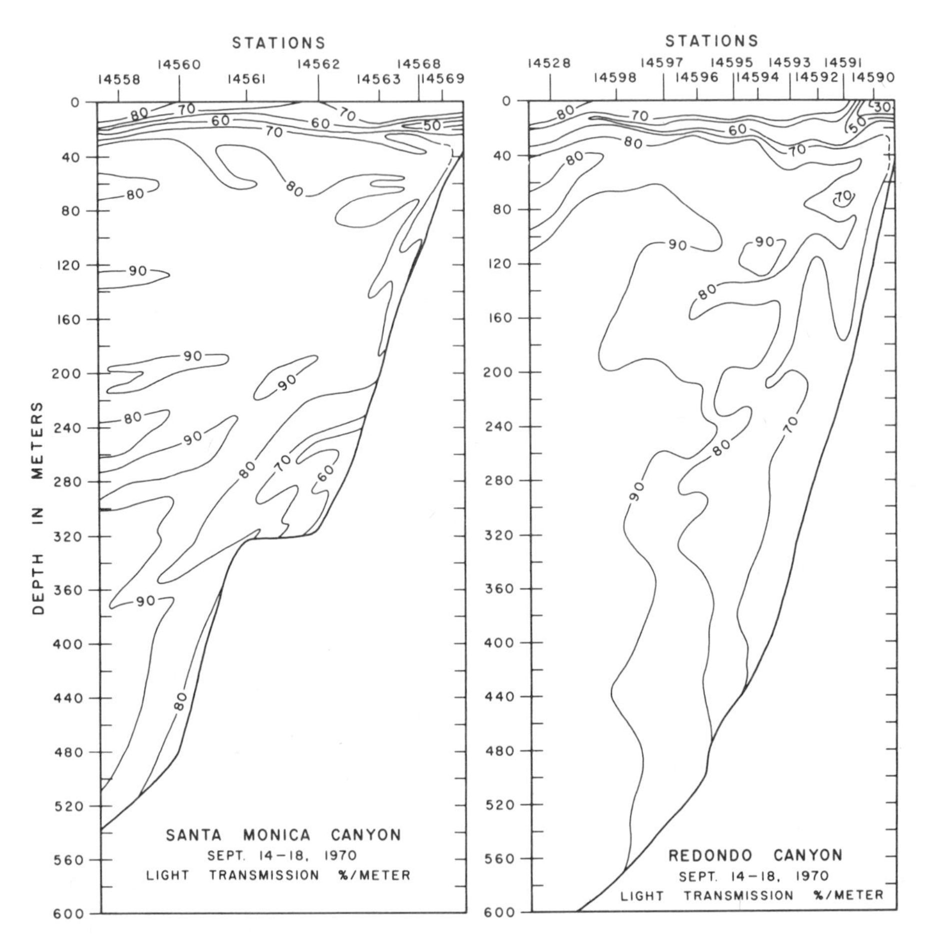

Figure 19
Cross section of light transmission values in axis of two submarine canyons. Seaward movement of suspended material is evident around 20 meters and near the bottom in both canyons.

discharge shows that the plume moves upward and toward the south until it reaches an equilibrium at a depth of approximately 20 meters. Wave activity and advective transport subsequently disperses suspended materials from the sewage discharge over the Santa Monica Shelf in a predominantly southeastward direction. Cross sections of light transmission values approximately normal to the coastline (Figs. 24 and 25) show that higher salinity water with less particulate material moves into the shelf region from the west and northwest. Internal waves also assist in broadening the depth interval of the sewage plume with increasing distance from the discharge. At relatively consistent geographic positions in the bay the net direction of particle transport is modified so that the transport is offshore and almost normal to the coastline. These zones of offshore transport also serve as conduits for mid-water transport of particulate material derived from nearshore mixing (Fig. 24-B).

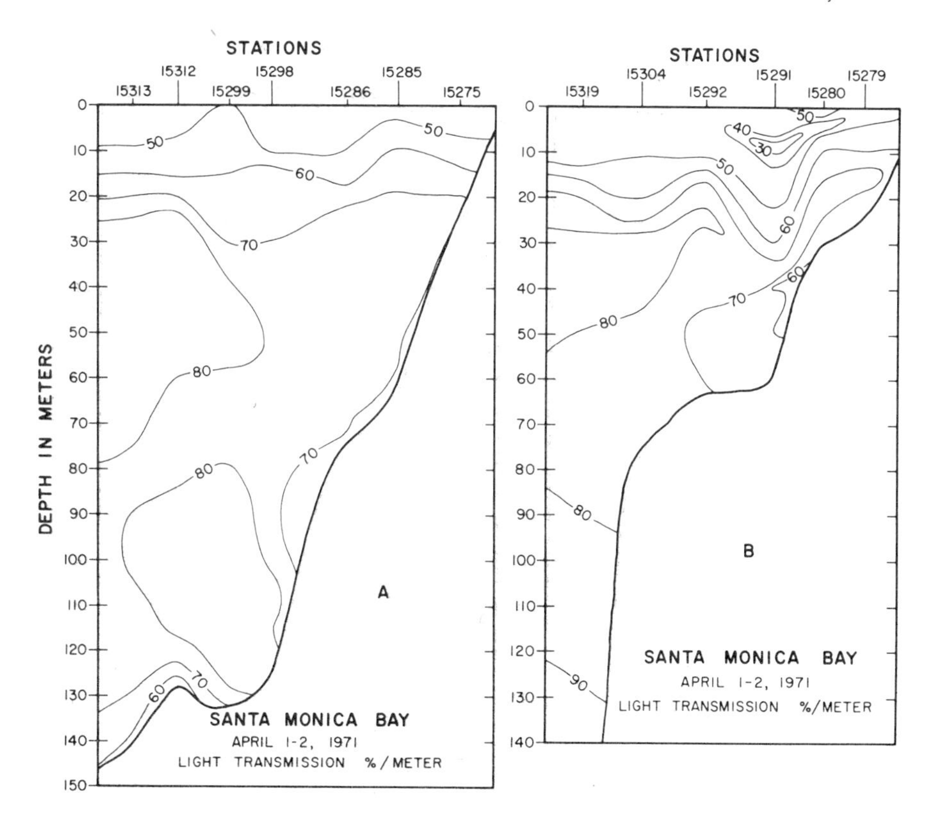

Figure 20
Cross section of light transmission values on both sides of Redondo Canyon. The lower values at about 10 meters on the south side of the canyon (b) represent material transported offshore from Palos Verdes peninsula (lower right corner of Fig. 18).

DISCUSSION

The east-west and north-south wave forms are interpreted as resulting from long-period waves and internal waves, respectively, striking the shelf edge. The position of these waves within the water column, during an entire cruise and also a series of cruises during more than a years time, also suggest the possibility of a standing wave with several nodes and anti-nodes. In the latter case, the distribution of particles from the sewage discharged in Santa Monica Bay could be strongly influenced by differential water motions induced by this wave action. If a standing wave exists in Santa Monica Bay, the horizontal water velocity should be greater at the nodes than at the anti-nodes (LaFond, 1962). Persistent standing waves presumably would then affect the deposition of sediments in Santa Monica Bay and produce a dominant alignment parallel to the wave forms. Consequently, substances such as toxic metals which are associated with fine-grained sewage particles would also tend to move

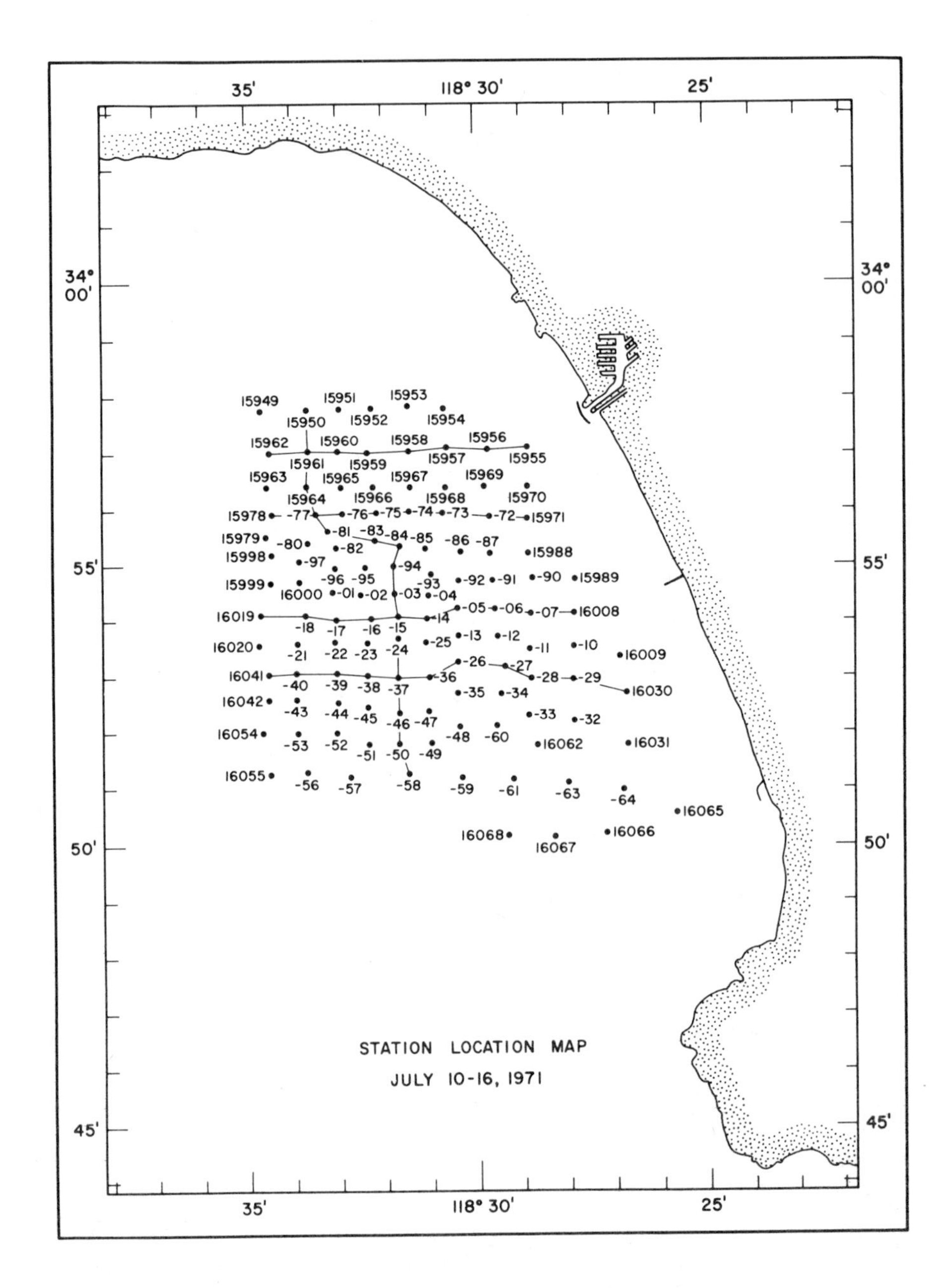

Figure 21
Station locations for July 10-16, 1971 cruise. Solid lines connect stations used in cross sections shown in Figs. 23-25.

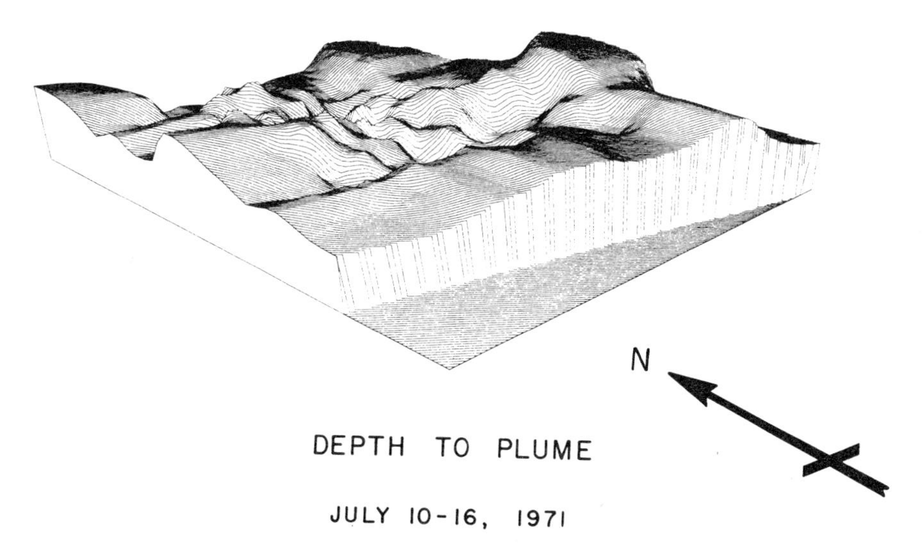

Figure 22
Three-dimensional depiction of depth to middle of sewage plume during July 10-16, 1971. View is from the southwest at a viewing angle of 30° from the horizontal. Pronounced east-west crests and secondary north-south crests in depth to middle of sewage plume are evident.

in a similar fashion, so that any potential impact on the benthic biota from this material would be greater where particles are deposited than in the intervening areas where particles remain in suspension or are eroded.

Surface water light transmission values were lower in May than during other months, and cross sections show that a larger portion of the sewage field ascended to the surface than, for example, during November. A reduction of light penetration caused by reduced density stratification in the water column and subsequent ascension of particles from the Hyperion outfalls could stress those organisms which normally inhabit the lower euphotic zone. Alternatively, the reduction in the depth of light penetration may be compensated for by migration of these organisms toward the water surface. The problem involves a sequence of possible interactions and is not especially straightforward because of the effects of nutrients and other substances such as heavy metals and chlorinated hydrocarbons which are introduced by sewage discharge. Understanding the situation will require a considerable amount of effort. However, light transmission studies provide a useful and relatively inexpensive technique for focusing work on those areas which promise to yield the largest amount of information.

The existence of several pathways of fine particle transport from the nearshore zone to the offshore area also suggest several intriguing possibilities that should be investigated. First, additional oceanographic investigations would provide greater insight into the mechanism(s) responsible for the observed patterns of particle transport. Additional light transmission measurements in the

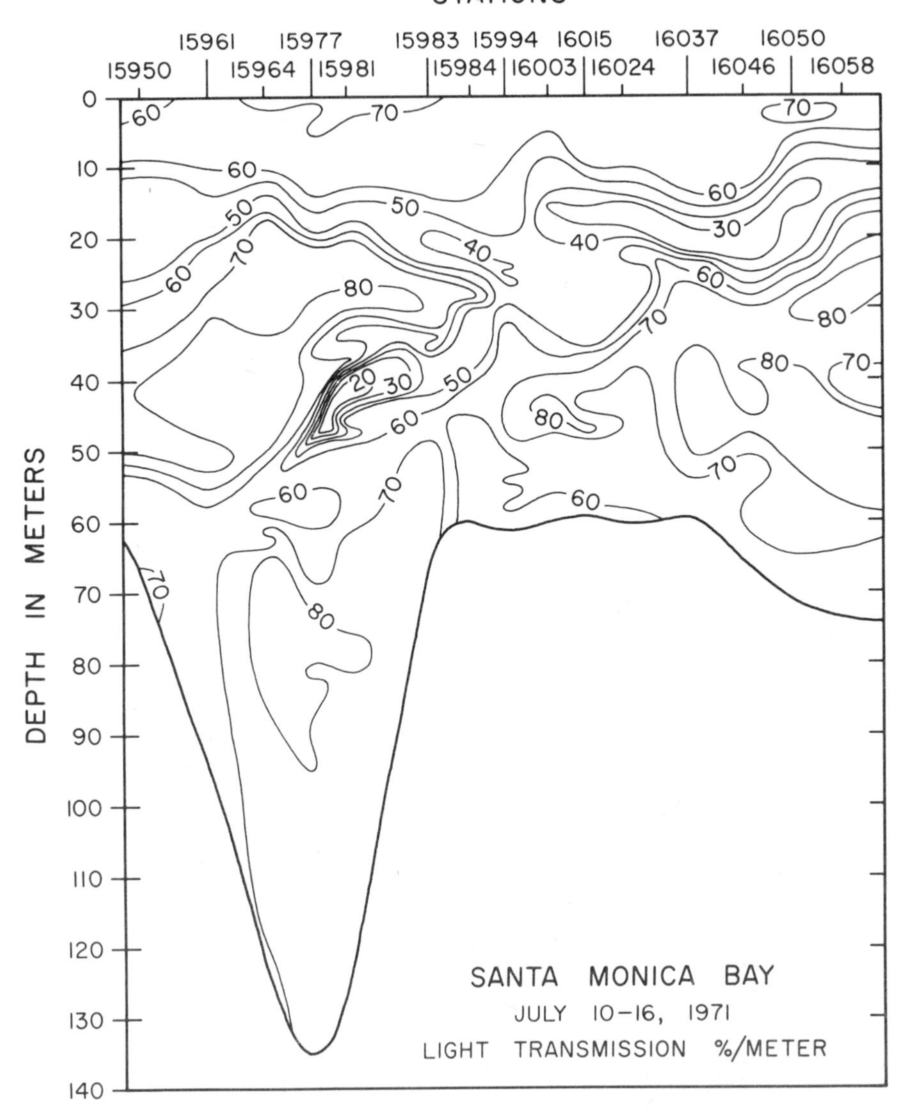

Figure 23
North-south cross section of light transmission values during July 10-16, 1971. Initial portion of sewage plume is evident near station 15981. Santa Monica Canyon is in the left part of the illustration.

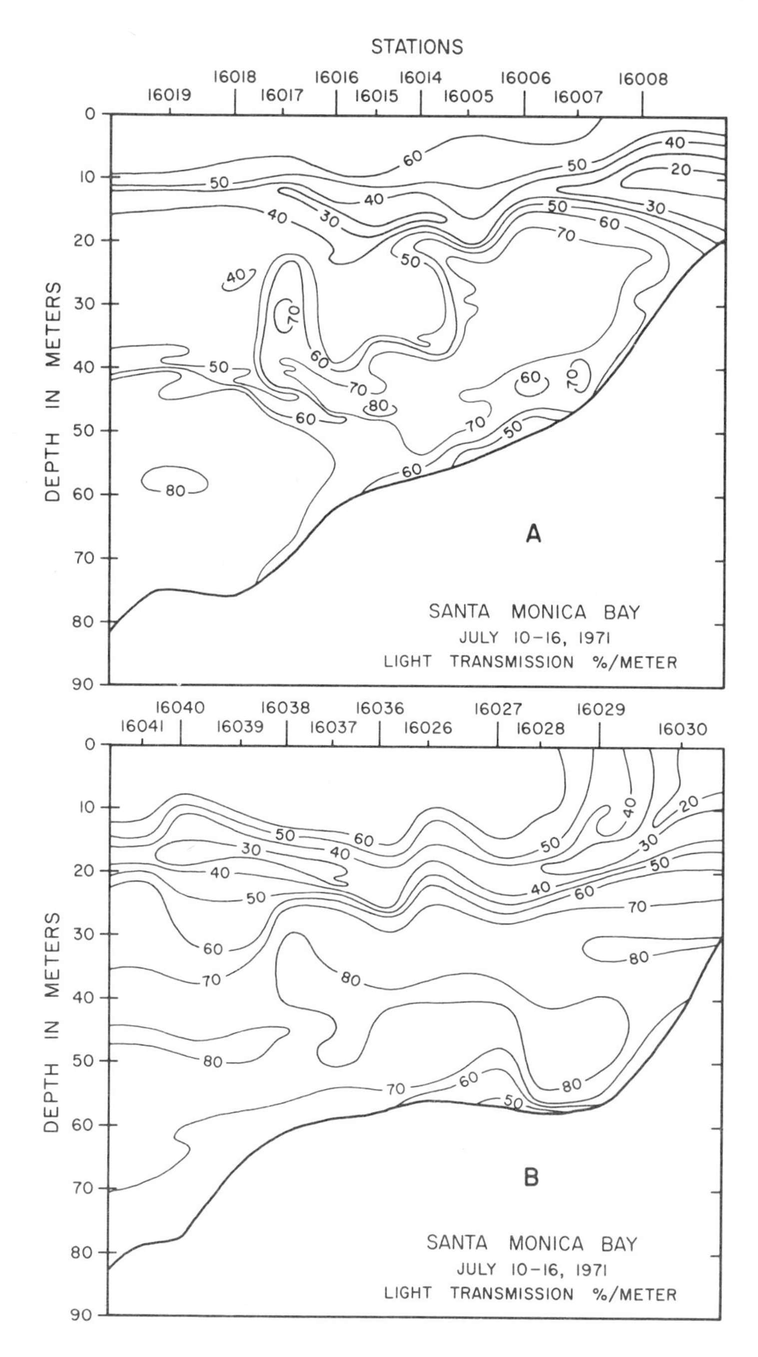

Figure 24
East-west cross sections of light transmission values south of the effluent outfall. The north-south crests in sewage plume outlined in Fig. 22 are evident in lower cross section at stations 16026 & 16040.

area seaward of the shelf also would provide a better indication of the possibility that much of the fine-grained sediments deposited in the offshore basins settle out from well-defined near-surface avenues of transport which are essentially normal to the coastline of this area. Finally, the alignment of the pathways of near-surface particle transport in Santa Monica Bay are of some consequence with respect to the interaction of suspended sediments and oil released from natural seeps near Redondo Canyon or oil that might be released in the area as a result of an accidental spill.

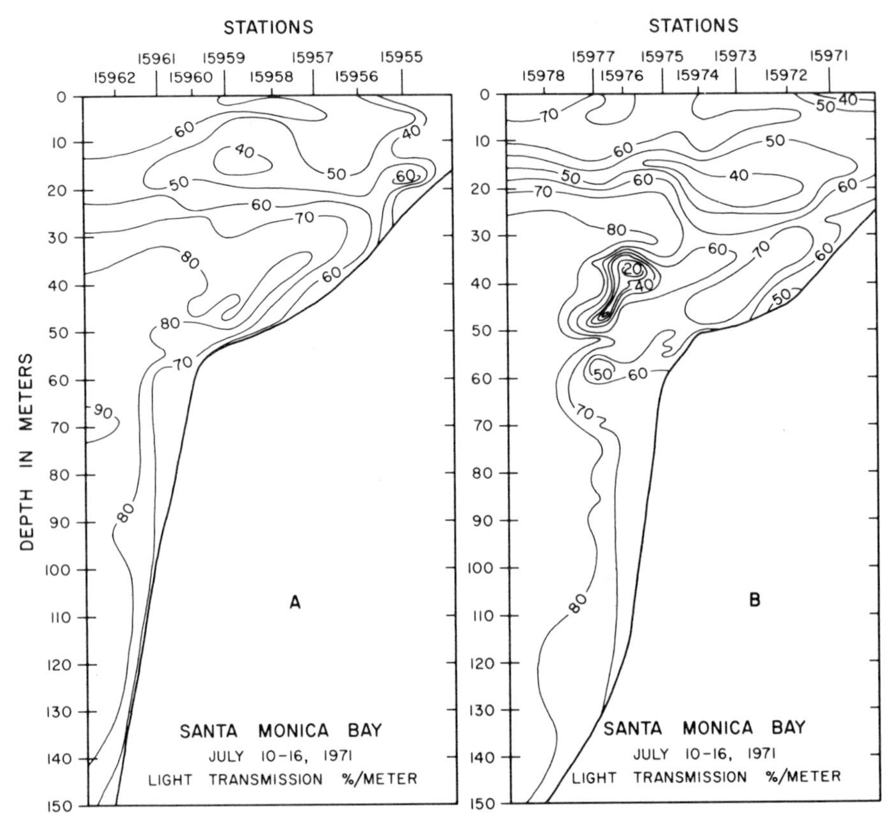

Figure 25
Cross sections of light transmission values north (a) and along outfall line (b) during July 10-16, 1971. The main sewage plume near the effluent outfall terminus is graphically outlined at a depth of 35-50 meters between stations 15975-15977.

ACKNOWLEDGEMENTS

Field work was carried out aboard the R/V Velero IV; ship time was funded by National Science Foundation Grant GB-8206. George Armstrong, Kendal Robinson, Ronald Zakrzewski and numerous students assisted in the collection and reduction of data. George Armstrong and Kendal Robinson drafted the figures and Joyce Barrow assisted in preparing the manuscript for publication. The computer graphics were produced by Robert Stearns and George Armstrong. Some financial support for this study was provided by the Environmental Geology Research Fund at the University of Southern California.

REFERENCES

Allan Hancock Foundation, 1965. An oceanographic and biological survey of the southern California mainland shelf. State of California, State Water Quality Control Board, Publ. 27, 231pp.

Anonymous, 1973. Color movie and slides presented by Los Angeles City Bureau of Sanitation at research conference on California ocean pollution. Lake Arrowhead, California, November 11-15, 1973.

Bandy, O.L., Ingle, J.C., Jr., and Resig, J.M., 1965. Foraminiferal trends, Hyperion outfall, California. Limnol. and Oceanog., 10: 314-332.

Beer, R.M., 1969. Suspended sediment over Redondo submarine canyon and vicinity, southern California. M.S. Thesis, Univ. of Southern California, Los Angeles, 94pp.

Drake, D.E., 1974. Distribution and transport of suspended particulate matter in submarine canyons off southern California. In: R.J. Gibbs (Editor), Suspended Solids in Water. Plenum Press, New York, pp. 133-153.

Drake, D.E. and Gorsline, D.S., 1973. Distribution and transport of suspended particulate matter in Hueneme, Redondo, Newport, and La Jolla submarine canyons, California. Geol. Soc. Am. Bull., 84: 3949-3968.

Drake, D.E., Kolpack, R.L. and Fischer, P.J., 1972. Sediment transport on the Santa Barbara-Oxnard shelf, Santa Barbara Channel, California. In: D. Swift, D. Duane and O. Pilkey (Editors), Shelf Sediment Transport: Process and Pattern. Dowden, Hutchinson & Ross, Stroudsburg, Pa., pp. 307-332.

Emery, K.O., 1952. Continental shelf sediments of southern California. Geol. Soc. Am. Bull., 63:1105-1108.

Emery, K.O., 1960. The Sea Off Southern California, John Wiley & Sons, Inc., New York, 366pp.

Føyn, E., 1971. Municipal wastes. In: D.W. Hood (Editor), Impingement of Man on the Oceans. Wiley-Interscience, New York, pp. 445-459.

Gorsline, D.S. and Emery, K.O., 1959. Turbidity current deposits in San Pedro and Santa Monica basins off southern California. Geol. Soc. Am. Bull., 70:279-290.

Gross, M.G., 1972. Geologic aspects of waste solids and marine waste deposits, New York Metropolitan region. Geol. Soc. Am. Bull., 83:3163-3176.

Haner, B.E., 1969. Geomorphology and sedimentary characteristics of Redondo submarine fan, southern California. M.S. Thesis, Univ. Southern California, Los Angeles, 84pp.

Hartman, O., 1956. Contributions to a biological survey of Santa Monica Bay, California. Final Report to Hyperion Engineers, Inc. Geology Dept., Univ. Southern California, 161pp.

Hartman, O., 1960. The benthonic fauna of southern California in shallow depths and possible effects of wastes on the marine biota. In: E.A. Pearson (Editor), Waste Disposal in the Marine Environment. Pergamon Press, New York, pp. 57-81.

Kerr, A.R., 1938. Littoral erosion and deposition of Santa Monica Bay. M.S. Thesis, Univ. California at Los Angeles, 49pp.

Ketchum, B.H. (Editor), 1972. The Water's Edge: Critical Problems of the Coastal Zone. Massachusetts Institute of Technology Press, Cambridge, Mass., 393pp.

LaFond, E.C., 1962. Internal waves. In: M.N. Hill (Editor), The Sea, 1, Pt. 1. Interscience Publishers, New York, pp. 731-751.

LeBlond, P.H. and Mysak, L.A., 1977. Trapped coastal waves and their role in shelf dynamics. In: E. Goldberg, I. McCave, J. O'Brien and J. Steele (Editors), The Sea, 6, Interscience Publishers, New York, pp. 459-495.

Merlini, M., 1971. Heavy-metal contamination. In: D.W. Hood (Editor), Impingement of Man on the Oceans. Wiley-Interscience, New York, pp. 461-486.

Munk, W.H., 1962. Long ocean waves. In: M.N. Hill (Editor), The Sea, 1. Interscience Publishers, New York, pp. 647-699.

Munk, W., Snodgrass, F., and Carrier, G., 1956. Edge waves on the continental shelf. Science, 123:127-132.

National Academy of Sciences, 1975. Assessing Potential Ocean Pollutants. National Academy of Sciences, Washington, D.C., 438pp.

Palmer, H.D., 1977. The use of submersibles in the study of ocean waste disposal. In: R. Geyer (Editor), Submersibles and their use in oceanography and ocean engineering, Elsevier Oceanography Series, no. 17. Elsevier Publishing Company, New York, pp. 317-334.

Petzold, T.J. and Austin, R.W., 1968. An underwater transmissometer for ocean survey work. Scripps Inst. Oceanography, Tech. Rept. Ref. 68-9, 5pp.

Pomeroy, R., 1960. The empirical approach for determining the required length of an outfall. In: E.A. Pearson (Editor), Waste Disposal in the Marine Environment. Pergamon Press, New York, pp. 268-278.

Richards, F.A., 1971. Anoxic versus oxic environments. In: D.W. Hood (Editor), Impingement of Man on the Oceans. Wiley-Interscience, New York, pp. 201-217.

Risebrough, R.W., 1971. Chlorinated hydrocarbons. In: D.W. Hood (Editor), Impingement of Man on the Oceans. Wiley-Interscience, New York, pp. 259-286.

Rodolfo, K.S., 1964. Suspended sediment in southern California waters. M.S. Thesis, Univ. Southern California, Los Angeles, 91pp.

Shepard, F.P. and Macdonald, G.A., 1938. Sediments of Santa Monica Bay, California. Amer. Assoc. Petrol. Geol. Bull., 22:201-216.

Shepard, F.P. and Emery, K.O., 1941. Submarine topography off the California coast: Canyons and tectonic interpretations. Geol. Soc. Am. Special Paper 31, 171pp.

Southern California Coastal Water Research Project (SCCWRP), 1973. Ecology of the southern California bight: Implications for water quality management. SCCWRP, El Segundo, California 531pp.

Stevenson, R.E., Tibby, R.B. and Gorsline, D.S., 1956. The oceanography of Santa Monica Bay, California. Final Report to Hyperion Engineers, Inc. Geology Dept., Univ. Southern California, 268pp.

Tibby, R.B., 1960. Inshore circulation patterns and the oceanic disposal of waste. In: E.A. Pearson (Editor), Waste Disposal in the Marine Environment. Pergamon Press, New York, pp. 296-327.

Troxell, H.C. and others, 1942. Floods of March, 1938 in southern California. U.S. Geol. Survey, Water Supply Paper 844, 399pp.

Tyler, J.E., Austin, R.W. and Petzold, T.J., 1974. Beam transmissometers for oceanographic measurements. In: R.J. Gibbs (Editor), Suspended Solids in Water. Plenum Press, New York, pp. 51-59.

Wildharber, J.L., 1966. Suspended sediment over the continental shelf off southern California. M.S. Thesis, Univ. Southern California, Los Angeles, 159pp.

Dredged Material, Ocean Disposal, and the Regulatory Maze

David D. Smith

ABSTRACT

Dredged material disposed of at sea each year totals about 50 to 75 x $10^6 m^3$, which is roughly 15 to 20 percent of the total sediment dredged in the U.S. annually. These volumes are discharged at some 125 designated ocean disposal sites, of which about 60 percent are located within the three mile limit.

Disposal of dredged material requires a permit issued by the Corps of Engineers in accordance with Environmental Protection Agency regulations and criteria. Obtaining this permit requires maneuvering through a complex procedural network (i.e. the regulatory maze) comprised in part by the numerous and varied criteria that govern the review and processing of the permit application. For a dredging project of even moderate size, negotiating this maze and obtaining the permit may require one to two years, including preparation and review of an Environmental Impact Statement.

The regulations governing ocean disposal also prescribe additional sets of criteria for evaluating (a) the dredged material to be disposed of, (b) the site proposed for disposal use, and (c) the environmental effects of the proposed disposal activity. Thus, literally dozens of criteria must be considered by the Corps in reaching a decision as to whether or not a permit for ocean disposal should be issued. Unfortunately, the governing regulations contain no recommended framework or standard methodology within which these many criteria can be applied--a major omission that requires resolution. Until such a framework or methodology is established, the decisions reached using the criteria must be recognized as highly subjective and potentially vulnerable to litigation.

INTRODUCTION

Dredging and dredged material[1] disposal are two of the most rigorously regulated construction activities in the U.S. today (Smith, 1975a, 1976; Boerger and Cheney, 1976; Smith and Graham, 1976). Since the early 1970's, a progressively more complex, legal, regulatory, and procedural framework has been developed by Federal agencies to control disposal of dredged material and, in particular, ocean disposal.

For the last three years, dredged material has comprised about 85 to 95 percent of all U.S. wastes dumped at sea. Thus it is not surprising that more than half of the papers presented in this volume deal with this "waste" sediment. The focus of this paper, however, is on the institutional aspects of ocean disposal rather than the scientific. Although a paper whose theme primarily is legal and regulatory may, at first, seem out of place in a scientific symposium, as will become evident, the geologic and other environmental aspects of oceanic waste disposal are, to a large degree, dependent on and controlled by the requirements and provisions of the legal framework.

Within the broader category of ocean waste disposal, the writer has selected dredged material as a representative example with which to illustrate how ocean disposal regulations function. As will be developed in following sections, disposal of dredged material at sea requires a permit issued by the Corps of Engineers (Corps) in accordance with Environmental Protection Agency (EPA) guidelines and criteria. To obtain this permit, the prospective disposer prepares and submits an application to the Corps, which then evaluates the proposed disposal and, by means of a complex set of procedural steps, arrives at a decision to issue or deny the permit.

This paper examines the legal basis on which the permit process rests, the procedural framework for review and processing of permit applications, and some of the various sets of criteria used in evaluating (a) the permit application, (b) the dredged material to be disposed of, (c) the site proposed for disposal use, (d) the environmental effects of the proposed disposal activity.

PRESENT STATUS OF OCEAN DISPOSAL

In order to develop and maintain the navigable waterways of the United States, the Corps is responsible for dredging on the order of 275 to $350 \times 10^6 m^3$ of sediment each year at a cost of over $200 million. In addition, in some years privately sponsored dredging projects may add another 25 to $50 \times 10^6 m^3$ to the total volume dredged.

Disposal practices for dredged material vary widely, depending upon a number of factors, including (a) the type of dredge in use,

---

[1]Dredged material is the term used in various Federal regulations for the mix of sediment and water generated as waste by a dredging operation. According to Section 227.13 of the EPA 1977 Ocean Dumping Regulations and Criteria, "Dredged materials are bottom sediments or materials that have been dredged or excavated from the navigable waters of the United States,... . Dredged material consists primarily of natural sediments or materials which may be contaminated by municipal or industrial wastes or by runoff from terrestrial sources such as agricultural lands."

(b) the category of work (maintenance versus new), (c) the geographic region, (d) various environmental and geographic characteristics of the area being dredged, and (e) various institutional and economic factors. Examples of some of these variations are presented in Boyd et al. (1972), Gren (1976), Mohr (1976), and Pequegnat et al. (1978).

Of the 300 to $400 \times 10^6 m^3$ total volume dredged annually in the U.S., about 50 to $75 \times 10^6 m^3$ (or about 15 to 20 percent are disposed of in the ocean.[2] Based on data provided by the Corps (USACE, 1977), volumes of dredged material disposed at sea for one four year period 1973-1976 are as follows:

Volume ($m^3 \times 10^6$)			
1973	1974	1975	1976
51	75	67	50

In all, these volume figures comprise about 85 to 95 percent of the total annual volume of U.S. wastes dumped in the ocean.

REGIONAL DISTRIBUTION

Volumes

The regional distribution of volumes of dredged material disposed at sea in U.S. waters in the three year period 1974-1976 is presented in Table 1a. As evident, the Gulf Coast, on the average accounts for more than half of the volume.

Disposal Sites

In a 1970 study of ocean disposal of solid waste, Smith and Brown (1971) reported some 160 active disposal sites for dredged material in U.S. coastal waters (not including sites in coastal bays, sounds, or estuaries). Of the 160 sites, 60 percent were located within the three mile (5.6km) limit in water depths less than about 30m, and over 90 percent were located along the Atlantic and Gulf Coasts.

As pointed out in Boyd et al. (1972), the principal criterion in the selection of these sites (many of which had been in use for 40 years or more) was disposal cost. Locations selected were usually those closest to the dredging project which offered reasonable assurance that the disposal material would neither return directly to the dredged channel, nor restrict navigation.

In 1977, EPA (1977a) promulgated a list of 127 interim disposal sites for dredged material; many of these are identical to sites listed by Smith and Brown (1971). The regional distribution of these sites is summarized in Table 1b. Of the 127, about 60 percent are located within the three mile (5.6km) limit, and all are within the 12 mile (22.2km) limit. Regionally, about 40 percent are along the Gulf Coast and about 30 percent each along the Atlantic and Pacific

---

[2] By way of comparison, about 500×10^6 tons (about $285 \times 10^6 m^3$, assuming a bulk density of 1.6) of suspended sediment from the conterminous United States reach the ocean each year via stream runoff (Curtis et al., 1973).

TABLE 1

REGIONAL DISTRIBUTION OF DISPOSAL VOLUMES AND SITES

Table 1a — Volumes by Region and Year

	1974	1975	1976	3 Year Average
Atlantic	19%	28%	35%	28%
Gulf	66%	50%	49%	55%
Pacific	15%	22%	16%	17%

Source: U.S. Army Corps of Engineers (USACE, 1977, and personal communications)

Table 1b — Disposal Sites by Region

(Sites designated as interim in EPA 1977 Regulations)

			Number Used in 1976
Atlantic	37	(29%)	28
Gulf	49	(39%)	20
Pacific	37	(29%)	24
Other	4	(3%)	-0-
Total	127 Sites		72 Sites (57%)

Source: Environmental Protection Agency (USEPA, 1977a)

Coasts. Less than 60 percent of the 127 sites were used in 1976; sites not used in 1976 may be used in subsequent years in connection with periodic maintenance dredging requirements.

FUTURE TRENDS

Various lines of scientific and technical reasoning favor ocean disposal of dredged material (see, among others, Smith, 1971, 1973; Emery, 1972; Bascom, 1974; Pequegnat et al., 1978; and Nichols and Fass, this volume). Further, with growing public awareness of the value of estuaries and coastal wetlands, it is logical to expect increasing pressure for dredged material disposal at sea (Boyd et al., 1972; Gross, 1972; U.S. Congress, 1976; Pequegnat et al., 1978). Environmental restrictions on terrestrial and estuarine sites warrant consideration of additional disposal sites in the ocean, particularly if appropriate care is taken to ensure that chemically polluted sediments are handled in such a way as to prevent significant adverse effects on the marine ecosystem.

LEGAL FRAMEWORK

GENERAL

Environmental and legal controls on dredging projects (Smith, 1975a; Boerger and Cheney, 1976) stem from regulatory policies specified in various Federal and state laws. The specific requirements of the Federal laws (and the complicated regulatory procedures for implementing them) generally are set forth as detailed regulations and guidelines published in the Federal Register (USACE, 1975; Smith, 1976; USEPA, 1977a). At the state level, the detailed regulations typically are issued by the appropriate regulatory boards and commisions. Generally, the pertinent regulatory procedures require issuance of permits for specific acts such as dredging, discharge of dredged material, placement of fill, etc.

HISTORICAL

At the Federal level, responsibility for primary regulatory control of dredging projects is vested in the Secretary of the Army and is exercised by the Corps of Engineers. Federal regulatory control of dredging, construction, and related actions in U.S. navigable waters dates back to the River and Harbor Act of 1899, which requires that a "Work in Navigable Waters" permit be obtained from the Corps for virtually all structures or work in U.S. navigable waters.

More recently, Section 404 of the 1972 Amendments to the Federal Water Pollution Control Act (P.L. 92-500) authorizes the Corps to issue permits for discharging dredged or fill material in navigable waters at specified disposal sites. Selection of the disposal sites is to be made in accordance with guidelines developed by the environmental Protection Agency and the Corps (see Smith, 1976).

For disposal of dredged material at sea, P.L. 92-532, the 1972 Marine Protection, Research, and Sanctuaries Act (MPRSA) provides for

Corps regulations and EPA criteria for discharge of dredged material into the territorial sea, the contiguous zone, and the ocean.

In 1975, the Corps issued regulations prescribing the policy, practice, and procedures for administering Corps permits for activities (including dredged material disposal) in navigable waters or the ocean, per the requirements of the three laws cited above. In 1977, EPA issued the Final Revision of Regulations and Criteria - Ocean Dumping: 40 CFR 220-229 (see USEPA, 1977a) which set forth provisions governing ocean disposal of dredged material, as called for in MPRSA.

The reader interested in a more detailed discussion of these laws, and the various regulations and guidelines that implement them, is referred to Black (1976), Bradley (1976), Hollis (1976), Kamlet (1976), Smith (1976), Webb and Holmes (1976).

In addition to these laws and regulations which deal specifically with dredging, the National Environmental Policy Act (NEPA) requires preparation of an environmental impact statement for all projects carried out by the Federal Government or involving Federal funds. In practice, NEPA's requirement for environmental statements has been interpreted broadly to include industrial projects which require Federal regulatory approval by agencies such as the Nuclear Regulatory Commission, the Corps of Engineers, the Federal Power Commission, and others. Further, certain states (for example, California) have enacted similar environmental protective legislation which requires formal assessment of a proposed project's environmental impacts. Because the environmental impacts of dredging and disposal are potentially significant, these activities and their impacts generally must be addressed at length in preparing an environmental impact statement for a project.

As a result of these laws and the administrative procedures stemming from them, regulatory agencies and public interest environmental groups have substantially greater influence on dredging projects than on most terrestrial projects of equivalent size.

MARINE PROTECTION, RESEARCH, AND SANCTUARIES ACT

The intent of the 1972 Marine Protection, Research, and Sanctuaries Act (MPRSA) is to regulate ocean disposal to the extent that such disposal "...will not unreasonably degrade or endanger human health, welfare, or amenities, or the marine environment, ecological systems, or economic potentialities" (USEPA, 1977b). Under the provisions of MPRSA, it was left to EPA to set criteria under which ocean dumping could be permitted without unreasonable degradation.

This expressed intent of MPRSA was reiterated in 1976 in Congressional reports developed from oversight hearings on MPRSA. As stated in EPA's (1977b) summary of the hearings:

> ...the clearly stated intent of the Congress is to permit ocean dumping as an acceptable alternative means of waste disposal under strict regulation as long as there is not 'unreasonable degradation.'
>
> Within the limits set by the factors required to be considered in setting criteria, it was left to EPA to determine what should be regarded as 'unreasonable degradation' and to establish specific criteria to insure that the impacts of dumping do not reach or exceed this level.

EPA (1977b) goes on to state:

> In developing the proposed criteria and in establishing the level of impact which would be regarded as 'unreasonable degradation,' EPA has attempted to establish a regulatory framework which would permit use of the ocean as an acceptable alternative for waste disposal without creating permanent damage to any part of the ocean bottom by inert material.

The key point is that the objective of the Congress and MPRSA was not, and is not, to terminate ocean disposal of dredged material.

OCEAN DUMPING CRITERIA

By means of a series of standard Federal rule-making procedures, EPA has over the past four years issued and revised a set of regulations governing ocean dumping (see USEPA, 1977a). Among the key elements of these regulations are the criteria for (1) the evaluation of permit applications for ocean dumping of materials, and (2) the selection and management of disposal sites for ocean dumping.

The criteria that govern evaluation of applications for ocean dumping include a series of criteria on environmental impact, as well as additional criteria concerning (a) the need for ocean dumping, (b) impact on esthetic, recreational, and economic values, and (c) impact on other uses of the ocean. These criteria are summarized in Table 2.

Subsequent sections of this paper address how some of the criteria are applied in the permit application review process and in selection of disposal sites.

PROCEDURAL FRAMEWORK

The legal framework just described would appear to be reasonably straight forward. In simplest terms, the law specifies that (1) a Dredged Material Permit is required in order to dispose of dredged material in the ocean, (2) the Corps is the agency responsible for deciding whether or not a permit should be issued, and (3) the Corps' decision will be guided in part by criteria established by EPA.

Experience has shown, however, that obtaining this Corps permit for disposal of dredged material can be an extremely complicated matter that requires moving through an intricate series of steps which comprise the permit application processing and review procedures. This interconnecting network of requirements, criteria, and administrative procedures make up, what amounts to, a regulatory maze for dredged material disposal. The intricacy of this maze may be perceived most readily by examination of Figure 1, which is a flow diagram summarizing the more important sequential steps involved in maneuvering through the regulatory requirements to obtain (or be denied) a Dredged Material Permit.

The complexity of the Corps' maze-like network stems primarily from the language and requirements of (a) the Congressional legislation cited earlier, and (b) regulations and guidelines which have been developed by the Corps and EPA to implement these laws. In

TABLE 2

EPA CRITERIA FOR THE EVALUATION OF PERMIT APPLICATIONS FOR OCEAN DUMPING OF MATERIALS

Subpart A - General

Sections

227.1 - Applicability
227.2 - Materials which satisfy the environmental impact criteria of Subpart B
227.3 - Materials which do not satisfy the environmental impact criteria of Subpart B

Subpart B - Environmental Impact

227.4 - Criteria for evaluating environmental impact
227.5 - Prohibited materials
227.6 - Constituents prohibited as other than trace contaminants
227.7 - Limits for specific wastes or waste constituents
227.8 - Limitations on disposal rates of toxic wastes
227.9 - Limitations on quantities of waste materials
227.10- Hazards to fishing, navigation, shorelines or beaches
227.11- Containerized wastes
227.12- Insoluble wastes
227.13- Dredged materials

Subpart C - Need for Ocean Dumping

227.14- Criteria for evaluating the need for ocean dumping and alternatives to ocean dumping
227.15- Factors considered
227.16- Basis for determination of need for ocean dumping

Subpart D - Impact on Esthetic, Recreational and Economic Values

227.17- Basis for determination
227.18- Factors considered
227.19- Assessment of impact

Subpart E - Impact on Other Uses of the Ocean

227.20- Basis for consideration
227.21- Uses considered
227.22- Assessment of impact

Subpart F - Special Requirements for Interim Permits

227.23 through 227.26 - Not applicable to dredged material

Subpart G - Definitions

227.27- Limiting permissible concentration (LPC)
227.28- Release zone
227.29- Initial mixing
227.30- High-level radioactive waste
227.31- Applicable marine water quality criteria
227.32- Liquid, suspended particulate, and solid phases of material

Source: USEPA, 1977a, Part 227.

particular, the requirements for substantial formal involvement of various Federal and state conservation agencies and the general public in the permit processing and review procedures has contributed to this complexity.

The flow diagram (Figure 1) summarizes the principal actions taken by (a) an applicant applying for a Dredged Material Permit, (b) the Corps in processing the application, (c) Federal and state conservation and/or regulatory agencies, as well as (d) the more important interactions among the various parties. For convenience, the diagram has been structured with the general sequence of actions progressing from left to right, and with the series of applicant, Corps, and other agency actions occupying the upper, center, and lower third of the diagram respectively.

As illustrated in Figure 1, a series of actions by the applicant, the Corps, and other agencies can be categorized as preliminary steps which culminate with the applicant's submission of the permit application to the Corps. The application then moves through a complicated, interconnected set of processing and review steps to reach the decision steps which result in issuance or denial of the Dredged Material Permit.

The primary path through the maze is delineated by the sequence of actions bearing the numbers 1 through 21 and connected by the heavy line. As many as another twenty-one actions may be required to complete the various subsidiary components of the secondary paths (these actions are numbered 1a, 1b, 2a, etc.). The more than 40 steps illustrated are representative of the sequential actions required of an applicant proposing a major private sector dredging project involving proposed ocean disposal. Time required to progress through the maze to a permit may vary widely depending on a number of factors (including degree of public support for or opposition to the project), but 12 to 18 months or more is not uncommon. As much as half to two-thirds of this period may involve preparation and review of the Environmental Impact Statement (see steps 6b and 8 in Figure 1).

On the other hand, an applicant for a project of smaller scope would follow a less complicated series of steps, and might be able to obtain the permit in just a few months, barring public opposition, and if no EIS were required. In the case of a Federally sponsored dredging project, the Corps generally does not issue a Dredged Material Permit, but virtually all the key processing and review steps shown in Figure 1 are addressed.

Preparation of the flow diagram is based on a detailed analysis of the pertinent Corps and EPA regulations, guidelines, and criteria (primarily 33 CFR 209.120, and 40 CFR 221-223, 225-228). The primary path through the maze, and the sequence of and interconnections between subsidiary actions within the secondary paths (as illustrated in the figure) represents the writer's best judgment of the preferred paths for the particular type of private sector projects for which the diagram was prepared. Obviously, somewhat different flow paths might be desirable for a project with different engineering, environmental, and economic characteristics.

Figure 1 (on following three pages)
Dredged Material Permit application processing and review procedure.

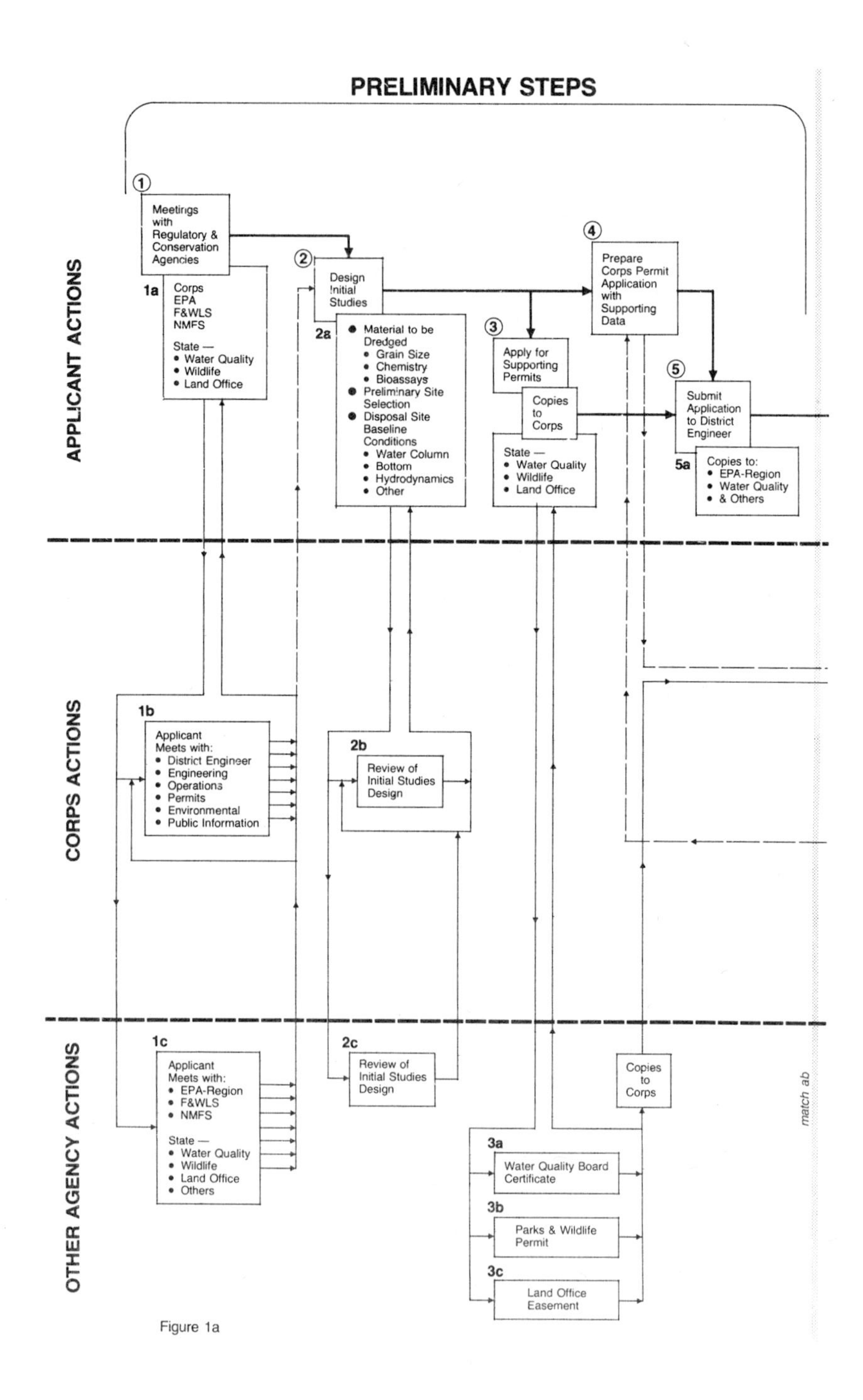

PRELIMINARY STEPS
APPLICANT ACTIONS
CORPS ACTIONS
OTHER AGENCY ACTIONS
1
Meetings with Regulatory & Conservation Agencies
1a
Corps
EPA
F&WLS
NMFS
State —
• Water Quality
• Wildlife
• Land Office
2
Design Initial Studies
2a
• Material to be Dredged
• Grain Size
• Chemistry
• Bioassays
• Preliminary Site Selection
• Disposal Site Baseline Conditions
• Water Column
• Bottom
• Hydrodynamics
• Other
3
Apply for Supporting Permits
Copies to Corps
State —
• Water Quality
• Wildlife
• Land Office
4
Prepare Corps Permit Application with Supporting Data
5
Submit Application to District Engineer
5a
Copies to:
• EPA-Region
• Water Quality
• & Others
1b
Applicant Meets with:
• District Engineer
• Engineering
• Operations
• Permits
• Environmental
• Public Information
2b
Review of Initial Studies Design
1c
Applicant Meets with:
• EPA-Region
• F&WLS
• NMFS
State —
• Water Quality
• Wildlife
• Land Office
• Others
2c
Review of Initial Studies Design
Copies to Corps
3a
Water Quality Board Certificate
3b
Parks & Wildlife Permit
3c
Land Office Easement
match ab

Figure 1a

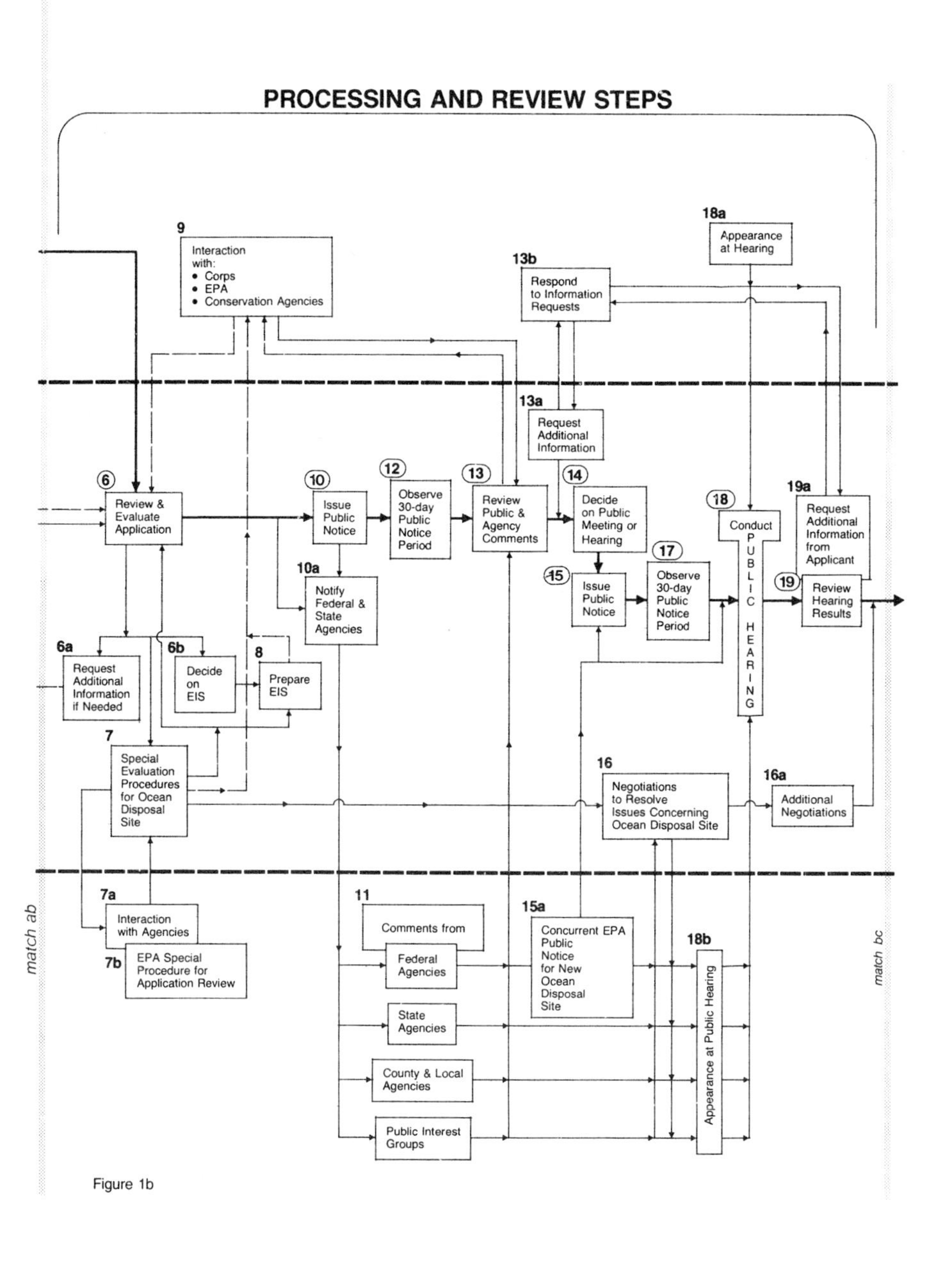

PROCESSING AND REVIEW STEPS
9
Interaction with: • Corps • EPA • Conservation Agencies
18a
Appearance at Hearing
13b
Respond to Information Requests
13a
Request Additional Information
6
Review & Evaluate Application
10
Issue Public Notice
12
Observe 30-day Public Notice Period
13
Review Public & Agency Comments
14
Decide on Public Meeting or Hearing
18
Conduct PUBLIC HEARING
19a
Request Additional Information from Applicant
19
Review Hearing Results
10a
Notify Federal & State Agencies
15
Issue Public Notice
17
Observe 30-day Public Notice Period
6a
Request Additional Information if Needed
6b
Decide on EIS
8
Prepare EIS
7
Special Evaluation Procedures for Ocean Disposal Site
16
Negotiations to Resolve Issues Concerning Ocean Disposal Site
16a
Additional Negotiations
7a
Interaction with Agencies
7b
EPA Special Procedure for Application Review
11
Comments from
Federal Agencies
State Agencies
County & Local Agencies
Public Interest Groups
15a
Concurrent EPA Public Notice for New Ocean Disposal Site
18b
Appearance at Public Hearing
match ab
match bc

Figure 1b

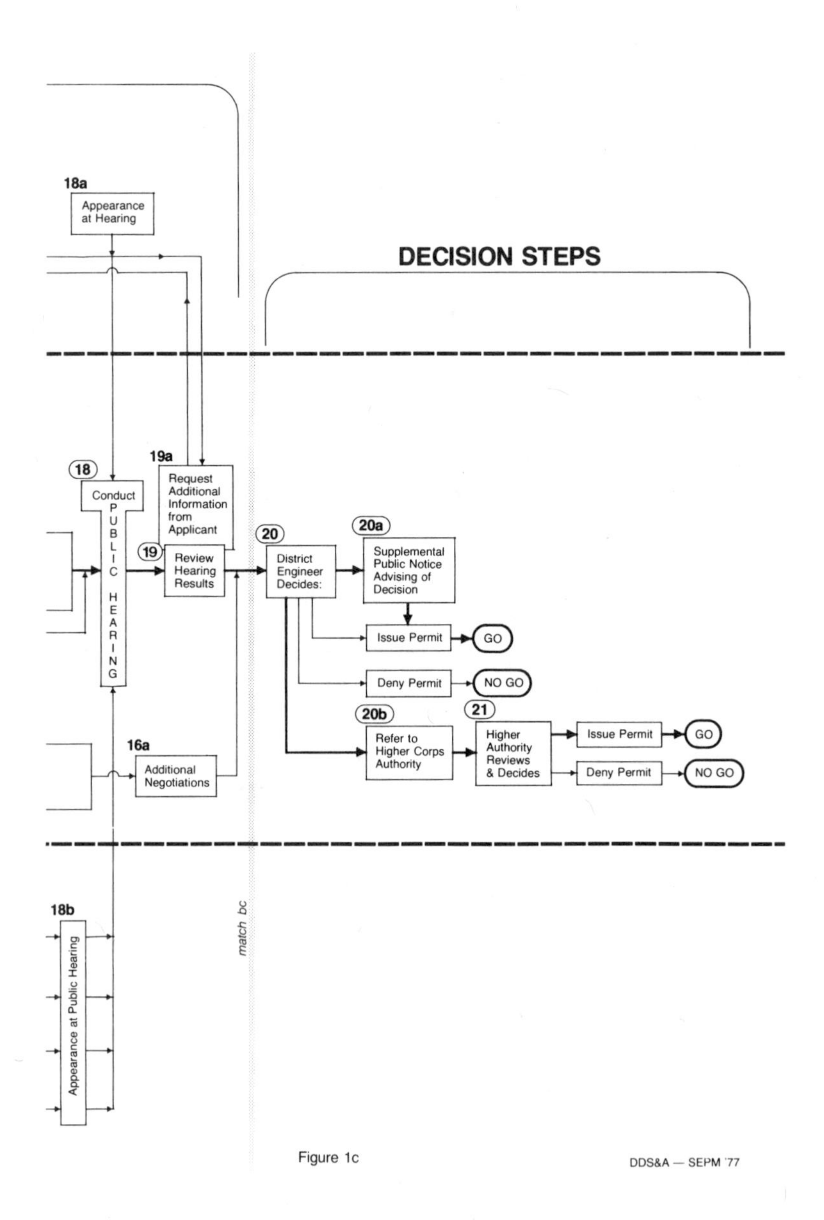

Figure 1c

DDS&A — SEPM '77

PERMIT APPLICATION EVALUATION CRITERIA

In evaluating and processing the Dredged Material Permit application, the Corps uses a series of evaluation criteria developed by EPA. These address (a) the acceptability of the materials for ocean disposal, (b) the probable environmental impact, (c) the probable impact on esthetic, recreational and economic values, (d) the probable impact on other uses of the ocean, and (e) the need for ocean disposal.

The responsibility for demonstrating that these criteria are satisfactorily complied with rests with the applicant. Therefore, the application must include a detailed treatment of each of the topics specified in the criteria.

The sequence in which the criteria are applied by the Corps in evaluating an application can be very important legally and technically. To provide guidance for Corps and EPA personnel in evaluating proposed ocean discharge of dredged material, and to ensure uniformity in application processing, EPA and the Corps jointly developed a flow diagram (reproduced here as Figure 2) setting forth the recommended sequence for the various testing and evaluation procedures required by the criteria. This diagram is part of the technical implementation manual (USEPA/Corps, 1977) which describes in detail how the evaluative procedures are to be applied.

As evident in Figure 2, this recommended sequence identifies some 20 steps through which it may be necessary to proceed in evaluating a proposed disposal. When viewed in terms of the overall permit processing flow diagram (Figure 1), the 20 steps of the EPA/Corps diagram (Figure 2) represent a detailed presentation of just three of the 21 major steps shown in Figure 1 (namely, steps 2, 6, and 7). Clearly, the regulatory maze is more complex than even Figure 1 conveys.

Although knowledge of the EPA/Corps recommended sequence (Figure 2) is helpful to the applicant in preparing an application, there may be more important reasons for the applicant to use a different sequence for some of the key steps. For example, based on the writer's experience, it is in the applicant's best interest to address the demonstration of need criteria first before proceeding with the other criteria.

Addressing the criteria for need first can save time and money because the requirements of the pertinent EPA regulations are such that unless need can be demonstrated convincingly, it is unlikely that the permit for ocean disposal would be granted. If expensive bioassay and other testing work is undertaken before the issue of need is examined and resolved, then substantial funds may be wasted.

DREDGED MATERIAL ACCEPTABILITY CRITERIA

Although the bulk of dredged material is natural sediment, some dredged material contains significant levels of pollutants derived from human activities. The potential effect of ocean disposal of dredged material on marine organisms and human uses of the ocean may range from unmeasurable to important (ODSSC, 1976; USEPA/Corps, 1977, p. 6; Pequegnat et al., 1978). These effects may differ at each disposal site and should be evaluated on a case by case basis. EPA regulations (USEPA, 1977a, p. 2478-79) provide criteria for such evaluations, with emphasis placed on direct assessment of biological impacts. Among these evaluative criteria are those which may be

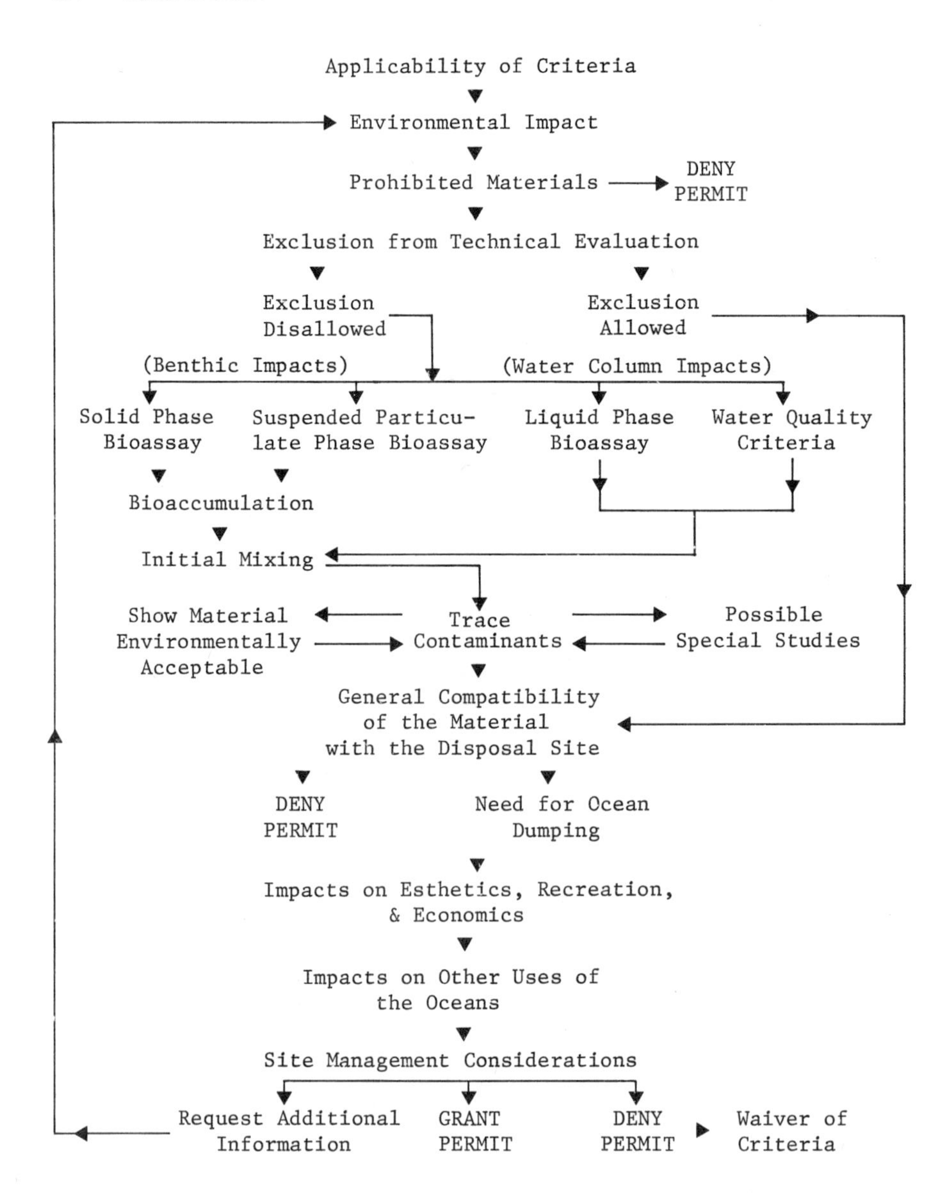

Figure 2. EPA/Corps flow diagram showing recommended sequence of testing and evaluation procedures (simplified from USEPA/Corps, 1977, Figure 1).

categorized as "acceptability criteria;" their purpose, in simplest terms, is to determine whether or not dredged material is acceptable for ocean disposal.

The acceptability criteria discussed here are used to (a) distinguish between dredged material which is likely to be free of pollutants and that which is not, and (b) determine the nature, concentrations, and availability of contaminants present in the polluted material.

EPA regulations recognize that dredged material may behave differently from other materials that may be ocean dumped, and treat dredged material as a special case of waste material containing liquid, suspended particulate, and solid phases. Nevertheless, the regulations provide that the test procedures applied to dredged material be the same as those applied to any multi-phase waste material. On the other hand, EPA and the Corps believe (USEPA, 1977b) that it is unnecessary to conduct expensive, time-consuming laboratory tests in evaluating every application for ocean disposal of dredged material. For this reason, the criteria dealing with dredged material include a straight forward screening procedure based in part on sedimentologic principles. Dredged material which meets the criteria of the screening procedure is judged to be environmentally acceptable for ocean disposal, and is excluded from further technical evaluation.

SCREENING PROCEDURE

The screening procedure includes (USEPA, 1977b, p. 90) an examination of the nature of the material to be dredged, the potential sources of man-made pollution in the geographical areas where dredging is to be done, and the characteristics of the area where the material will be dumped. In order to be judged environmentally acceptable for ocean disposal without further testing, dredged material must (as specified in 40 CFR 227.13(b)) meet any of the following criteria:

> (1) Dredged material is composed predominantly of sand, gravel, rock, or any other naturally occurring bottom material with particle sizes larger than silt, and the material is found in areas of high current or wave energy such as streams with large bed loads or coastal areas with shifting bars and channels; or
> (2) Dredged material is for beach nourishment or restoration and is composed predominantly of sand, gravel or shell with particulate sizes compatible with material on the receiving beaches; or
> (3) When: (i) The material proposed for dumping is substantially the same as the substrate at the proposed disposal site; and
> (ii) The site from which the material proposed for dumping is to be taken is far removed from historical sources of pollution so as to provide reasonable assurance that such material has not been contaminated by such pollution.

FURTHER TECHNICAL EVALUATION PROCEDURES

Dredged materials which cannot meet the screening procedure criteria are subject to testing of the liquid, suspended particulate, and solid phases using various bioassay procedures and chemical analyses. The appropriate technical procedures are specified in 40 CFR 227 and 228 and are explained in some detail in USEPA/Corps (1977).

To be judged environmentally acceptable for ocean disposal, the liquid phase, suspended particulate phase, and solid phase must meet a series of requirements dealing with: prohibited materials, trace contaminants, limiting permissible concentrations, water quality criteria, bioaccumulation potential, and compatibility with disposal site, among others (see USEPA, 1977a and USEPA/Corps, 1977, for a detailed listing of these requirements).

It is the intent of the regulations (USEPA, 1977a,) that:

> The [evaluative] criteria should be based, wherever possible, on impacts of dumped materials on marine ecosystems, and that these impacts could be measured best by bioassays rather than by relying on determination of total amounts of specific constituents present in a waste.
> ...thus, all criteria are based on ecosystem impact rather than on assumptions regarding allowable deviations from normal ambient values.

EPA (1977b) believes that this approach is consistent with the concept of "unreasonable degradation" specified in the basic law (MPRSA). Further, the use of bioassay results for regulatory purposes should provide EPA with direct measurements of the impact of dumping materials, so that it will no longer be necessary to infer damage indirectly through measurements related to normal ambient values.

SELECTION OF DISPOSAL SITE

One of the most important decisions in preparing a permit application for ocean disposal of dredged material is selecting the location of a disposal site. A judicious choice at the outset may substantially reduce opposition to the project by conservation agencies and environmental groups and, thus, materially reduce application processing time. Procedures for selecting a satisfactory disposal site may become quite involved. For example, in Figure 1, site selection involves steps 2, 2a, 2b, 7, 7b, 15, 16, and 16a. Thus, for a large project, a separate flow diagram should be prepared to define the sequence and interrelationships of the key steps in site selection. Further, for projects where a number of disposal alternatives are under consideration, some type of decision analysis procedure should be used (Smith, in preparation).

The EPA regulations require the applicant to propose use of a particular disposal site; either (a) an existing disposal site (i.e., one of the 127 sites already designated on an interim basis by EPA) or, alternatively, (b) a site not previously designated. This latter approach, in effect, amounts to requesting the District Engineer to designate a new site for disposal use by the applicant.

With regard to designation of disposal sites, Section 228.4 of the EPA regulations (USEPA, 1977a) specifies that:

> In those cases where a recommended disposal site has not been designated by the Administrator, or where it is not feasible to utilize a recommended disposal site that has been designated by the Administrator, the District Engineer shall, in consultation with EPA, select a site in accordance with the requirements of... [Sections] 228.5 and 228.6(a).

Sections 228.5 and 228.6 are, respectively, the general and specific criteria for selection of ocean disposal sites (reproduced here as Table 3 and discussed below).

If the applicant proposes use of a new disposal site (which is more or less typical of large private sector projects), the District Engineer's decision to authorize use of the proposed site will be made in consultation with EPA and in accordance with these various general and specific criteria. In addition, a disposal site designation study probably will be required, as will preparation of an environmental assessment, the results of which will be used by the District Engineer and the EPA in deciding whether a full scale Environmental Impact Statement should be prepared before authorizing the new site. When the substantial time and costs involved in these steps become evident, many applicants abandon the proposal for a new site designation and opt for the use of an already designated site.

SITE SELECTION CRITERIA

EPA's criteria for site selection (Table 3) are grouped in two categories: general and specific. The general criteria express site selection policy, and the specific criteria list the various factors (environmental, operational, etc.) to be considered.

General Criteria

Briefly, the general criteria (40 CFR 228.5) require the site location and size be selected so as to (a) minimize interference with other activities in the marine environment; (b) avoid areas of (i) existing fisheries or shellfisheries, or (ii) heavy commercial/ recreational navigation; (c) reduce temporary perturbations in water quality or other environmental conditions to normal ambient seawater levels or to undetectable contaminant concentrations or effects before reaching any beach, shoreline, marine sanctuary, fishery or shellfishery; (d) facilitate identification and control of any immediate adverse impacts, as well as effective monitoring and surveillance programs to prevent adverse long-range impacts. In addition, wherever feasible, the general criteria require selection of sites that have been used historically, or are located beyond the edge of the continental shelf.

Specific Criteria

A major provision of EPA's specific criteria requires that a "disposal site evaluation and/or designation study" be carried out,

TABLE 3
SPECIFIC CRITERIA FOR SELECTION OF OCEAN DISPOSAL SITES
(40 CFR 228.6)

In the selection of disposal sites, in addition to other necessary or appropriate factors determined by the EPA Administrator, the following factors will be considered:

(1) Geographical position, depth of water, bottom topography and distance from coast;

(2) Location in relation to breeding, spawning, nursery, feeding, or passage areas of living resources in adult or juvenile phases;

(3) Location in relation to beaches and other amenity areas;

(4) Types and quantities of wastes proposed to be disposed of, and proposed methods of release, including methods of packing the waste, if any;

(5) Feasibility of surveillance and monitoring;

(6) Dispersal, horizontal transport and vertical mixing characteristics of the area, including prevailing current direction and velocity, if any;

(7) Existence and effects of current and previous discharges and dumping in the area (including cumulative effects);

(8) Interference with shipping, fishing, recreation, mineral extraction, desalination, fish and shellfish culture, areas of special scientific importance and other legitimate uses of the ocean;

(9) The existing water quality and ecology of the site as determined by available data or by trend assessment or baseline surveys;

(10) Potentiality for the development or recruitment of nuisance species in the disposal site;

(11) Existence at or in close proximity to the site of any significant natural or cultural features of historical importance.

Source: USEPA, 1977a, p. 2483.

and that this study make use of the eleven criteria presented in Table 3, as well as such other factors as EPA determines to be appropriate or necessary. The majority of the eleven criteria specified are environmental in character.

These criteria also require that (a) the results of this study be presented as an environmental assessment of the impact of the proposed use of the site for disposal, and (b) this environmental assessment in turn be presented in support of the site designation action by the Corps and EPA.

The organization and degree of completeness of the environmental factors included in the specific criteria warrant examination and discussion (see Smith, 1975b), but such is outside the scope of this paper.

DISCUSSION

As this paper demonstrates, in order to obtain a permit for ocean disposal of dredged material, an applicant must move through a complex procedural network (i.e. the regulatory maze). Some of the major components of the maze are the numerous and varied criteria that must be complied with in the preparation, review, and processing of the permit application. Briefly, these criteria deal with (a) the acceptability of dredged material for ocean disposal, (b) assessment of environmental, esthetic, recreational, economic and other types of impacts, and (c) selection of disposal sites.

The EPA and Corps regulations and evaluative criteria on which this paper focuses have been the subject of extensive study, technical discussion, public hearings, and letters of comment in the course of the standard Federal rule-making procedure by which they were developed and subsequently revised (see USEPA, 1977b).

In addition, the regulations and criteria have been the subject of litigation by the National Wildlife Federation (see Kamlet, in EPA 1977B, vol. II), and the revised criteria were reviewed in a two day technical workshop. For the sum and substance of the workshop discussions, the reader is referred to EPA, 1977b, Appendix G.

Despite the numerous previous comments, to the writer's knowledge at least one fundamental omission has not been addressed--this is the absence of any provisions on <u>how</u> the numerous sets of criteria should be applied. Simply put, although there are several dozens of criteria to be considered, the regulations contain no recommended framework or standard methodology within which the criteria can be applied. Further, there are no provisions for development of such a framework or methodology. This is a major omission that should be remedied because, as evident from Tables 2 and 3 above, the criteria "laundry lists" are long and diverse, and many of the criteria or factors are not readily comparable (Smith, 1975b). The need for at least prescribing some type of rank ordering and/or comparative weighting techniques seems evident.

The importance of specifying a standard evaluative methodology within which the criteria are to be applied can hardly be overstated. Without a standard methodology, the results of an evaluation using this number of criteria must perforce by highly subjective and judgmental in character, both on the part of the applicant and on the part of the review agencies (EPA, the Corps, and the various Federal and state conservation agencies). Without a standard methodology, the applicant knows only the factors to be considered but not their relative importance or by what means the factors should be assessed, much less compared.

When this type of situation obtains in the regulatory arena, there is a significant potential for controversy, delay, and, eventually, unsound regulatory decisions. Even worse, the potential seems high for legal actions by either an applicant or an intervenor on the grounds, in part, that the manner in which the criteria were applied in reaching a particular decision was, in fact, capricious and arbitrary.

ACKNOWLEDGMENTS

This paper is based on detailed study of Corps and EPA regulations for dredged material disposal over the last several years in the course of the work on several large private sector dredging projects for Dames & Moore, Woodward-Clyde Consultants, the El Paso Company, Pacific Lighting, and Rohr Industries. The support received from these firms, and the cooperation of the Corps of Engineers and the Environmental Protection Agency are gratefully acknowledged.

In particular, the writer wishes to thank Messrs. D. Woodford and I. C. Macfarlane of Dames & Moore, Dr. G. A. Robilliard and Mr. C. Batra of Woodward-Clyde, Dr. J. M. Craig and Mr. M. C. Richards of El Paso, Drs. R. T. Saucier, R. M. Engler, and R. Peddicord of the Corps, Messrs. T. A. Wastler, T. W. Musser, and W. S. Davis of EPA, LCDR R. B. Franks of USCG, Dr. W. E. Pequegnat of TerEco, and Ms. K. F. Graham of DDS&A.

REFERENCES

1. Bascom, W. 1974. The disposal of waste in the ocean. *Scientific American,* 231:2:18-25.
2. Black, W. L. 1976. Waters of the United States. In: P. A. Krenkel, J. Harrison and J. C. Burdick III (Editors). *Dredging and Its Environmental Effects.* American Society of Civil Engineers, New York, pp. 10-38.
3. Boerger, F. C. and M. H. Cheney. 1976. Economic impact of dredging regulations. In: P. A. Krenkel, J. Harrison and J. C. Burdick III (Editors). *Dredging and Its Environmental Effects.* American Society of Civil Engineers, New York, pp. 408-417.
4. Boyd, M. B., R. T. Saucier, J. W. Keeley, R. L. Montgomery, R. D. Brown, D. B. Mathis, and C. J. Guice. 1972. *Disposal of Dredge Spoil: Problem Identification and Assessment and Research Program Development.* U.S. Army Waterways Experiment Station, Vicksburg, H-72-8, 121 p.
5. Bradley, J. H. 1976. Dredging and the Environmental Policy Act of 1969. In: P. A. Krenkel, J. Harrison and J. C. Burdick III (Editors). *Dredging and Its Environmental Effects.* American Society of Civil Engineers, New York, pp. 39-48.
6. Curtis, W. F., J. K. Culbertson, and E. B. Chase. 1973. Fluvial-sediment discharge to the oceans from the conterminous United States. U.S. Geological Survey, Circular 670, 17 p.

7. Emery, K. O. 1972. Statement on desirability of disposal of waste on the sea floor. As cited in: V. L. Andrelinunas, and C. G. Hard. 1972. Dredging disposal: real or imaginary dilemma? *Water Spectrum,* 4:16-21.
8. Gren, G. G. 1976. Hydraulic dredges, including boosters. In: P. A. Krenkel, J. Harrison and J. C. Burdick III (Editors). *Dredging and Its Environmental Effects.* American Society of Civil Engineers, New York, pp. 115-124.
9. Gross, M. G. 1972. Geologic aspects of waste solids and marine waste deposits, New York metropolitan region. *Geological Society of America Bulletin,* 83:3163-3176.
10. Hollis, C. W. 1976. Legislative impacts on dredging: general regulatory functions. In: P. A. Krenkel, J. Harrison and J. C. Burdick III (Editors). *Dredging and Its Environmental Effects.* American Society of Civil Engineers, New York, pp. 1-9.
11. Kamlet, K. S. 1976. Impact of Public Law 92-532 on dredging and disposal. In: P. A. Krenkel, J. Harrison and J. C. Burdick III (Editors). *Dredging and Its Environmental Effects.* American Society of Civil Engineers, New York, pp. 49-82.
12. Mohr, A. W. 1976. Mechanical dredges. In: P. A. Krenkel, J. Harrison and J. C. Burdick III (Editors). *Dredging and Its Environmental Effects.* American Society of Civil Engineers, New York, pp. 125-138.
13. Ocean Disposal Study Steering Committee (ODSSC). 1976. *Disposal in the Marine Environment-An Oceanographic Assessment.* National Academy of Sciences, Washington, D. C., 76 p.
14. Pequegnat, W. E., D. D. Smith, R. M. Darnell and R. O. Reid. 1978. *An Assessment of the Potential Impacts of Dredged Material Disposal in the Open Ocean.* U.S. Army Waterways Experiment Station, Vicksburg, D-78-2, 660 p.
15. Smith, D. D. 1971. Testimony on ocean disposal of dredged material before U.S. Senate Subcommittee on Oceans and Atmosphere, April 21, 1971. U.S. Government Printing Office, Washington D. C., 92-11:205, 214-218.
16. Smith, D. D. 1973. Marine disposal of selected solid wastes--a major beneficial use of ocean space. American Institute of Chemical Engineers, New York, 68:122:132-136.
17. Smith, D. D. 1975a. Disposal of dredged material--a key environmental consideration in the construction of major nearshore and coastal marine facilities. American Society of Mechanical Engineers, New York, 75-WA/Pet-7, 7 p.
18. Smith, D. D. 1975b. Dredged material disposal guidelines. *Science*, 189:8.
19. Smith, D. D. 1976. New federal regulations for dredged and fill material. *Environmental Science & Technology,* 10:328-333.
20. Smith, D. D. (in preparation). The use of decision analysis techniques in selection of dredged material disposal sites.
21. Smith, D. D. and R. P. Brown. 1971. *Ocean Disposal of Barge-delivered Liquid and Solid Wastes from U.S. Coastal Cities.* U.S. Environmental Protection Agency, Washington, D. D., SW-19c, 119 p.
22. Smith, D. D. and K. F. Graham. 1976. The effects of institutional constraints on dredging projects: San Diego Bay, a case history. In: *Dredging: Environmental Effects & Technology.* WODCON Association, San Pedro, pp. 119-141.

23. U.S. Army Corps of Engineers (USACE), 1975. Permits for activities in navigable waters or ocean waters. *Federal Register,* 40:144:31320 31343.
24. U.S. Army Corps of Engineers (USACE). 1977. *1976 Report to Congress on Administration of Ocean Dumping Activities.* U.S. Government Printing Office, Washington, D. C., 0-730-462/1597, 69 p.
25. U.S. Congress. 1976. Report of oversight hearings, House Committee on Merchant Marine and Fisheries, fiscal year 1977. U.S. Government Printing Office, Washington, D. C., 94-1047.
26. U.S. Environmental Protection Agency (USEPA). 1977a. Ocean dumping, final revision of regulations and criteria. 40 CFR SubChapter H, Parts 220-229. *Federal Register,* 42:7:2462-2490.
27. U.S. Environmental Protection Agency (USEPA). 1977b. *Proposed Revisions to Ocean Dumping Criteria, Final Environmental Impact Statement.* U.S. Environmental Protection Agency, Washington, D. C., 2 vols., 201 pp. + 8 appendices.
28. U.S. Environmental Protection Agency/Corps of Engineers. 1977. *Ecological Evaluation of Proposed Discharge of Dredged Material in Ocean Waters.* U.S. Army Waterways Experiment Station, Vicksburg, 19 pp. + 8 appendices.
29. Webb, D. H. and A. P. Holmes, Jr. 1976. Legislative impact of P.L. 92-500 on dredging. In: P. A. Krenkel, J. Harrison and J. C. Burdick III (Editors). *Dredging and Its Environmental Effects.* American Society of Civil Engineers, New York, pp. 83-114.

Index

Acid site, 53, 61
Agriculture, 2
Agglomerated particles, 172–173, 176, 182
Ambrose light, 52, 63
Anthropogenic solids, 51
Annapolis, Maryland, 142
Artifacts, human, 3
Asymmetrical mound (ridge), 106
Atlantic
  Beach, N. Y., 89
  Bight, Middle, 185–201
  Coast, 148, 149, 151, 185–201, 243, 244
  Highland, N. J., 56

Bacteria, 114
  coliform, 186, 192
  populations, 126
Baltimore, Md., Harbor, 131, 133–135, 137–140, 143
  channels, 134
Bar-Channel, system, 97
Bathymetric
  charts, 51
  surveys, 112
Bay, entrance, 126
Bedload transport, 41
Benthic
  animals, 114, 115
    activity, 126
  biomass, 141
  communities, 114
  organisms, 5, 140
    resettlement of, 140

Bight, apex gyre, 88
Bio-aggregation, 109
Biochemical, oxygen demand (BOD), 186, 192
Biological mixing. *See* Bioturbation
Biota, 136, 137
  benthic, 140, 141, 143, 165, 167–170, 173, 182, 200, 201, 207, 233
  diatoms, 137
  nekton, 166
  plankton, 137, 140, 166, 200
Bioturbation, 104, 109, 115, 126, 168, 169, 176, 182, 201, 202
  rate of, 126
Blue Plains, wastewater treatment plant, 138
Blue Ridge, 99
Brooklyn Bridge, 63
Buzzards Bay, Mass., 114

Cadmium, 139
Carbon. *See* Organic carbon
Cascade Mountains, 23
Castle Hill, 57, 61, 63
Cellar dirt, 1, 53, 61
Charleston Harbor, South Carolina, 149–150
Chesapeake and Delaware Canal, 134
Chesapeake Bay, 115, 131–143. *See also specific tributaries*
  Northern, 133, 137
Chlorite, 138

Cholera Banks, 59
Christianensen Basin, 73, 76, 88, 90
Chromium, 139
Clays, 10
  minerals, 138
Coal mining and washing, 133
Coastal plain, 99
Coastwise transport, sediment, 82
Coliform bacteria, 186, 192
Coney Island, New York, 59
Consolidation, 112
Continental shelf
  nearshore, 149, 158, 159
  Oregon and Washington, 18
  outer, 164, 186, 220
  sediments, 136, 166, 167, 207, 208
    source, 136
Copper, 139
Cornfield Point, Connecticut, 109
Corps of Engineers. *See* U.S. Corps of Engineers
Craney Island, Virginia, disposal area, 152–153
Crisfield, 136
  Harbor, Maryland, 135
Critical erosion velocity, 114
Currents. *See* Shelf currents
  bottom, 21, 24, 40, 46, 166, 172, 182, 187, 190, 191
  thermohaline, 10
  tidal. *See* Tidal currents
  volocity, 116
  wind driven, 76

Delaware Bay, Delaware, 2, 137, 138, 149, 150, 152, 154, 164
Deltaic areas, 12
Density current, 110
Derrick stone, 61
Diamond Hill, 53, 57, 63
Diapatra sp., 104
Diatom production, 137
Diked disposal, 131, 138, 150–153
Disposal site. *See specific sites*
  containment capacity of, 123
  location, 123
  sedimentary, environmental, 124
Dissolved oxygen (DO), 186, 192
Dredge spoil, 53, 75, 137–139, 242–244, 253–255
  dumpsite, 89. *See also specific sites*
Dredged material, 9
  disposal, long-term effects, 131
  materials, strategies for disposal, 141
Dredging, 132–143, 147–160, 242–246
  engines, 1
Drift, littoral, 3

Eatons Neck, New York, 110
Elutriate test, 141
Engineering, coastal, 1
Environmental Protection Agency. *See* U.S. Environmental Protection Agency
Erosion, 133
  cropland, 3
  estuary bed, 137
  resistance, 109
  resuspension, rates, 131
  shell, 5
Estuaries, 12
  as sediment sinks, 10
  circulation, 116, 140, 143, 151–153, 155–158, 160

Factor analysis, 29, 46
Fine sediment budget, 90
Finrot, 5
Fire Island inlet, 91
Fish, 140
  eggs, 140
Floatables, 212, 215

Fluctuation spectrum, 116
Fluid mud, 139
Fly ash, 1

Garbage, 61
Geological Survey. *See* U.S. Geological Survey
Georgia, 99
Grain-size analyses, 53
Gravels, artifacts, 5
Great Britain, 114
Great South Bays, 89
Gulf of Mexico, 2, 243, 244

Hampton Roads, Virginia, 1, 133–135, 137, 142. *See also* Norfolk, Virginia
Harbors, 9, 12, 131
Hempstead Bay, 89
Holocene, 55
  deposits, 100
Hopper dredges, 99
Hudson Channel, 4, 58, 61
  buried, 59
Hudson Shelf Valley, 5, 76
Human pathogens, 5
Hydrocarbons, 5

Illite, 138
Industrial
  discharge, 136
  wastes, 136, 142, 143, 164, 165, 171–173, 176
Interstitial water, 109
Itai-itai (cadmium poisoning), 2

Jamaica Bay, 89

Kaoline, 138

Landfill, 133, 138, 141, 148
Langmuir circulation, 176
Larvae, 140
Lead, 4, 139
London, convention of 1973, 133
Long Beach, California, 55
Long Island, New York, 55, 89
  Sound, 109, 110, 112, 114–116
  south shore, 76
Lower Bay, 56, 57

Magnetic ratio, 46
Maintenance dredging, 2, 100
Management
  permitting, 242–260
  regulatory, 141–143, 245–260
  testing, 141, 254–256
Manganese, 139
Marcus Hook, New Jersey, 138
Marina del Ray, California, 215
Marine organisms, 136
Marshes, 12
Mechanical, stability of waste deposits, 112
Mercury, 139
Methods of disposal, 124
Microtopography, 77, 166–168, 172–174, 176, 182, 201, 202
Minamata disease, 2
Mineral-organic aggregates, 114
Mining, 2
Models
  circulation, 190
  mathematical, 185–201
  waste dispersion, 191–200
Mud, 61

Navigation channels, 132
Nepheloid transport, 13
New England, southern, 55
New Haven, Connecticut, 109, 110, 112
  disposal site, 116
New Orleans, Louisiana, 12
New York
  Bay, lower, 56

Bight, 1, 45, 53, 55–57, 66, 148
Bight, apex gyre, 88
City of, 52
Harbor, 74
New Jersey metro region, 3
Nickel, 139
Nitrogen, dissolved, 140
Norfolk, Virginia, 131, 152, 153, 333, 435, 437, 438, 443
Northeast Pacific, 19
North Pacific High, 19, 21
Nutrients, 140, 148, 235

Ocean dumping
criteria, 247, 248, 253, 255. *See also* Management
site selection, 133, 140, 256–258
One-dimensional model of deposit, 112
One-man stone, 61
Open water, disposal sites, 109, 110
Organic carbon, 4, 5, 138, 165, 174, 182

Pacific Coast, 244
Palos Verdes, California, 228
Panama City, Florida, 14
Particulate wastes, 109
retention, 109
Philadelphia, Pennsylvania, 186
Phosphate, 171
dissolved, 140
Photosynthesis, phytoplankton, 140
Piedmont Province, 56, 99
Ports, 1. *See also specific sites*
facilities, 132
Portsmouth, Virginia, 143
Postdepositional, sedimentological response, 47
Primary productivity, 140
Pycnocline, 189

Raritan Bay, 56, 59
Redondo Canyon, 208, 215, 220, 221, 225, 228, 236
Relict sediments, 100
Resuspension, 114
Rip-currents, 81
Ripples, 166–169, 172–174, 176, 182, 201, 202. *See also* Microtopography
River
Columbia, poll, 18, 29
Columbia, sediments, 46
Delaware, 197
discharge, 82
Elizabeth, Maryland, 137–139
Elizabeth, Southern branch, 135
Hampton Roads, Elizabeth, 136
Hudson, 56, 89, 150
James, Virginia, 134–136, 138–140, 153, 155
Mississippi, 11
Patuxent, 138
Potomac, 136, 138, 139
Rappahannock, 134
Raritan, 56
Saint Lawrence, 11
sediment discharge, 131
Susquehanna, 133, 136, 138, 197
Thomas, England, 156–159
York, 134, 136
Riverborne, solids, 137
Rockaway, 55
Beach, 90

Salinity, 189, 209, 211, 212, 230
San Francisco Bay, 18
Sandwave-like mud deposits, 77
Sandy Hook, 55, 59
Santa Monica Bay, California, 205–236
Santa Monica Canyon, 208, 211, 212, 220, 225, 228, 234

Savannah, Georgia, 99
  navigational channel, 97
Scotland Light, 62
Sea Bright, New Jersey, 90
Sea level, 142
Sediments
  accumulation, natural rate of, 123
  anthropogenic, 73
  chemistry, 176, 180–182
  composition, 53
  compressibility of, 112
  mechanical properties of, 114
  movement, frequency of, 41
  permeability of, 112
  relict, 208
  stratigraphy, 53
  strength, 124
  suspended, 115, 140, 149, 174, 175, 215, 220, 225, 228, 233, 236. *See also* Turbidity
    clay, 138, 143, 155, 176–179
    sand, 138, 155, 165, 167, 173, 176–179, 190
    silt, 140, 143, 155, 176–179
  transport, 39
  water interface, 114, 126
Sedimentary regime, 109
Sedimentation
  estuaries, 136–140, 151, 157–160
  precipitates, 172
  rate, 109
Seismic reflection profiling, 51, 52
Self-consolidation, 109
Settling rates, 195–201
Severn estuary, 14
Sewage
  discharge, 136
  effluents, 136, 138, 142, 164–171, 176, 186, 191, 194–198, 205–236
    sludge, 51, 61
    dumpsite, 53, 73, 74
    front, 73
    particles, 75
Sheet, sweepings, 61
Shelf, currents
  bottom, 165, 166, 169, 172–174, 182, 186–189, 201
  surface, 170
Shoreline erosion, 136, 137
Shrewsbury Rocks, 59
Sidon, 1
Sludge, 68
Silver, 4
Slumping, 112
Soil erosion, 133, 141, 142
Solomons, Maryland, 142
Southern Atlantic Region, 2
Southern Piedmont, 3
Species fluctuation, 141
Spoil banks, 152, 153, 155
Steam ashes, 61
Storms, 194, 301
  effects of, 116, 118
  Northeastern, 89
  recurrence intervals, 118
Strain rate, 114
Strait of Juan de Fuca, 18
Submersibles, 164–182
Suburbanization, 137
Subway rocks, 63
Supervisor, of New York Harbor, 59
Suspended
  load transport, 45
  material fall, velocity of, 115
  sedimentation, 137, 174, 175, 186, 192, 194, 195, 209

Television, 165
Temperature, 165, 166, 171, 173, 209, 211, 212

Thermocline, 166, 170, 171, 173, 174, 182, 189, 194, 199, 220, 225, 228
Tidal
  currents, 141–143, 148, 152, 153, 155, 160, 186–189, 198, 201, 209
  prism, 21
  resuspension, 109
  stream, 109, 116
Tides
  diurnal, 21
  semi-diurnal, 21
"Tin Can" grounds, 59
Tobacco farming, 133
Toxic substances, 141, 143, 148, 207, 253–255
  chlorinated hydrocarbons, 233
  metals, 138, 139, 143, 207, 231, 233
Trace metals, 186, 192
Transmissometry, 165, 174, 207–217, 220–223, 225–231, 233–235
Transportation, 114
Tripod, instrumented, 24
Turbidity, 1, 28, 140, 141, 148, 159, 160, 171–175, 182, 209, 211–213, 215, 217, 220, 224, 225
  maximum, 12, 80, 131, 136–138, 142, 143
  transparency, 215, 218, 219
Turbulent flow, 115
Tyre, 1

U.S. Army Corps of Engineers, 148, 242–257, 259
U.S. Environmental Protection Agency (EPA), 164, 174, 186, 242, 243, 245–249, 253–259
U.S. Geological Survey (USGS), 174
Urban runoff, 136, 141, 142

Verrazano Narrows, 57
Vibratory cores, 52
Volcanic material, 23
Volumes of waste materials, New York Bight, 68

Wards Island Treatment Plant, N. Y., 90
Washington, D. C., 133, 137, 142, 143
Waste. *See specific types*
  chemical, 63
    composition, 4
  cohesionless, 112
  deposit, 109
    armored surface, 121
    formation, 121
    surface, slope on, 123
  disposal, water depths for, 123
Water
  discolored, 5
  level, data, 118
Waterborne, transport, 131
Waves, 142, 143, 148, 153, 167, 215, 230, 231
  affected zone, 109
  edge, 10
  internal, 10, 176, 225, 229–231
  sand, 24
  standing, 231
  surface, 10
Wetland, disposal sites, 131
Willapa Canyon, Wash., 23
Winds, 24
  mean, 22

Yusho (PCB poisoning), 2

Zinc, 139
Zooplankton, 140

About the Editors

HAROLD D. PALMER presently serves as Technical Coordinator for Marine Services with Dames & Moore in Washington, D. C. Dr. Palmer received his B.S. in Geology from Oregon State University and his M.S. and Ph.D. from the University of Southern California. He is a Fellow of the Geological Society of America and has served as both chairman and secretary of the American Society of Civil Engineers, Committee of Hydrographic and Oceanographic Surveying and Charting. Currently, he serves on the evening faculty at The Johns Hopkins University.

M. GRANT GROSS is Director of the Chesapeake Bay Institute, The Johns Hopkins University, in Baltimore, Maryland. His research on marine waste disposal problems and processes spans nearly twenty years and includes studies of dredging and disposal problems in the Chesapeake Bay, New York Harbor, and Long Island Sound. Earlier work was concerned with radioactive wastes in the Columbia River. He has published more than ninety scientific papers, three textbooks in oceanography, and five other books on coastal zone planning and waste disposal.

Dr. Gross received his A.B. in Geology from Princeton University and his M.S. and Ph.D. from the California Institute of Technology and has been on the faculties of the University of Washington, State University of New York at Stony Brook, and The Johns Hopkins University. He is a consultant to numerous federal agencies and foreign governments.